Encyclopedia of
Genetics

Encyclopedia of
Genetics

Volume I
Aggression – Heredity and Environment

Editor
Jeffrey A. Knight
Mount Holyoke College

Project Editor
Robert McClenaghan

Salem Press, Inc.
Pasadena, California
Hackensack, New Jersey

Executive Editor: Dawn P. Dawson
Managing Editor: Christina J. Moose
Project Editor: Robert McClenaghan
Copy Editor: Douglas Long
Acquisitions Editor: Mark Rehn
Production Editor: Joyce I. Buchea
Photograph Editor: Karrie Hyatt
Design and Layout: James Hutson

Illustrations by: *Electronic Illustrators Group,* 24, 33, 94-95, 102, 119, 159-160, 170, 175, 205, 211, 323, 334, 369, 380, 384-385, 390, 404, 438, 444; *Hans & Cassidy, Inc.,* 3-4, 9, 11, 15, 18, 39, 43, 68, 71, 75, 132, 138, 141, 149, 167, 171, 260, 271, 300, 322, 340, 346, 358-359, 431, 506, 528

Library of Congress Cataloging-in-Publication Data

Encyclopedia of genetics / editor Jeffrey A. Knight; project editor Robert McClenaghan.
  p. cm.
Complete in 2 vols.
Includes bibliographical references and index.
  1. Genetics—Encyclopedias. I. Knight, Jeffrey A., 1948-   . II. McClenaghan, Robert, 1961-   .

QH427.E53 1999
576.5'03—dc21   98-31952
                                                        CIP
ISBN 0-89356-978-X (set)
ISBN 0-89356-979-8 (vol. 1)

Second Printing

PRINTED IN THE UNITED STATES OF AMERICA

Publisher's Note

The *Encyclopedia of Genetics* was created to provide the general reader with a thorough yet accessible overview of one of modern science's most vital and intriguing fields. Its two volumes and 172 entries survey this exciting and continually evolving discipline from a variety of perspectives, offering historical and technical background along with balanced discussion of recent discoveries and accomplishments. The encyclopedia also comprehends the broader social issues raised by modern genetics research and its far-reaching implications, which have brought developments in this once-obscure science out of the laboratories and into the headlines.

Genetics encompasses a variety of subdisciplines, and the entries in the *Encyclopedia of Genetics* are organized around eight related but distinct fields of study. Eleven essays examine the field of bacterial genetics; twenty-six treat topics in classical transmission genetics; four discuss developmental genetics; thirty explore the related areas of genetic engineering and biotechnology; sixty-five concern human genetics; eight survey immunogenetics; fifteen discuss molecular genetics; and nine center on population genetics.

The encyclopedia is further structured around a hierarchy of article lengths. Eighteen 3,500-word essays give in-depth treatment of overview topics and other centrally important subjects. Thirty-eight 2,500-word entries provide substantive discussion of major subtopics, while fifty-two 1,500-word articles and sixty 1,000-word entries explore subsidiary topics and support the longer essays with a wealth of background information. Four appendices of varying lengths provide convenient reference sources.

Each article begins with ready-reference top matter that provides a quick summary of essential information. A "Field of Study" line indicates the subdiscipline under discussion. A "Significance" section gives an overview of the article's most important points. A listing of "Key terms" defines important concepts and provides necessary technical background. Following the main body of the essay text, each entry concludes with a "See Also" cross-reference section, which directs readers to related articles within the encyclopedia, and an annotated "Further Reading" paragraph that points readers to illuminating discussions in other sources.

A series of appendices supplements the essays: A Time Line of milestones in genetics offers a chronological overview of the field's development; a Biographical Dictionary provides capsule summaries of the careers and accomplishments of seventy-five distinguished geneticists; a Glossary provides definitions of more than 350 commonly used terms and explications of important concepts; and a general Bibliography references important works in each field of study.

The articles in the *Encyclopedia of Genetics* are arranged alphabetically by title; an alphabetical list of contents appears at the beginning of each volume. To help readers locate topics of interest, a Category List and a Subject Index are included at the conclusion of volume 2.

Nearly two hundred photographs and illustrations amplify and enliven the text. Dozens of diagrams, charts, graphs, and tables illuminate complex passages.

Many hands went into the creation of this work. Special mention must be made of consulting editor Jeffrey A. Knight, who applied his broad knowledge of genetics to shaping the encyclopedia's contents. Thanks are also due to the many academicians, researchers, and other scholars who assisted in the production of individual articles. Each essay carries the name of its author, and all writers are listed at the beginning of volume 1. Their contributions to the creation of the *Encyclopedia of Genetics* are gratefully acknowledged.

Preface

The science of genetics, once the purview only of serious students and professionals, has in recent decades come of age and entered the mainstream of modern life. An unparalleled explosion of new discoveries, powerful new molecular techniques, and practical applications of theories and research findings has brought genetics and its related disciplines to the forefront of public consciousness. The successful cloning of "Dolly" the sheep has sparked widespread public interest and debate and raised new questions about the ethics of this and other genetic technologies. Gene therapy has made the transition from science fiction to reality and is used to treat serious diseases, and there is increasing demand for the newest health professionals, genetic counselors, at hospitals and medical centers around the world. As we celebrate the new millennium, it is perhaps worth noting that the young science of genetics celebrates its one hundredth birthday.

Among many other events of historical importance, the year 1900 marked the rediscovery of the Austrian monk Gregor Mendel's experimental work on the inheritance of traits in the garden pea. Mendel had published his results thirty-four years earlier, but his work attracted little attention and soon faded into obscurity. By the close of the nineteenth century, however, much had happened on the scientific front. Chromosomes had been discovered, and the cellular processes of mitosis and meiosis had been observed under the microscope. The physical bases for understanding Mendel's principles of inheritance had been established, and the great significance of his pioneering work could finally be appreciated. The so-called chromosome theory of heredity was born, and the age of transmission genetics had arrived.

The first great geneticist to emerge (and some would still call him the greatest of the twentieth century) was Thomas Hunt Morgan, who established his "fly laboratory" at Columbia University and began studying the principles of transmission genetics, using the fruit fly as a model organism. All the major principles of transmission genetics, including single and multifactorial inheritance, chromosome mapping, linkage and recombination, sex linkage, mutagenesis, and chromosomal aberrations, were first investigated by Morgan and his students.

The subdisciplines of bacterial and molecular genetics had their beginnings in the 1940's, when bacteria and their viruses became favored genetic systems for research because of their relative simplicity and the ease with which they could be grown and manipulated in the laboratory. In particular, the common intestinal bacterium *Escherichia coli* was studied intensely, and today far more is known about the biology of this single-celled organism than about any other living system. In 1952, James Watson and Francis Crick provided the molecular model for the chemical structure of DNA, the genetic material, and the next twenty years saw great progress in the understanding of the molecular nature of essential cellular processes such as DNA replication, protein synthesis, and the control of bacterial gene expression.

The 1970's witnessed the discovery of a unique class of enzymes known as restriction endonucleases, which set the stage for the development of the exciting new technology known by various names as cloning, genetic engineering, or recombinant DNA technology. Since that time, research has progressed rapidly on several fronts, with the development of genetic solutions to many practical problems in the fields of medicine, agriculture, plant and animal breeding, and environmental biology. With the help of the new technology, many of the essential questions in cell and molecular biology that were first addressed in bacteria and viruses in the 1950's and 1960's can now be effectively studied in practically any organism.

And what are the major problems remaining to be solved? No doubt there are many, some of which cannot even be articulated given the present state of scientific understanding. Two important questions, however, are drawing disproportionate shares of attention in the current sphere of basic research. One of these is the problem variously referred to as "the second genetic code" or "protein folding." Scientists know how a particular molecule of DNA, with a known sequence of nucleotide subunits, can cause the production of a particular unique protein composed of a known sequence of amino acid subunits. What is not understood, however, is the process by which that protein will spontaneously fold into a characteristic three-dimensional shape in which each amino acid interacts with other amino acids to produce a functional protein that has the proper pockets, ridges, holes, protuberances, and other features that it needs in order to be biologically active. If all the rules for protein folding were known, it would be possible to program a computer to create an instant three-dimensional picture of the protein resulting from any given sequence of amino acids. Such knowledge would have great applications, both for understanding the mechanisms of action of known proteins and for designing new drugs for therapeutic or industrial use.

The second "big question" at the forefront of experimental genetic inquiry relates to the control of gene expression in humans and other higher organisms. In other words, what factors come into play in turning on or turning off genes at the proper times, either during an individual cell cycle or during the developmental cycle of an organism? How is gene expression controlled differentially—that is, how are different sets of genes turned on or off in different tissues in the same organism at the same time? Many human genetic diseases are now known or suspected to be caused by errors in gene expression—that is, too much or too little of a particular protein is made in the critical tissues at the critical developmental times—so the answers to these and related questions are sure to suggest new possibilities for gene therapy or other treatments.

The purpose of these reference volumes is twofold. First, the editors seek to highlight some of the most exciting new advances and applications of genetic research, particularly in the fields of human medical genetics and agriculture. Second, we hope to provide a solid basis for understanding the fundamental principles of genetics as they have been developed over this first one hundred years, along with an appreciation of the historical context in which the most important discoveries were made. It is our hope that such an understanding and appreciation might help to inspire a new generation of geneticists who will continue to expand the boundaries of scientific knowledge well into the next millennium.   —*Jeffrey A. Knight*

Contributor List

Barbara J. Abraham
Hampton University

Linda R. Adkison
Mercer University

Richard Adler
University of Michigan, Dearborn

Oluwatoyin O. Akinwunmi
Muskingum College

P. Michele Arduengo
Morningside College

Carl L. Bankston III
University of Southwestern Louisiana

D. B. Benner
East Tennessee State University

Alvin K. Benson
Brigham Young University

Carol Bernstein
University of Arizona

R. L. Bernstein
San Francisco State University

Barbara Brennessel
Wheaton College

Fred Buchstein
John Carroll University

Paul R. Cabe
St. Olaf College

James J. Campanella
Lehigh University

Rebecca Cann
University of Hawaii, Manoa

J. Aaron Cassill
University of Texas, San Antonio

Kerry L. Cheesman
Capital University

Richard W. Cheney, Jr.
Christopher Newport University

Stacie R. Chismark
Heartland Community College

Jaime S. Colomé
California Polytechnic State University, San Luis Obispo

Jennifer Spies Davis
Shorter College

David K. Elliott
Northern Arizona University

Phillip A. Farber
Bloomsburg University of Pennsylvania

James L. Farmer
Brigham Young University

Linda E. Fisher
University of Michigan, Dearborn

Chet S. Fornari
DePauw University

Daniel R. Gallie
University of California, Riverside

W. W. Gearheart
Morris College

Soraya Ghayourmanesh
Independent Scholar

Sibdas Ghosh
University of Wisconsin, Whitewater

Sander Gliboff
Johns Hopkins University

D. R. Gossett
Independent Scholar

Daniel G. Graetzer
Independent Scholar

Randall K. Harris
William Carey College

Robert Haynes
Albany State College

Werner G. Heim
Colorado College

Carl W. Hoagstrom
Ohio Northern University

Austin L. Hughes
Pennsylvania State University

Karen E. Kalumuck
Exploratorium

Manjit S. Kang
Louisiana State University

Susan J. Karcher
Purdue University

Armand M. Karow
Xytex Corporation

Roger H. Kennett
Wheaton College

Stephen T. Kilpatrick
University of Pittsburgh, Johnstown

Jeffrey Knight
Mount Holyoke College

William R. Lamberson
University of Missouri, Columbia

Kate Lapczynski
Motlow State Community College

Craig S. Laufer
Hood College

Michael R. Lentz
University of North Florida

Doug McElroy
Western Kentucky University

Sarah Lea McGuire
Millsaps College

Nancy Farm Mannikko
Michigan Technological University

Sarah Crawford Martinelli
Southern Connecticut State University

Lee Anne Martínez
University of Southern Colorado

Grace D. Matzen
Molloy College

Ulrich Melcher
Oklahoma State University

Eli C. Minkoff
Bates College

Beth A. Montelone
Kansas State University

Nancy Morvillo
Florida Southern College

Donald J. Nash
Colorado State University

Bryan Ness
Pacific Union College

Henry R. Owen
Eastern Illinois University

Massimo Pigliucci
University of Tennessee, Knoxville

Frank E. Price
Hamilton College

James P. Prince
California State University, Fresno

Mary Beth Ridenhour
*State University of New York,
Potsdam*

Connie Rizzo
Pace University

James L. Robinson
University of Illinois

David Wijss Rudge
Iowa State University

Paul C. St. Amand
Kansas State University

Virginia L. Salmon
*Northeast State Technical
Community College*

Mary K. Sandford
*University of North Carolina,
Greensboro*

Matthew M. Schmidt
*State University of New York,
Empire State College*

Harold J. Schreier
*University of Maryland
Biotechnology Institute*

Tom E. Scola
*University of Wisconsin,
Whitewater*

Rose Secrest
Independent Scholar

Bonnie L. Seidel-Rogol
Plattsburgh State University

Nancy N. Shontz
Grand Valley State University

R. Baird Shuman
*University of Illinois,
Urbana-Champaign*

Sanford S. Singer
University of Dayton

Lisa Levin Sobczak
Independent Scholar

Jamalynne Stuck
Western Kentucky University

James N. Thompson, Jr.
University of Oklahoma

Leslie V. Tischauser
Prairie State College

Patricia G. Wheeler
Indiana University

Bradley R. A. Wilson
University of Cincinnati

James A. Wise
Hampton University

R. C. Woodruff
Bowling Green State University

Ming Y. Zheng
Houghton College

Contents

Encyclopedia of
Genetics

Aggression

Field of study: Human genetics

Significance: *The idea that some humans may be born with inherited or innate tendencies for aggressive and criminal behavior has long fascinated scientists. Accumulating evidence indicates that aggressive behavior is governed by the interaction of social, environmental, and genetic factors. If the nature of the roles of these various factors can be understood, it may be possible to develop biological treatments for individuals with aggressive behavioral disorders.*

Key terms

MONOAMINE OXIDASE (MAO): an enzyme that breaks down several of the brain's important chemical transmitters

TESTOSTERONE: the male sex hormone produced by the testis; testosterone is responsible for the development of male sex organs and secondary male characteristics, sperm production, and male behavior

Y CHROMOSOME: the sex chromosome that triggers the expression of male characteristics

Nature Versus Nurture

Individual people vary considerably in the amounts and types of aggressive behavior they exhibit. It has been relatively easy for researchers to establish that many individual differences in behavior are influenced by genetic factors. What has not been determined is the precise role that genes play in contributing to aggressive behavior, since social and environmental factors also play a major role.

Although the study of various genetic disorders has demonstrated that genetic factors are significant in certain specific cases of aggressive behavior, it is evident that genetic factors are probably not a major contributing force to high crime rates. Research has also demonstrated that a need exists in all cases to achieve an understanding of how heredity is related to aggressive behavior. Since the 1959 finding that Down syndrome is caused by the presence of one extra chromosome, numerous other chromosomal abnormalities have been discovered. Chromosomal abnormalities usually produce marked effects upon normal development and lead to many types of physical defects and mental retardation.

The XYY Syndrome

Individuals with XYY syndrome have an extra sex chromosome, making them XYY males instead of normal XY males. A 1965 study reported that 35 percent of the 197 inmates of a maximum-security prison in Scotland had the extra Y chromosome. A number of subsequent studies investigated whether XYY males were at risk for aggressive, antisocial, or criminal behavior. Early studies indicated that there was a definite association between the XYY chromosomal constitution and confinement in mental and penal institutions. In other words, the XYY combination shows up in greater proportions in confined populations than it does in the "normal" male population.

There do not appear to be any physical, behavioral, or physiological features that characterize all XYY individuals, and normal XYY males have been reported. Certain characteristics, however, do seem to have some degree of association with XYY males, including below-average intelligence, increased height, heart defects, neurological anomalies, and sexual development disorders. XYY males do not appear to be "supermales" with greatly elevated levels of aggressiveness; rather, they tend to be the more withdrawn or passive members of confined populations. It is possible that problems stemming from affected intelligence or nonspecific behavioral problems may lead to antisocial or criminal behavior.

Single Genes Associated with Aggression

There are a number of single-gene disorders in which some association with aggression has been noted. It should be emphasized that in none of these cases is aggressive behavior invariably associated with the gene. The single-gene disorders include Tourette's syndrome, male-limited precocious puberty, fragile X syndrome, deafness-hypogonadism syndrome, and Lesch-Nyhan syndrome.

One of the most interesting cases associated with aggression involved a group of violent men in a large Dutch family who lacked a gene for the production of monoamine oxidase (MAO). MAO is an enzyme that breaks down several of the brain's chemical transmitters and, in men who lacked this gene, produced excessive amounts of the chemical serotonin, which was thought to lead to the men's unprovoked, violent outbursts. It has also been noted that abnormal levels of serotonin have been found in some violent criminal offenders. It must be emphasized, however, that the MAO deficiency gene is very rare and certainly cannot account for a very great percentage of the incidents of aggressive or criminal behavior.

Other evidence linking genetic factors to aggression have come from family, twin, and adoption studies. A statement by John Paul Scott in his classic book *Aggression* (1958) still provides a good description of how genes may influence aggression:

> Genes plus training produce aggressiveness, environmental stimulation plus aggressiveness produce aggression. . . . The primary stimulation to fight must come from the outside. Heredity can enter the picture only in such ways as lowering or raising the threshold of stimulation or modifying the physical equipment for fighting.
>
> —*Donald J. Nash*

See Also: Behavior; Criminality; XYY Syndrome.

Further Reading: Dean Hamer and Peter Copeland, *Living with Our Genes* (1998), provides an interesting chapter on aggression, crime, and violence as well as other chapters on genes and behavior. The so-called aggression gene is discussed in "The Bad Seed: Amid Controversy, Scientists Hunt for the 'Aggression' Gene," *Omni* 17 (February, 1995), by Jeff Goldberg. The relationship among genes, aggressiveness, and testosterone is presented by Robert Sapolsky in "Testosterone Rules," *Discover* 18 (1997).

Aging

Field of study: Human genetics
Significance: *Genes control the rate of aging. Un-*

derstanding which genes are involved in the aging process and how they act may ultimately allow scientists to promote the function of genes that slow the rate of aging.

Key terms

GENE: a specific segment of the long, double-stranded molecule deoxyribonucleic acid (DNA); a gene usually encodes information for making a specific protein

DEOXYRIBONUCLEIC ACID (DNA): the hereditary material composed of subunits called nucleotides joined end to end; a typical gene consists of a sequence of more than one thousand nucleotides

DNA DAMAGE: damage caused to nucleotides within genes by other metabolic products and by chemicals or agents that come from outside the body (such as X rays or by-products in the air)

What Aging Is and How It Is Controlled

During their life spans, organisms undergo changes that occur in a more or less predictable sequence. The changes that lead to an enhanced level of function are often referred to as "development," while the changes that lead to a decrease in functional ability are often referred to as "aging."

Metabolism, the set of chemical reactions occurring in the body, uses oxygen to turn food into useful energy and into components of the body. Unavoidably, metabolism produces some by-products that are not useful and some that are harmful. Humans breathe out the carbon dioxide that is produced by metabolism and dispose of urea in their urine. Another by-product is hydrogen peroxide (commonly used as a liquid disinfectant), which interacts with superoxide radicals (a special form of oxygen inside cells) to form a molecule called the hydroxy radical. The hydroxy radical is an oxidative free radical. A free radical is a molecule that reacts very quickly to join with and damage or break down other molecules.

Cells of the human body contain proteins, lipids, carbohydrates, and nucleic acids. The nucleic acids include deoxyribonucleic acid (DNA) and the similar molecule ribonucleic acid (RNA). Within each cell of the body, RNA carries instructions from the DNA for the formation of proteins. Many proteins are enzymes that promote the chemical reactions required by the cell, including the formation of lipids and carbohydrates. Proteins, lipids, carbohydrates, and nucleic acids are all susceptible to damage by free radicals. Proteins, lipids, and carbohydrates that sustain such damage are generally degraded by enzymes and then replaced. For instance, proteins in the liver of a mouse last only about three days, on average, before they are replaced. On the other hand, damage can accumulate in DNA, which does not turn over. DNA damages accumulate particularly in nondividing cells such as those of the brain and the muscles. When unrepaired damages occur in the DNA of a brain, muscle, or other long-lived cell, a gene containing the damaged DNA can no longer make its corresponding RNA. That particular cell will no longer be able to make the protein coded by that particular RNA molecule. The increase in unrepaired DNA damages per cell causes a general decline in the function of these long-lived cells, which leads to aging. While there are some other theories of aging, the DNA damage theory is probably the best at explaining most observed facts.

Normal mammalian metabolism gives rise to oxidative free radical damages to DNA at an average rate of about ten thousand damages per cell per day in humans and about eighty thousand per cell per day in rats. Although repair processes can remove most of these DNA damages, a significant portion (about eighty per cell per day in the liver of the rat) remain unrepaired in some tissues and accumulate with age. The number of DNA damages in a cell is determined by the balance between the production of damages and their repair. In comparisons among humans, monkeys, rats, and mice, a shorter life span correlates with a higher frequency of unrepaired oxidative DNA damage. Thus it has been hypothesized that accumulated DNA damage, particularly oxidative DNA damage, is the cause of aging.

Genes Affecting Aging and Longevity

When an inherited defective form (mutant form) of a gene is expressed in a human, that person can have a group of characteristic symp-

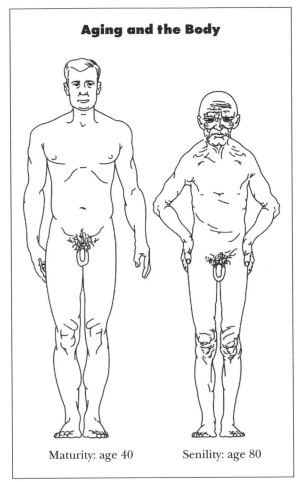

Aging and the Body

Maturity: age 40   Senility: age 80

Among the most obvious effects of aging are reduced body mass, a "shrinking" of height with loss of bone mass, sagging and wrinkling skin, and graying or loss of hair.

toms known as a genetic syndrome. Many genetic syndromes have been identified, and some of them include features of premature aging. Cells of individuals with premature aging types of genetic syndromes often have a deficiency in DNA repair ability or an excess of DNA damage. One example of a genetic syndrome associated with premature aging is Cockayne's syndrome (CS). The cells of individuals with CS are deficient in their ability to repair DNA damage. Individuals with CS suffer from premature neurological degeneration, cataracts, diabetes, and osteoporosis. A second example is ataxia telangiectasia (AT), in which cells are also unable to repair DNA normally. AT causes premature neurodegeneration, dia-

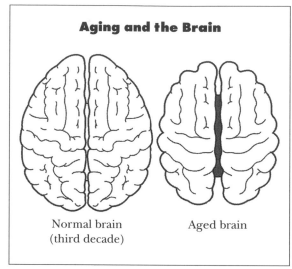

Aging and the Brain

Normal brain
(third decade)

Aged brain

The human brain shrinks with age as nerve cells are lost and brain tissue atrophies.

betes, and graying and loss of hair.

Animals with different life spans have been tested for their ability to carry out repair of DNA damage, including, in order of increasing life span, shrews, mice, rats, hamsters, cows, elephants, and humans. These animals exhibited an increasing ability to carry out DNA repair in the same order as increasing longevity. When thirteen mammals with different life spans were tested for the effectiveness of a specific DNA repair enzyme known as "PARP," the activity of this enzyme also increased along with increased longevity of the species.

When mutant animals with extended longevity are examined, they are often found to have increased defenses against free radicals. One type of nematode worm (*Caenorhabditis elegans*) has a mutation that causes increased levels of the enzymes catalase and superoxide dismutase (SOD), which increases its life span by about 50 percent. Catalase causes the breakdown of hydrogen peroxide to water and oxygen, while SOD causes the breakdown of the superoxide radical to hydrogen peroxide (from there, catalase breaks down the hydrogen peroxide). *Drosophila melanogaster*, a mutant fruit fly with a life span about 30 percent longer than that of a normal fruit fly, has increased levels of catalase and SOD as well as increased levels of another enzyme, xanthine

dehydrogenase (XDH), which protects against free radicals.

In a number of animal species, a calorie-restricted diet has been found to increase longevity. The average or maximum life spans of mice, rats, fish, rotifers, insects, and protozoa have all been shown to increase when their food intake was reduced by about 40 to 60 percent. In the case of mice, food restriction was shown to raise the level of DNA repair ability in their cells.

Impact and Applications

A number of studies with nematode worms, fruit flies, and mice suggest that coordinate increases in enzymes that protect against oxidative free radical damage (for example, catalase, SOD, and XDH) or that carry out DNA repair may slow the rate of aging without degrading normal abilities. By the late 1990's, methods were developed for inserting genes or modified forms of genes into cells. Although ethically questionable, such methods could, in principle, be applied to fertilized human egg cells. This approach might be used to increase expression of genes that decrease DNA damage and hence slow the rate of aging. However, there may be special dangers in attempting to increase human life span in this way. Any harmful side effects that might show up in such a genetically engineered human would affect not only that individual but also the descendants of that individual. It may be that slowing the aging process by putting more energy into assuring longevity may impose the price of reducing energy put into other functions such as mental capacity.

Genetic engineering technology has not been promising as a method for slowing the rate of aging of currently living individuals. Genetic engineering would require changing the levels of key enzymes in a substantial fraction of the long-lived cells of the body such as the trillion or so neurons in the brain. On the other hand, a calorie-restricted diet has substantial scientific support for potentially reducing the rate of aging in existing human individuals. Other dietary changes such as increased levels or combinations of vitamins or dietary supplements may be found that increase the activity of DNA repair enzymes, increase catalase and SOD production, or otherwise decrease DNA damage. Such dietary changes, if identified, might slow the aging process for currently living humans.

—*Carol Bernstein*

See Also: Cloning: Ethical Issues; Developmental Genetics; DNA Repair; Gene Regulation: Bacteria; In Vitro Fertilization and Embryo Transfer.

Further Reading: *Biology of Aging: Observations and Principles* (1991) by Robert Arking provides an overview of human aging for the general reader. Ricki L. Rusting, in "Why Do We Age?" *Scientific American* 267 (December, 1992), summarizes the changes that occur with aging and the roles of oxidants and free radicals. A more in-depth discussion of the importance of free radicals in aging is presented in *Free Radicals in Aging* (1993), edited by Byung Pal Yu. Carol Bernstein and Harris Bernstein, in *Aging, Sex, and DNA Repair* (1991), present an overview of aging in many organisms and explain how aging is overcome in the immortal germ line.

Albinism

Field of study: Human genetics
Significance: *Albinism is a direct result of decreased or nonexistent pigmentation of the skin, hair, and eyes. Albino humans are susceptible to sunburns and skin cancer, while albino animals lack the ability to adjust to environments in which nonalbino animals thrive.*

Key terms

ALBINISM: the absence of pigment such as melanin in eyes, skin, hair, scales, or feathers

MELANISM: the opposite of albinism, a condition that leads to the overproduction of melanin

PHOTOPHOBIA: a condition, often observed in albinos, in which sunlight is painful to the eyes

PIEBALDISM: a condition involving the patchy absence of skin pigment seen in partial albinos

Occurrence and Symptoms

Tyrosine, an amino acid, is normally converted by the body to a variety of pigments called melanins, which give an organism its characteristic colors in areas such as the skin, the hair, and the eyes. Albinism results when the body is unable to produce melanin because of defects in the metabolism of tyrosine caused by the lack or inactive presence of the enzyme tyrosinase. Those with albinism may thus be divided into two subgroups: the tyrosinase-negative (lack of tyrosinase) and the tyrosinase-positive (inactive presence of tyrosinase). The most serious case is that of complete albinism or tyrosinase-negative oculocutaneous albinism, in which there is a total absence of pigment. People with this condition have white hair, colorless skin, red irises, and serious vision defects. The red irises are caused by the lack of pigmentation in the eye retina and subsequent light reflection from the blood present in the retina. These people also display rapid eye movements (nystagmus) and suffer from photophobia, decreased visual acuity, and, in the long run, functional blindness. People with this disorder sunburn easily, since their skin does not tan. Partial albinos have a condition known as piebaldism, characterized by the patchy absence of skin pigment in places such as the hair, the forehead, the elbows, and the knees.

Several complex diseases are associated with albinism. The Waardenberg syndrome is identified with the presence of a white forelock (a lock of hair that grows on the forehead) or the absence of pigment in one or both irises, the Chediak-Higashi syndrome is characterized by a partial lack of pigmentation of the skin, and Tuberous sclerosis patients have only small, localized depigmented areas. A more serious case is the Hermansky-Pudlak syndrome, a disorder that also includes bleeding.

Ocular albinism is inherited and involves the lack of melanin only in the eye while the rest of the body shows normal or near-normal coloration. The condition reduces visual acuity from 20/60 to 20/400, with African Americans occasionally showing acuity as good as 20/25. Other problems include strabismus (crossed eyes or "lazy eye"), sensitivity to brightness, and nystagmus. The color of the iris may be any of the

An albino girl. (Ben Klaffke)

normal colors, but an optician can easily detect the condition by shining a light from the side of the eye. In ocular albinos, the light shines through the iris because of the absence of the light-absorbing pigment. Children with this condition have difficulty reading what is on a blackboard unless they are very close to it. Surgery and the application of optical aids appear to have had positive results in correcting such problems.

Albinism has long been studied in humans and captive animals. It has also been detected among wildlife, but such animals have little chance of survival because of natural selection, especially weaker animals that cannot develop their camouflage colors and therefore protect themselves from predators. Animals in which albinism has been recorded include deer, giraffes, squirrels, frogs, parrots, robins, turtles, trout, and lobsters. Partial albinism has also been reported in wildlife. In other cases, such as the black panther of Asia, too much melanin is formed and the disorder is called melanism. Albinism has also been observed in plants, but their life span rarely goes beyond the sprouting period, since starch, the main factor in plant metabolism and growth, is not able to form.

Impact and Applications

Albinism appears in various forms and may be passed to offspring through autosomal recessive, autosomal dominant, or X-linked modes of inheritance. In the autosomal recessive case, both parents of a child with autosomal recessive albinism carry the gene and transfer it equally to their sons and daughters. When both parents have it, there is a one in four chance that the child will inherit the condition. On the other hand, X-linked albinism occurs almost exclusively in males, and mothers who carry the gene will pass it on 50 percent of the time.

Albinism has not been found to affect expected life span among humans but has clearly modified these persons' lifestyles. Treatment of the disease involves reduction of the discomfort the sun creates. Thus photophobia may be relieved by sunglasses that filter ultraviolet light, while sunburn may be reduced by the use of sun protection factor (SPF) sunscreens and by covering the skin with clothing. Since albinism is a basically inherited condition, genetic counseling is of great value to individuals with a family history of albinism. Dopa reaction tests in hair bulbs of fetuses and electron microscopy of fetal skin have been successfully applied to the prenatal diagnosis of albinism.

—Soraya Ghayourmanesh

See Also: Hereditary Diseases; Inborn Errors of Metabolism; Natural Selection.

Further Reading: A thorough review on albinism was written and published by C. J. Witkop, Jr., in *Clinical Dermatology* 7 (1989). "Albinism" by the same author and his coworkers appears in *The Metabolic Basis of Inherited Disease* (1989), edited by Charles R. Scriver. The hair bulb tyrosinase test is described in *The Journal of the American Academy of Dermatology* 24 (1991) by R. Gershoni-Baruch et al.

Alcoholism

Field of study: Human genetics
Significance: *Since the discovery of the three-dimensional shape of deoxyribonucleic acid (DNA) in the 1950's, scientists have been describing genes for specific functions at an ever-increasing pace. Although tentative links have been established between certain genetic factors and the tendency to abuse alcohol, methodological problems have prevented the definitive identification of any particular gene as predisposing people to alcoholism.*

Key terms

DEOXYRIBONUCLEIC ACID (DNA): a macromolecule containing the genetic code or hereditary blueprint

GENE: a unit of heredity; a segment of DNA that contains the instructions to build a protein

PROTEIN: a macromolecule made of amino acids used for the function and structure of cells

ENZYME: a special protein that controls metabolism by regulating chemical reactions in cells

BEHAVIOR: action resulting from a combination of environmental and genetic factors

Fundamentals of Genetics

A gene is a sequence of approximately one thousand nitrogenous bases (adenine, guanine, cytosine, and thymine) along a stretch of a deoxyribonucleic acid (DNA) molecule. A single gene provides a cell with the instructions on how to make a specific protein. In other words, genes contain a code for building proteins. Each human cell uses the approximately 100,000 genes with which it is provided to make the approximately 50,000 proteins it needs to function. Each human has approximately 3 billion bases and about 100,000 genes, which means that 95 percent or so of the human genome does not code for proteins. Much of this 95 percent of human DNA is thought to be "junk," or not involved in a cell's daily activity, but much of this noncoding DNA is involved in the direct control and functioning of the genes themselves. One of the most common misconceptions about genes is that they have absolute control over a cell; actually, genes are ultimately under the control of other noncoding regions of DNA called "regulatory sequences." Likewise, these regulatory sequences often answer to cellular or environmental factors.

A mutation is simply a change in the original base sequence of a gene that may or may not alter the final protein to be built. Mutations occur in nature either by random or in response to radiation or chemical influences. Theoretically, no two organisms are genetically equivalent (with the exception of identical twins). These slight variations in genes occur from individual to individual, and there is thought to be considerable variation among humans in the 95 percent of the DNA that is noncoding. It makes sense that most of the variation among humans would be in the "junk" DNA, since the same basic set of genes are required, with modest variations, to build a functional human being.

Genes and Behavior

To investigate what links, if any, exist between an organism's genetic makeup and its behavior, it is important to remember that a gene makes a protein and that a protein may cause a specific or general biochemical response in a cell. The behavior of an animal is under the combined influence of genes and its environment. A good example is the mating seasons in many animals. As the number of daylight hours gradually increases toward spring and summer, a critical day length is reached that signals the release of hormones, which results in increased sexual activity and the beginning of the mating season. The production and activity of hormones involves genes or gene products. Until the critical number of daylight hours is reached, genes will not be activated and sexual behavior will not increase.

Each brain cell (neuron) that makes up the intricate networks and circuits throughout the cerebrum (80 percent of the human brain) has protein receptors that respond to specific signaling molecules (chemoreceptors). The production of the receptors and signaling molecules used for any type of brain activity is directly tied to genes. A slightly different gene may lead to a slightly different signaling molecule or receptor and thus a slightly different cell neuron response. A larger difference between genes may lead to a larger difference between signaling molecules or receptors and therefore a larger variation in cell response.

Since behavior involves the response of neurons and neuron networks to specific signals and since the response of neurons is caused by the interaction between a signaler and a receptor built by specific genes, the genetic link seems very straightforward: input, signal, response, behavior. However, when the slight variations between genes is added to the considerable variation among noncoding or regulatory sequences of DNA, the genetic connection to behavior becomes much less direct. Since a gene is under the control of one to several regulatory sequences that, in turn, may be under the control of various environmental inputs, the amount of genetic variation between organisms is compounded by two other critical factors: the environmental variations under which each brain develops and the daily environmental variations to which each brain is exposed.

A convenient way to think of genetics and behavior is to consider that genes simply allow humans to respond to a specific stimulus by

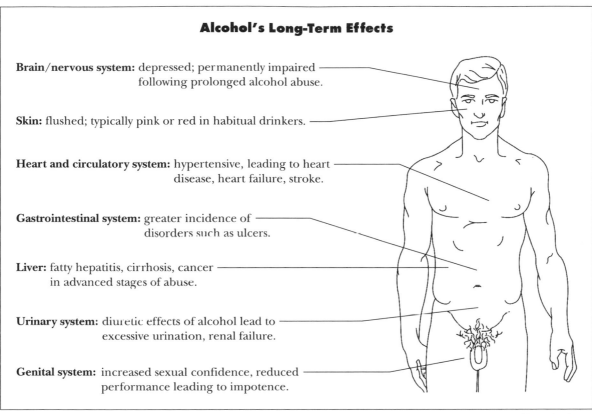

Alcohol's Long-Term Effects

Brain/nervous system: depressed; permanently impaired following prolonged alcohol abuse.

Skin: flushed; typically pink or red in habitual drinkers.

Heart and circulatory system: hypertensive, leading to heart disease, heart failure, stroke.

Gastrointestinal system: greater incidence of disorders such as ulcers.

Liver: fatty hepatitis, cirrhosis, cancer in advanced stages of abuse.

Urinary system: diuretic effects of alcohol lead to excessive urination, renal failure.

Genital system: increased sexual confidence, reduced performance leading to impotence.

The abuse of alcohol has extensive and serious effects on the body; unarrested, alcohol abuse will lead inevitably to death.

building the pathway required for a response, while behavior is the degree and manner of the response. With variations possible in genes, gene regulators, and the final cellular response, it is virtually impossible to disconnect the "nature versus nurture" tie that ultimately controls human behavior. Genes are simply the tools by which the environment shapes and reshapes human behavior. Although there is a direct correlation between genes and protein (if the gene is changed, the protein is also changed), there is no direct correlation between genes and behavior: Changing the gene does not necessarily change the behavior.

Alcoholism and Genetics

Years of studies have given little insight into what role genetics may play in the development of alcoholism. Some studies have shown that there may be a mutant form of a gene in the brain related to clearing alcohol from between brain cells. Alcoholics may have the mutant

gene that does not clear alcohol away as quickly as the normal gene; therefore, alcoholics may slowly become dependent on having a certain level of alcohol in the fluid between brain cells.

Part of the difficulty in genetic studies of humans is that scientists cannot artificially breed humans to produce specific offspring as they can with plants and animals. Studies involving sets of identical twins, fraternal twins, and adopted children have indicated that alcoholism may be anywhere from 40 to 60 percent genetic. This means that genes probably account for about one-half of an alcoholic's predisposition to drink excessively, but the environment plays an equally important role. Molecular genetic studies have identified some specific genes that may or may not play a role in the predisposition to drink, or more likely, in how people respond to alcohol in their blood. One renowned molecular geneticist has pointed out that it is difficult, and maybe impossible, to determine if one becomes an alcoholic because of genes or because of the exposure to alcohol use by one's family. One approach to the problem of genetics and alcoholism is to ask whether alcoholics have a mutant gene or set of genes that causes them to either crave alcohol or respond to alcohol in an abnormal way. Based on what is known about genetics, either (or perhaps both) scenario is possible, and the genes currently under investigation may prove to be related to this question. Another possibility is that alcoholics may simply have a mental or emotional disorder that causes them to seek or become attracted to antisocial or harmful behavior, which could also be caused by mutant genes.

Is there a genetic link to alcoholism? If so, how strong is this link? Could alcoholism be a behavior that people simply choose, even though they know it can lead to physical and mental damage, or is there an underlying gene or set of genes that makes it more difficult for some people to resist behaviors that they know will result in harm to themselves or others? These questions will only be answered when researchers gain a better understanding of human behavior and genetics.

—*W. W. Gearheart*

See Also: Classical Transmission Genetics; Eugenics; Genetic Code; Heredity and Environment.

Further Reading: *The Genetics of Alcoholism* (1995), edited by Henri Begleiter and Benjamin Kissin, is a thorough overview of the topic. *Alcohol and the Family* (1990), edited by R. Lorraine Collins et al., discusses the genetics of alcoholism in the context of the disease's effect on families. Cleamond D. Heskelson's "Hereditary Predisposition for Alcohol Abuse," in *Diagnosis of Alcohol Abuse* (1989), edited by Ronald R. Watson, provides much useful data on genetic markers that may be linked to alcohol abuse.

Allergies

Field of study: Immunogenetics
Significance: *In economically developed countries, allergies are responsible for a large portion of illnesses and medical expenses. Many allergies have genetic components and thus tend to "run" in families; the identification of such hereditary factors can help in diagnosis and in family planning. Moreover, research into the causes of allergies may lead to a more precise understanding of how the immune system functions. This may lead ultimately to the development of better drugs to treat allergies.*

Key terms

HYPERSENSITIVITY: an exaggerated response of the immune system to an antigen beyond what is considered "normal"; a synonym for allergy

IMMUNE SYSTEM: the defense mechanism of the body against foreign matter (bacteria, viruses, and parasites); it is composed of different types of cells and chemical substances

ANTIGEN: any substance that, when injected into the body, causes antibody formation that reacts specifically to that substance; also known as an allergen or an immunogen

ANTIBODY: a protein made by the body in response to an antigen; antibodies or immunoglobulins are specific for each antigen

The Basic Information About Allergies

Sneezing, sniffling, and wheezing are the symptoms most often associated with allergies. Allergies, or hypersensitivities, are the human body's exaggerated response to a foreign substance such as pollen. Hypersensitivity reactions can be immediate (hay fever) or delayed (contact dermatitis—for example, a reaction to latex or poison ivy) depending upon the body's immune reaction to the antigen.

Essentially, there are three stages of an allergic reaction. The first stage causes no symptoms. It is the immune system's initial contact with the antigen. The cells of the immune system react to the antigen by producing IgE antibodies that attach to mast cells and eosinophils (two cell types of the immune system) that are circulating in the blood. When the same antigen is encountered a second time and attaches to two adjacent IgE antibodies on a mast cell, the mast cell is said to be "activated." During this second stage, the mast cell releases chemical substances (such as histamines, prostaglandins, and leukotrienes) that are responsible for many of the common allergic symptoms. The third and final stage of an allergic reaction is the prolonged immune activity caused by the chemical substances released by cells of the immune system. This prolonged or late-phase reaction can cause the immune system to continue to react and cause tissue damage.

Based on varying responses to antigens, researchers Peter Gell and Robert Coombs have classified allergies into four types: I (anaphylaxis), II (cytotoxic), III (immune complex), and IV (cell-mediated). Type I hypersensitivity—anaphylaxis, from the Greek *ana* (against) and *phylaxis* (protection), or "the opposite of protected"—can be further divided into either systemic or local response. Systemic anaphylaxis is the whole body's response to an antigen such as a bee sting. Because of the amount of chemical substances released by the cells of the immune system, the body reacts immediately by a drop in blood pressure (leading to shock), difficulty in breathing, and swelling of the airways. If not treated immediately, anaphylactic shock can be fatal. Localized anaphylactic reactions (atopy) are the most familiar of the hypersensitivities. The symptoms are dependent upon the route the antigen uses to enter the body. For airborne antigens such as house dust, pollens, and animal dander, symptoms may include hay fever (itchy eyes, runny nose, sneezing, and coughing) or bronchial asthma (wheezing, coughing, and difficulty breathing). Other atopic symptoms may include hives, itchy skin, and diarrhea. Food allergies are also examples of an atopic reaction.

Type II (cytotoxic) hypersensitivity reaction involves the binding of an antigen and antibody complex to a cell that destroys the target cell. Examples of this type of hypersensitivity are incompatible blood groups (giving type B blood to a person who has type A blood), hemolytic anemia (destruction of red blood cells), and hemolytic disease of a newborn (the mother produces antibodies against the fetus based on a protein found in the blood).

Type III (immune complex) hypersensitivity reaction involves the depositing of immune complexes (an antigen bound to an antibody) on the walls of blood vessels, causing inflammation and tissue damage. Glomerulonephritis, inflammation of the blood vessels in the kidneys, is a type III hypersensitivity reaction. This disease is believed to be a reaction to a particular bacterial infection.

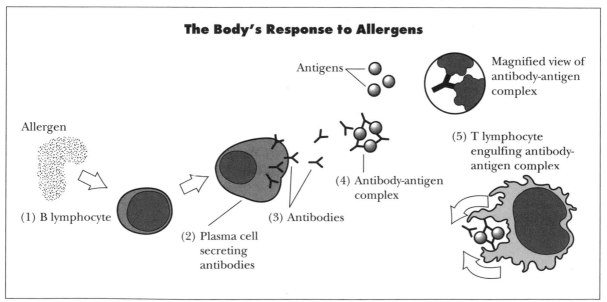

The Body's Response to Allergens

Antigens

Magnified view of antibody-antigen complex

Allergen

(5) T lymphocyte engulfing antibody-antigen complex

(4) Antibody-antigen complex

(1) B lymphocyte

(3) Antibodies

(2) Plasma cell secreting antibodies

An allergic reaction is caused when foreign material, or antigens, enter the immune system, which produces B lymphocytes (1) that cause blood plasma cells to secrete antibodies (2). The antibodies (3) link with antigens to form antigen-antibody complexes (4), which then are engulfed and destroyed by a T lymphocyte (5).

Contact dermatitis (for example, a reaction to poison ivy, latex, cosmetics, or jewelry) is a common example of the last category, type IV (cell-mediated) hypersensitivity. T cells (a cell type of the immune system) initially react with the antigen; upon a second exposure to the antigen, clones of the same T cell release chemical factors that cause a reaction to the antigen. This release of chemicals results in a red, itchy rash or hives. The reaction to *Mycobacterium tuberculosis* (the bacteria that causes tuberculosis) is a type IV hypersensitivity. The immune system has also been known to attack itself and cause disease. These disorders (autoimmune disorders), such as multiple sclerosis, juvenile diabetes, and systemic lupus erythematosus, are only beginning to be understood.

Impact and Applications

Treatment of allergies may include avoidance of the antigen, use of antihistamines (drugs that block the release of histamine from mast cells) and anti-inflammatories (steroids), and desensitization (allergy shots). Efforts by scientists to learn how the immune system functions and why it overreacts to antigens will lead to the development of better, less toxic drugs to combat allergies and their symptoms.

—*Mary Beth Ridenhour*

See Also: Antibodies; Autoimmune Disorders; Immunogenetics.

Further Reading: Benjamin A. Pierce's *Family Genetic Sourcebook* (1990) and Doris Teichler Zallen's *Does It Run in the Family? A Consumer's Guide to DNA Testing for Genetic Disorders* (1997) discuss genetic screening for allergies and many other hereditary ailments. *The Metabolic Basis of Inherited Disease* (1989), edited by Charles R. Scriver, is a technical introduction to the processes by which diseases are inherited. *Microbiology: An Introduction* (1998), by Gerard Tortora et al., gives a general overview of the human immune system and allergies. Lawrence Lichtenstein explains allergies in "Allergy and the Immune System," *Scientific American* 269 (September, 1993), while Lawrence Steinman describes what happens when the immune system develops an autoimmune disorder that causes it to see the body as foreign and attack itself in "Autoimmune Disease," *Scientific American* 269 (September, 1993).

Altruism

Field of study: Population genetics
Significance: *Altruism is behavior that enhances the fitness of a beneficiary at the expense of the actor's own fitness. Theorists have had difficulty explaining how altruistic behavior could have evolved by natural selection. The problem can be resolved with an expanded concept of fitness in which an altruistic individual can pass its own genes to the next generation by benefiting close relatives.*

Key terms

EUSOCIALITY: a social system in which some members of a colony (nonreproductive castes) do not reproduce but help raise offspring of other colony members (reproductive castes)

HAPLODIPLOIDY: a system of sex determination in which males are haploid and raised from unfertilized eggs whereas females are diploid and raised from fertilized eggs

INDIVIDUAL FITNESS: the extent to which an individual contributes genes to future generations through its own offspring

INCLUSIVE FITNESS: the extent to which an individual contributes genes to future generations through both its own offspring and the offspring of close relatives

The Problem of Altruism

In an evolutionary sense, "altruism" refers to any behavior that decreases the individual fitness of the actor while enhancing the individual fitness of one or more beneficiaries. The most striking examples of altruistic behavior in the animal kingdom are found in the eusocial (truly social) insects such as termites, ants, and many species of bees and wasps. In these insect species, sterile, nonreproductive castes (such as the "workers" in a honeybee hive) work to help raise the offspring of the reproductive caste (the "queen" in a honeybee hive). Natural selection will favor any gene that enhances the

The altruistic behaviors of honeybees and some other animal species pose puzzles for geneticists. (Ben Klaffke)

reproductive success of the individual bearing it, causing that gene to increase in frequency. However, it is problematic how any gene that predisposes its bearer to altruism could ever have evolved as a result of natural selection. English naturalist Charles Darwin realized the problem posed for his theory of natural selection by the social insects: "[W]ith the working ant we have an insect differing greatly from its parents, yet absolutely sterile; so that it could never have transmitted successively acquired modifications of structure or instinct to its progeny. It may well be asked how is it possible to reconcile this case with the theory of natural selection?"

Hamilton's Solution

The most widely accepted explanation of how altruistic behavior can evolve was developed in the 1960's by evolutionary geneticist William D. Hamilton. Hamilton reasoned that an individual can pass its genes to the next generation in two different ways: by producing offspring of its own or by helping closely related individuals raise offspring. An individual can spread its own genes through the offspring of close relatives because relatives are likely to share genes that are identical by descent from a common recent ancestor. Thus Hamilton distinguished between individual fitness (fitness derived from the individual's own offspring) and inclusive fitness (fitness derived both from the individual's own offspring and from nondescendant relatives). Natural selection occurring through inclusive fitness is called kin selection.

The coefficient of relatedness (r) measures the probability that a gene is identical by descent in two individuals. For example, in outbred diploids, r between parent and offspring is $\frac{1}{2}$; r between full siblings is $\frac{1}{2}$; r between half siblings is $\frac{1}{2}$; and r between first cousins is $\frac{1}{8}$. Hamilton used the coefficient of relatedness to specify the conditions under which a mutant gene predisposing its bearer to altruism would increase in frequency in a population: Where c

is the cost to the altruist (in terms of individual fitness) and b is the benefit to the recipient (again in terms of individual fitness), the gene will increase when $b/c > 1/r$. This condition is known as "Hamilton's rule."

Impact and Applications

Eusociality has evolved in only two orders of insects: Isoptera (termites) and Hymenoptera (including bees, wasps, and ants). However, eusociality has evolved several times independently in the Hymenoptera. The Hymenoptera have an unusual system of sex determination called haplodiploidy. Male Hymenoptera are haploid, being hatched from unfertilized eggs, whereas females are diploid and hatch from fertilized eggs. In this system, r between full sisters is $3/4$. Therefore, a female is more closely related to her sisters than she would be to her own offspring. In eusocial Hymenoptera, workers are always sterile females, and it makes good sense in terms of inclusive fitness for such females to help raise sisters as queens rather than producing their own offspring. Hamilton argued that this system of sex determination has predisposed the Hymenoptera toward evolving eusocial behavior.

So far, the genes controlling altruistic behavior have not been identified in any species; indeed, like most behavioral traits, altruistic behaviors are probably controlled by several loci. In the honeybee, there is evidence for heritable variation in some aspects of altruistic behavior. When a honeybee worker stings a mammalian intruder into the hive, the worker itself dies; thus aggressive defense of the hive is the ultimate act of altruism. It is known that genetically distinct strains of honeybee can differ markedly with respect to their aggressiveness toward intruders. The most dramatic example of such differences involves the so-called "Africanized" honeybees. These highly aggressive bees of African origin were introduced to Brazil in 1956, and in subsequent decades their offspring spread throughout the New World tropics, reaching the southern United States in the 1980's.

—*Austin L. Hughes*

See Also: Evolutionary Biology; Natural Selection; Sociobiology.

Further Reading: *The Ant and the Peacock* (1991), by Helena Cronin, provides a lively history of evolutionary biologists' attempts to explain the evolution of altruism. Edward O. Wilson's 1971 classic *The Insect Societies* remains the best general account of insect social behavior. Bert Holldobler and Wilson have together written a comprehensive description of ant societies in *The Ants* (1990).

Alzheimer's Disease

Field of study: Human genetics
Significance: *Alzheimer's disease is the leading cause of dementia in the elderly. Determining the underlying genetic factors of Alzheimer's disease will enable more accurate diagnosis, better understanding of the process, and more effective treatment of the disease.*

Key terms

DEOXYRIBONUCLEIC ACID (DNA): the genetic material that serves as the "instructions" for the growth and development of most organisms

MUTATION: any change in a DNA molecule that can be passed from one generation to another

GENE: a unit of a DNA molecule that codes for the production of a specific molecule such as a protein

Alzheimer's Disease Pathology

Alzheimer's disease (AD) is the most prevalent form of dementia in people more than forty years old. Although AD was first described by German neurologist Alois Alzheimer in 1907, the biological mechanisms of the disease are still poorly understood. AD is a slowly progressing neurodegenerative disorder that begins with subtle problems with short-term memory and gradually worsens to include loss of memory, loss of intellectual abilities, and loss of language skills. From the time of onset to death (typically from five to twenty years), the AD patient loses virtually all cognitive function. Studies of brain tissue from AD patients reveal two common features of the disease at the cellular level: the formation of tangles of proteins within neurons (specialized cells within the brain) and the presence of amyloid plaques in the brain. Amyloid is associated with the blood vessels of the brain; in AD patients, a short, "sticky" form of the protein creates aggregates or plaques within nervous tissue.

Alzheimer's disease can be divided into two major categories: early-onset AD and late-onset AD. Early-onset AD occurs prior to age sixty (as early as age thirty) and is inherited in an autosomal dominant fashion with 50 percent of children of an affected parent developing the disease. Late-onset AD includes those cases occurring after age sixty. No clear inheritance pattern is seen with late-onset cases; they are likely to result from a combination of genetic and environmental factors. All AD cases show the pathology of amyloid plaques and tangles regardless of the genetic and environmental background of the patient, indicating that a variety of biological disruptions may lead to Alzheimer's disease.

The study of AD is problematic because the behaviors associated with it are also associated with many other problems in the elderly such as overmedication and depression. Therefore, scientists have turned to the analysis of pedigrees and genetic mapping in an attempt to understand more about the events involved in AD pathogenesis. Genetic mapping involves locating a gene that is associated with a particular condition on a chromosome. Once a candidate gene is identified, scientists look for mutations that might be responsible for disrupting the gene's function. The normal function of genes containing mutations that cause a particular disease can be studied in a variety of ways, leading to an understanding of the disease process in question. Scientists have determined that mutations in any of several different genes can contribute to AD pathology.

The Amyloid Beta Precursor Gene

The first genetic alteration shown to be associated with AD was trisomy 21, the genetic defect causing Down syndrome, a condition in which a person's cells contain three copies of the human chromosome 21 instead of the normal two copies. The amyloid precursor protein gene, from which the sticky amyloid beta (Ab)

product is made, is located on chromosome 21. Down syndrome patients develop amyloid plaques that are similar to those in AD-affected brains. The amyloid precursor protein that is encoded by this gene can be made in many different forms by the cell, but only the Ab product, a truncated form of the protein, is associated with the AD plaques. Research of AD in non-trisomic 21 individuals led to the discovery of mutations in the amyloid precursor gene in people suffering from familial (inherited) Alzheimer's disease, indicating that these mutations can cause AD.

Several functions have been proposed for the amyloid precursor protein. One hypothesis suggests that it participates in wound healing and vascular injury repair, a second hypothesis links the protein with the adherence of neurons to their surrounding substrate, and a third proposes that the protein has growth-promoting activity. Disruption of any one of these normal activities could result in the neuronal cell loss and neurodegeneration associ-

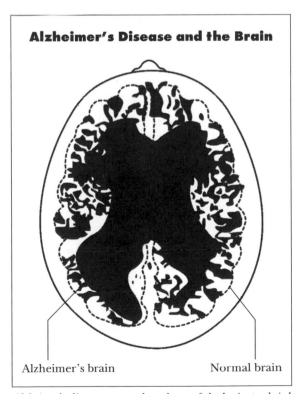

Alzheimer's Disease and the Brain

Alzheimer's brain   Normal brain

Alzheimer's disease causes the volume of the brain to shrink substantially.

ated with AD. The Ab product of the precursor protein, which accumulates in the plaques of AD brains, is present in small quantities in normal tissue. However, large amounts of Ab can induce neuronal cells to undergo a process called apoptosis or genetically programmed cell death. One reason for the cell loss seen in AD brain tissue may be that the accumulated Ab protein in the plaques causes the neurons in the brain to die by apoptosis.

The Apolipoprotein E Gene

A second gene that seems to be associated with an increased risk for Alzheimer's disease is the apolipoprotein E (ApoE) gene located on chromosome 19 in humans. Apolipoproteins transport fat-soluble molecules such as cholesterol through the blood stream. Although no mutations in this gene have been shown to directly cause AD, certain variants (polymorphisms) of this gene are associated with an increased risk for developing AD. Three different polymorphisms are common in human populations: e2, e3, and e4. Each person possesses two copies of the ApoE gene and may have any combination of these polymorphisms. People with two copies of the e4 form of the ApoE gene appear to be at highest risk for developing AD. The ApoE protein interacts with the pathogenic Ab product and associates with the tangles that are characteristic of AD. ApoE normally functions in neuronal injury response, transporting cholesterol (a molecule absolutely necessary for cell membrane integrity) to the injured tissue. Because of the normal role in chaperoning fat-soluble molecules such as cholesterol and the Ab protein, certain forms of ApoE that have increased affinity for these types of molecules may result in increased Ab deposition and thus contribute to AD pathology in predisposed individuals.

The Presenilin Genes

In June of 1995, researchers collaborating with Peter St. George-Hyslop at the University of Toronto announced the discovery of the presenilin-1 gene that, when mutated, results in familial early-onset Alzheimer's disease. The human gene implicated in early-onset AD is similar to two other genes (spe-4 and sel-12) cloned from the round worm, *Caenorhabditis elegans,* a model genetic system used by scientists to study fundamental processes of development. After 1995, a third presenilin gene was identified in *C. elegans*; a second human presenilin was cloned, and presenilins were cloned and sequenced from rats, mice, fruit flies, and frogs. Although these presenilin genes are found in many different organisms, suggesting a fundamental biological role for these proteins, the gene sequence reveals very little about their normal function in cells or how they might be important in Alzheimer's disease. In mice, the proteins produced by these genes are necessary for early development of the central nervous and skeletal systems because mice lacking normal presenilin genes show gross disruption of the central nervous system, spinal column, and ribs and die as embryos. The presenilins are thought possibly to be involved in the transport of proteins within cells. In addition to the two human presenilins, a third gene, presumably on chromosome 12 of humans, may also contribute to late-onset AD risk.

Impact and Applications

The advances in AD genetics have dramatic implications for the understanding and treatment of this devastating disease. Presenilin genes have been found in every animal model system examined, and their ubiquitous nature suggests a role in biological processes shared by all these organisms. Studying the roles of these genes in model systems is easier and less expensive than human genetics and will allow a more rapid understanding of their normal function and of how mutations in these genes cause AD. Also, genetic testing of the genes associated with AD will provide a powerful diagnostic tool, and specific genetic targets will enable the development of more effective pharmaceuticals for the disease.

—*P. Michele Arduengo*

See Also: Classical Transmission Genetics; Developmental Genetics; Down Syndrome; Human Genetics.

Further Reading: *Hannah's Heirs: The Quest for the Genetic Origins of Alzheimer's Disease* (1993), by Daniel A. Pollen, gives a wonderful account of the personal side of the science behind the identification of the presenilin gene locus by Peter St. George-Hyslop. Dennis J. Selkoe provides an easy-to-read summary of the genes implicated in AD in "Alzheimer's Disease: Genotypes, Phenotype, and Treatments," *Science* 275 (January 31, 1997). An interesting study of the verbal skills and cognitive abilities of elderly nuns is described by Sarah Richardson in "Alzheimer's Begins at 20," *Discover* (January, 1997). Jean Marx provides excellent commentary on the proposed roles of the presenilins in AD in "Dissecting How Presenilins Function and Malfunction," *Science* 274 (December 13, 1996).

Amniocentesis and Chorionic Villus Sampling

Field of study: Human genetics

Significance: *Amniocentesis and chorionic villus sampling continue to assume ever-increasing roles as prenatal diagnostic tools. The number of genetic disorders that can be detected prenatally has widened the possibilities for more effective genetic counseling. Parents who are at risk for having a child with a birth defect have more opportunities of having a healthy child.*

Key terms

AMNIOCENTESIS: a procedure for removing amniotic fluid and fetal cells around the sixteenth week of pregnancy in order to obtain information about the chromosomes and genetic makeup of the fetus

AMNIOTIC FLUID: the fluid in which the fetus is immersed during pregnancy

CHORIONIC VILLUS SAMPLING (CVS): a procedure usually carried out between the ninth and twelfth weeks of pregnancy to obtain fetal cells from placental tissue

PRENATAL TESTING: testing that is done during pregnancy to examine the chromosomes or genes of a fetus to detect the presence or absence of a genetic disorder

Prenatal Testing

The goal of prenatal testing is to provide at-risk families with information about the chances of having a child with a specific genetic disorder or birth defect. Only a small minority of such disorders can now be detected, but the list continues to grow.

The primary techniques for prenatal testing include amniocentesis, chorionic villus sampling (CVS), ultrasonography (in which high-frequency sound waves are used to "view" the fetus and obtain information about its position and structure), fetoscopy (a procedure that utilizes a fiber-optic instrument to obtain a direct image of the fetus), fetal blood sampling (in which blood cells of the fetus are obtained by inserting a needle directly into the umbilical cord), and screening for alpha fetoprotein (a fetal protein found in amniotic fluid, high levels of which may indicate the presence of neural tube defects).

Amniocentesis and chorionic villus sampling are not recommended for every pregnancy. Although the two procedures are relatively safe, they do not carry a zero risk factor and are not likely to be employed unless the risk of the procedure is lower than the risk factor for a birth defect in a specific pregnancy. The general risk for having a child with a significant birth defect is about 2 to 3 percent. Among the factors that indicate an increased risk of having a child with a birth defect are maternal age (the incidence of chromosomal defects in children increases sharply in pregnant women over age thirty-five), a previous child with a known chromosomal or genetic disorder (assuming the genetic disorder can be detected prenatally), previous problems with spontaneous abortions or miscarriages, a history of a structural chromosomal abnormality in one of the parents, a previous child with a neural tube defect, and a closely related couple.

Amniocentesis

Amniocentesis has been used safely and widely since 1967 and is used more often than other methods of prenatal testing. The procedure is usually performed on an outpatient basis between the fourteenth and eighteenth week of gestation. By this stage in the pregnancy, the volume of amniotic fluid is large enough to get an adequate sample. Also, it

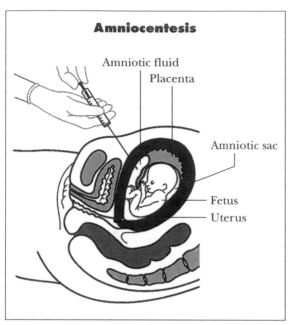

Amniocentesis

Amniotic fluid

Placenta

Amniotic sac

Fetus

Uterus

Removal and analysis of fluid from the amniotic sac that surrounds a fetus during gestation can be used to rule out or confirm the presence of serious birth defects or genetic diseases.

allows sufficient time for testing to be done in the laboratory, minimizing complications if a therapeutic abortion is performed.

The skin of the abdomen is scrubbed, and a topical anesthetic may be applied. The exact location of the placenta and fetus are found by ultrasound. A long, thin needle is inserted through the abdominal wall into the amniotic sac that encloses the fetus. A small amount of amniotic fluid is withdrawn. This fluid contains cells that have been sloughed off by the fetus. There are not enough cells to analyze, so cells must be grown in the laboratory to produce a sufficient number for testing. A variety of chromosomal analyses and biochemical tests can then be carried out. Any sort of numerical chromosomal abnormality, such as Turner's syndrome and Down syndrome, can be detected. Structural abnormalities of chromosomes, missing or extra pieces, also can be detected. Cri du chat syndrome is one such genetic disorder. Biochemical and deoxyribonucleic acid (DNA) analyses can be carried out on the cells, and some specific genetic disorders can be detected in this manner. The biochemical assays are used to detect low levels of

particular enzymes involved in specific biochemical defects. Although most genetic disorders cannot be diagnosed in this manner, the list is rapidly growing. Examples of some inborn errors are galactosemia, Hurler's syndrome, and Lesch-Nyhan syndrome.

It is also possible to analyze DNA directly in cells obtained from amniocentesis. Using specific genetic probes, it is possible to identify mutant genes associated with specific genetic diseases such as sickle-cell anemia, cystic fibrosis, hemophilia A, and Duchenne muscular dystrophy. This approach is not yet possible for many genes since it is necessary to know the DNA sequence of the gene involved.

Chorionic Villus Sampling

Although amniocentesis has been a successful prenatal testing procedure, it does present some disadvantages. Perhaps the major disadvantage involves the need to perform it during the sixteenth week of pregnancy, which provides a fairly narrow window in which cells can be grown in culture, tests can be carried out, and procedures can be replicated, if necessary. If procedures run past the nineteenth or twentieth week of pregnancy, the physical and psychological complications associated with a late termination of pregnancy rise considerably. The technique of chorionic villus sampling addresses some of these problems.

Like amniocentesis, chorionic villus sampling can be performed on an outpatient basis. After cleansing the vagina and cervix with an antiseptic, the physician uses ultrasound to guide the insertion of a catheter (a small, thin tube) through the cervix into the uterus. The catheter is placed in contact with the placenta, where the chorionic villi are located, and gentle suction is used to remove a small sample. Villus cells are produced by the fetus and comprise one of the outer layers of the placenta. Of major importance is the fact that this small sample of tissue contains millions of cells that can be used immediately for testing (recall that cells obtained during amniocentesis must be grown for a week or longer before testing can be done). This means that chromosomal analyses and some biochemical tests can be performed and results given to the patient before she leaves the physician's office. Chorionic villus sampling is usually performed between the ninth and twelfth weeks of pregnancy so that complete results are likely to be reported one month earlier than is usually the case with amniocentesis. A termination of pregnancy after chorionic villus sampling is expected to have fewer complications than a termination performed at a later stage in pregnancy.

Impact and Applications

Both amniocentesis and chorionic villus sampling provide significant information to couples at risk for having a child with a genetic disorder or other type of birth defect. It is estimated that approximately one-half of the women over the age of thirty-five who are pregnant utilize amniocentesis or chorionic villus sampling. Thousands of women undergo some form of prenatal genetic testing each year, and studies have determined the overall safety of the procedures. Amniocentesis increases the risk of spontaneous abortion only about 0.5 percent above the overall general risk. Chorionic villus sampling probably carries an increased risk of miscarriage of 1 to 2 percent above the overall general risk. It should be kept in mind, however, that these two techniques are usually not performed unless there is some additional risk already present in a specific pregnancy.

The techniques of amniocentesis and chorionic villus sampling may assure parents at risk of having a child with a genetic disorder that their child will be born without the disorder. Results may also be such that parents must be told that their child will definitely have a certain disorder. However, even if a couple elects to continue with a pregnancy, useful information is provided about the nature of the disorder and about treatments that might be used after birth to equip the parents to better prepare for raising a child with a birth defect. It is to be expected that further developments in techniques will dramatically improve means of prenatal testing.

—Donald J. Nash

See Also: Genetic Counseling; Genetic Screening; Genetic Testing; Prenatal Diagnosis.

Further Reading: *Human Heredity* (1998), by Michael R. Cummings, is a good introduction to human genetics for nonspecialists. *The Family Genetic Sourcebook* (1990), by Benjamin A. Pierce, gives basic information about many genetic disorders and has a good chapter on genetic counseling and prenatal testing. Doris Teichler Zallen's book, *Does It Run in the Family?* (1997), provides much information about the biological, social, and ethical issues involved in genetic testing.

Antibodies

Field of study: Immunogenetics
Significance: *Antibodies provide the main line of defense (immunity) in all vertebrates against infections caused by bacteria, fungi, viruses, or other foreign agents. Antibodies are used as therapeutic agents to prevent specific diseases and to identify the presence of antigens in a wide range of diagnostic procedures. Large quantities of antibodies have also been produced in plants for use in human and plant immunotherapy. Because of their importance to human and animal health, antibodies are widely studied by geneticists seeking improved methods of antibody production.*

Key terms

B CELLS: a class of white blood cells (lymphocytes) derived from bone marrow responsible for antibody-directed immunity

B MEMORY CELLS: descendants of activated B cells that are long-lived and that synthesize large amounts of antibodies in response to a subsequent exposure to the antigen, thus playing an important role in secondary immunity

HELPER T CELLS: a class of white blood cells (lymphocytes) derived from bone marrow that prompts the production of antibodies by B cells in the presence of an antigen

LYMPHOCYTES: types of white blood cells (including B cells and T cells) that provide immunity

PLASMA CELLS: descendants of activated B cells that synthesize and secrete a single antibody type in large quantities and also play an important role in primary immunity

Antibody Structure

Antibodies are made up of a class of proteins called immunoglobulins (Ig) produced by plasma cells (descendants of activated B cells) in response to a specific foreign molecule known as an antigen. Most antigens are also proteins or proteins combined with sugars. Antibodies recognize, bind to, and inactivate antigens that have been introduced into an organism by various pathogens such as bacteria, fungi, and viruses.

The simplest form of antibody molecule is a *Y*-shaped structure with two identical, long polypeptides (substances made up of many amino acids joined by chemical bonds) referred to as "heavy chains" and two identical, short polypeptides referred to as "light chains." These chains are held together by chemical bonds. The lower portion of each chain has a constant region made up of similar amino acids in all antibody molecules, even among different species. The remaining upper portion of each chain, known as the "variable region," differs in its amino acid sequence between antibodies. The three-dimensional shape of the tips of the variable region (antigen-binding site) allows for the recognition and binding of target molecules (antigens). The high-affinity binding between antibody and antigen results from a combination of hydrophobic, ionic, and van der Waals forces. Antigen binding sites have specific points of attachment on the antigen that are referred as "epitopes" or "antigenic determinants."

Antibody Diversity

There are five classes of antibodies (IgG, IgM, IgD, IgA, and IgE), with each having a distinct structure, size, and function (see table below). IgG is the principal immunoglobulin and constitutes up to 80 percent of all antibodies in the serum.

The human body can manufacture a limitless number of antibodies, each of which can bind to a different antigen; however, human genomes have a limited number of genes that code for antibodies. It has been proposed that random recombination of deoxyribonucleic acid (DNA) segments is responsible for antibody variability. For example, one class of genes (encoding light chain) contains three regions: the L-V (leader-variable) region (in which each variable region is separated by a leader sequence), the J (joining) region, and the C (constant) region. In the embryonic B cells, each gene consists of from one hundred

---

Classes, Locations, and Functions of Antibodies

Class	Location	Functions
IgG	Blood plasma, tissue fluid, fetuses	Produces primary and secondary immune responses; protects against bacteria, viruses, and toxins; passes through the placenta and enters fetal bloodstream, thus providing protection to fetuses.
IgM	Blood plasma	Acts as a B-cell surface receptor for antigens; fights bacteria in primary immune response; powerful agglutinating agent; includes anti-A and anti-B antibodies.
IgD	Surface of B cells	Prompts B cells to make antibodies (especially in infants).
IgA	Saliva, milk, urine, tears, respiratory and digestive systems	Protects surface linings of epithelial cells, digestive, respiratory, and urinary systems.
IgE	In secretion with IgA, skin, tonsils, respiratory and digestive systems	Acts as receptor for antigens causing mast cells (often found in connective tissues surrounding blood vessels), to secrete allergy mediators; excessive production causes allergic reactions (including hay fever and asthma).

to three hundred L-V regions, approximately six J regions, and one C region. These segments are widely separated on the chromosome. As the B cells mature, one of the L-V regions is randomly joined to one of the J regions and the adjacent C region by a recombination event. The remaining segments are cut from the chromosome and subsequently destroyed, resulting in a fusion gene encoding a specific light chain of an antibody. In mature B cells, this gene is then transcribed and translated into polypeptides that form a light chain of an antibody molecule. Genes for the other class of light chains as well as heavy chains are also made up of regions that undergo recombination during B-cell maturation. These random recombination events in each B cell during maturation lead to the production of billions of different antibody molecules. Each B cell has, however, been genetically programmed to produce only one of the many possible variants of the same antibody.

Production of Antibodies: Immune Response

Immunity is a state of bodily resistance brought about by the production of antibodies against an invasion by a foreign agent (antigen). The immune response is mediated by white blood cells known as lymphocytes that are made in the bone marrow. There are two types of lymphocytes: T cells, which are formed when lymphocytes migrate to the thymus gland, circulate in the blood, and become associated with lymph nodes and the spleen; and B cells, which are formed in bone marrow and move directly to the circulatory and the lymph systems. B cells are genetically programmed to produce antibodies. Each B cell synthesizes and secretes only one type of antibody, which has the ability to recognize with high affinity a discrete region (epitope or antigenic determinant) of an antigen. Generally, an antigen has several different epitopes, and each B cell produces a set of different antibodies corresponding to one of the many epitopes of the same antigen. All of the antibodies in this set, referred to as "polyclonal" antibodies, react with the same antigen.

The immune system is more effective at con-

trolling infections than the nonspecific defense response and has three characteristic responses to antigens: diverse (effectively neutralizes or destroys various foreign invaders, whether they are microbes, chemicals, dust, or pollen), specific (effectively differentiates between harmful and harmless antigens), and anamnestic (which has a memory component that remembers and responds faster to a subsequent encounter with an antigen). The primary immune response involves the first combat with antigens, while the secondary immune response includes the memory component of a first assault. This is why humans get some diseases only once (such as chicken pox); other infections (such as cold and influenza) often recur because the causative viruses mutate, thus presenting a different antigenic face to the immune system each season.

An antibody-mediated immune response involves several stages: detection of antigens, activation of helper T cells, and antibody production by B cells. White blood cells known as "macrophages" continuously wander through the circulatory system and the interstitial spaces between cells searching for antigen molecules. Once an antigen is encountered, the invading molecule is engulfed and ingested by a macrophage. Helper T cells become activated by coming in contact with the antigen on the macrophage. In turn, an activated helper T cell identifies and activates a B cell. The activated T cells release cytokines (a biochemical) that prompt the activated B cell to divide. Immediately, the activated B cell generates two types of daughter cells: plasma cells (which synthesize and release approximately two thousand to twenty thousand antibody molecules per second into the bloodstream during its life span of four to five days) and B memory cells (which have a life span of a few months to a year). The B memory cells are the component of the immune memory system that, in response to a second exposure to the same type of antigen, produces antibodies in larger quantities and at faster rates over a longer time frame than the primary immune response. A similar cascade of events occurs when a macrophage presents an antigen directly to a B cell.

Polyclonal and Monoclonal Antibodies

Plasma cells originating from different B cells manufacture distinct antibody molecules because each B cell was presented with a specific portion of the same antigen by a helper T cell or macrophage. Thus a set of polyclonal antibodies is released in response to an invasion by a foreign agent. Each member of this group of polyclonal antibodies will launch the assault against the foreign agent by recognizing different epitopes of the same antigen. The polyclonal nature of antibodies has been well recognized in the medical field.

In the case of multiple myeloma (a type of cancer), one B cell out of billions in the body proliferates in an uncontrolled manner. Eventually, this event compromises the total population of B cells of the body. The immune system will produce huge amounts of IgG originating from the same B cell, which recognizes only one specific epitope of an antigen; therefore, this person's immune system produces a set of antibodies referred to as "monoclonal" antibodies. Monoclonal antibodies form a population of identical antibodies that all recognize and are specific for one epitope on an antigen. Thus, someone with this condition may suffer frequent bacterial infections because of a lack of antibody diversity. Indeed, a bacterium whose antigens do not match the antibodies manufactured by the overabundant monoclonal B cells has a selective advantage.

The high-affinity binding capacity of antibodies with antigens has been employed in both therapeutic and diagnostic procedures. It is, however, unfortunate that the effectiveness of commercial preparations of polyclonal antibodies varies widely from batch to batch. In some instances of immunization, certain epitopes of a particular antigen are strong stimulators of antibody-producing cells, whereas at other times, the immune system responds more vigorously to different epitopes of the same antigen. Thus one batch of polyclonal antibodies may have a low level of antibody molecules directed against a major epitope and not be as effective as the previous batch. Consequently, it is desirable to produce a cell line that will produce monoclonal antibodies with a high affinity for a specific epitope on the antigen for commercial use. Such a cell line would provide a consistent and continual supply of identical (monoclonal) antibodies. Monoclonal antibodies can be produced by hybridoma cells, which are generated by the fusion of cancerous B cells and normal spleen cells obtained from mice immunized with a specific antigen. After initial selection of hybridoma clones, monoclonal antibody production is maintained in culture. In addition, the hybridoma cells can be injected into mice to induce tumors that, in turn, will release large quantities of fluid containing the antibody. This fluid containing monoclonal antibodies can be collected periodically and may be used immediately or stored for future use. Various systems used to produce monoclonal antibodies include cultured lymphoid cell lines, yeast cells, *Trichoderma reese* (ascomycetes), insect cells, *Escherichia coli*, and monkey and Chinese hamster ovary cells. Transgenic plants and plant cell cultures have been explored as potential systems for antibody expression.

Impact and Applications

The high-affinity binding capacity of antibodies may be used to inactivate antigens in vivo (within a living organism). The binding property of antibodies may also be employed in many therapeutic and diagnostic applications. In addition, it is a very effective tool in both immunological isolation and detection methods.

Monoclonal antibodies may outnumber all other products being explored by various biotechnology-oriented companies for the treatment and prevention of disease. For example, many strategies for the treatment of cancerous tumors as well as for the inhibition of human immunodeficiency virus (HIV) replication are based on the use of monoclonal antibodies. HIV is a retrovirus (genetic material is ribonucleic acid, or RNA) that causes acquired immunodeficiency syndrome (AIDS). Advances in plant biotechnology have made it possible to use transgenic plants to produce monoclonal antibodies on a large scale for therapeutic or diagnostic use. Indeed, one of the most promising applications of plant-produced antibodies in immunotherapy is in passive immunization (for example, against *Streptococcus mutans,* the most common cause of tooth decay). Large doses of the antibody are required in multiple applications for passive immunotherapy to be effective. Transgenic antibody-producing plants may be one source that can supply huge quantities of antibodies in a safe and cost-effective manner. It has been demonstrated that a hybrid IgA-IgG molecule produced by transgenic plants prevented colonization of *Streptococcus mutans* in culture, which appears to be how the antibody prevents colonization of this bacterium in vivo.

It has been estimated that antibodies expressed in soybeans at a level of 1 percent of total protein may cost approximately one hundred dollars per kilogram of antibody, which is relatively inexpensive in comparison with the cost of traditional antibiotics. Transgenic plants have also been used as bioreactors for the large-scale production of antibodies with no extensive purification schemes. In fact, antibodies have been expressed in transgenic tobacco roots and then accumulated in tobacco seeds. If this technology could be employed to obtain stable accumulation of antibodies in more edible plant organs such as potato tubers, it could potentially allow for long-term storage as well as a safe and easy vector of delivery of specific antibodies for immunotherapeutic applications. In addition, plant-produced antibodies may be more desirable for human use than microbial-produced antibodies because plant-produced antibodies undergo eukaryotic rather than the prokaryotic (bacterial) post-translational modifications. Human glycosylation (a biochemical process whereby sugars are attached onto the protein) is more closely related to that of plants than that of bacteria.

The potential use of antibody expression in plants for altering existing biochemical pathways has also been demonstrated. For example, germination mediated by phytochrome (a biochemical produced by plants) has been altered by utilizing plant-produced antibodies. In addition, antibodies expressed in plants have been successfully used to immunize host plants against pathogenic infection; for example, tobacco plants have already been immunized with antibodies against viral attack. This approach has great potential to replace the traditional methods (use of chemicals) in controlling pathogens.

—Sibdas Ghosh
—Tom E. Scola

See Also: Hybridomas and Monoclonal Antibodies; Immunogenetics; Synthetic Antibodies.

Further Reading: *Antibodies: A Laboratory Manual* (1988), edited by Ed Harlow and David Lane, provides a detailed account of different methods involved in production and application of antibodies. Mathew D. Smith, "Antibody Production in Plants," *Biotechnology Advances* 14 (1996), summarizes production and applications of plant-produced antibodies. *Molecular Biotechnology: Principles and Applications of Recombinant DNA* (1994), edited by Bernard R. Glick and Jack J. Pasternak, discusses structure and function of antibodies as well as the role of biotechnology in the use of antibodies.

Archaebacteria

Field of study: Bacterial genetics
Significance: *Archaebacteria are a diverse group of microorganisms found in environments once considered to be inhospitable to life. Archaebacteria are prokaryotes, but biochemical and genetic studies have indicated that these organisms may be closely related to an ancestor that gave rise to both bacteria and eukaryotes. Archaebacteria thus provide some insight into evolutionary processes, and their unique biological properties have yielded numerous substances for biotechnological applications.*

Key terms

ENZYME: a molecule, usually protein, composed of amino acids that assists and accelerates cellular reactions without itself being altered by the reactions

EUKARYOTE: a complex cell type characterized by having many internal compartments, each enveloped by membranes, including the nucleus, which harbors the cell's chromosomes

PROKARYOTE: a simple cell type characterized by the absence of a nucleus and any other internal, membrane-bound compartments

RIBONUCLEIC ACID (RNA): a form of genetic material consisting of units known as ribonucleotides used for the synthesis of proteins

RIBOSOME: a structure within the cell used to produce proteins

Archaebacteria: A Third "Kingdom" of Life

For many years, biologists have categorized living organisms into two groups based on their cellular organization and complexity. The single-celled organisms whose chromosomes are not compartmentalized inside a nucleus have been classified as prokaryotic bacteria. All other organisms containing a nucleus have been classified as eukaryotes. In the late 1970's, however, studies on a unique group of microorganisms led investigators to believe that the two-domain classification ignored the tremendous diversity of the prokaryotes. These microorganisms, called archaebacteria by molecular biologist Carl Woese and his colleagues in 1977, live in habitats with extremes in heat, pressure, and salinity, and many are able to utilize sulfur and molecular hydrogen as part of their growth process. Like all prokaryotes, archaebacteria do not have a nucleus. However, in their biochemistry and the structure and composition of certain large molecules, they are as different from prokaryotes as they are from eukaryotes. Woese and his colleagues analyzed and compared specific molecules of ribonucleic acid (RNA) present within the ribosome in all organisms (ribosomal RNA). Their findings suggested that all extant life is composed of three distinct groups of organisms: the eukaryotes and two different prokaryotes (bacteria and archaebacteria). In 1990, Woese and others recommended the replacement of a two-domain view of life with a new tripartite scheme based on three kingdoms or domains: Bacteria (eubacteria or "true" bacteria), Archaea (archaebacteria), and Eucarya (eukaryotes). After 1990, the three-domain classification became the subject of considerable debate, and, as a consequence, both old and new terminology began to be used in scientific and popular literature.

Characteristics of Archaebacteria

Generally, the size and shape of archaebacterial cells are similar to those of bacteria. They are single-celled microscopic organisms that, in some cases, are motile and may be found in chains or clusters. They multiply in the same manner as bacteria—that is, via binary fission, budding, or fragmentation. Like bacteria, archaebacterial chromosomes are circular, indicating the absence of breaks or discontinuities, and many archaebacterial genes are organized in the same fashion as those found in bacteria. On the other hand, the specific chemical composition of archaebacterial carbohydrate, lipid, and protein polymers that form the structural barrier to the external environment is uniquely archaebacterial and is quite different from the composition of those structures typically found in either bacteria or eukaryotes.

The characteristics that define the archaebacterial class of organisms and make it distinct from both bacteria and eukaryotes can be seen in the composition of large molecules. For example, the molecule involved in protein synthesis, namely the ribosome, resembles the bacterial ribosome in shape and composition and is distinct from that present in eukaryotes. However, the RNA component of the ribosome as well as certain RNAs used by the ribosome during the protein synthesis process (transfer RNAs) contain specific characteristics that are unique to archaebacteria. On the other hand, the enzyme utilized by the archaebacteria in the production of RNA, namely RNA polymerase, is quite different from the enzyme

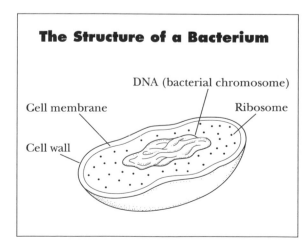

The Structure of a Bacterium

DNA (bacterial chromosome)

Cell membrane

Ribosome

Cell wall

The extreme salinity of the Dead Sea that allows bathers to float easily also provides an ideal environment for archaebacteria. (Warren Lieb/Archive Photos)

found in bacteria. Bacterial RNA polymerase is composed of four major proteins; in archaebacteria, RNA polymerase consists of as many as ten proteins and is remarkably similar to the enzyme found in eukaryotes. In fact, archaebacterial RNA polymerase is so similar to the eukaryotic enzyme that combining certain proteins from both archaebacterial and eukaryotic sources results in a functional enzyme, a manipulation that is not possible with any bacterial RNA polymerase proteins.

Among species of the archaebacteria, there are a variety of metabolic processes that differ greatly from the better-known metabolic routes of bacteria and eukaryotes. Many of the archaebacterial pathways utilized to convert food sources to energy and building blocks for growth involve enzymes having biological activities not found in any other biological systems. In some cases, the enzymes are novel and require the involvement of unusual metals such as tungsten. While a requirement for metals in the activity of many bacterial and eukaryotic enzymes is ubiquitous, the utilization of tungsten appears to be unique to archaebacteria.

Archaebacterial Diversity

A fascinating feature of archaebacteria (and a major reason for their discovery) is the diversity of the environments in which they thrive. Archaebacteria encompass three basic identifiable types: methanogenic (methane forming), sulfur dependent (requiring sulfur for growth), and halophilic (salt loving).

The methanogenic archaebacteria are found in strictly anaerobic environments. That is, they have absolutely no tolerance for oxygen—trace amounts inhibit growth, and too much is lethal. To obtain energy for growth, these archaebacteria carry out a process called methanogenesis, which involves the conversion of carbon dioxide to methane gas. Microorganisms that produce methane have been known for centuries. In 1776, Italian scientist

Alessandro Volta demonstrated that air generated from bogs, streams, and lakes whose sediments are rich in decaying vegetation can be ignited. It is now known that the microorganisms responsible for generating this "marsh gas" are methanogenic archaebacteria. Because methanogens require an oxygen-free environment for growth, they are found only where carbon dioxide and hydrogen are available and oxygen has been excluded. Thus methanogens thrive in stagnant water, sewage treatment plants, and the rumen of cattle and other ruminants as well as the intestinal tracts of animals. Methanogens are also found in hot springs and the deep ocean.

The sulfur-dependent archaebacteria are generally thermophilic (heat loving) and live in environments ranging in temperature from 55 degrees Celsius (131 degrees Fahrenheit) to greater than the boiling point of water, as high as 113 degrees Celsius (235 degrees Fahrenheit). They have been detected near hot sulfur springs, in sulfur-laden mud at the base of volcanoes, and near very hot deep-sea hydrothermal vents where hot water is emitted at much higher temperatures. Species that can utilize oxygen as well as those that have no tolerance for oxygen are known. In addition, many thermophilic archaebacteria are found in environments that are extremely acidic.

The halophilic archaebacteria require extremely high concentrations of salt for survival, and some grow readily in saturated brine. Halophiles grow in salty habitats along ocean borders and inland waters such as the Dead Sea between Israel and Jordan and the Great Salt Lake in Utah. The red color observed in salt evaporation ponds is caused by halophiles. Interestingly, some halophilic archaebacteria have the ability to carry out photosynthesis (the harvesting of light to provide energy for growth) using a pigment that is remarkably like one associated with vision.

Impact and Applications

The extreme conditions in which archaebacteria are found suggest that these organisms have adapted to environments thought to exist three to four billion years ago during early life on Earth. Thus the archaebacteria may be con-sidered a window into the past, and, in the early 1980's, investigators began studying them in order to shed light on the processes involved in evolution as well as relationships among these organisms, bacteria, and eukaryotes. In order to survive in their unique environments, archaebacteria possess molecules that withstand heat, acids, and salt, characteristics that are tailor-made for applications in molecular biology and biotechnology. Such applications have included the use of heat-stable enzymes for analyses used in genetic fingerprinting and cancer detection; the use of halophilic pigments for holographic applications, optical signal processing, and photoelectric devices; and the use of methanogenesis as an alternative fuel source. The understanding that the archaebacteria are a unique domain of organisms has opened up new areas of both basic and applied research and has held promise for allowing scientists to develop new applications to solve difficult problems in the extended biological world.

—Harold J. Schreier

See Also: Bacterial Genetics and Bacteria Structure; Biotechnology; Methane-Producing Bacteria; Polymerase Chain Reaction.

Further Reading: Michael T. Madigan et al., *Brock Biology of Microorganisms* (1997), includes several excellent chapters on the ecology and biology of archaebacteria. An overview of halophilic microorganisms can be found in Russel H. Vreeland et al., *The Biology of Halophilic Bacteria* (1993). James G. Ferry, *Methanogenesis: Ecology, Physiology, Biochemistry, and Genetics*, 1993, provides an excellent overview of methanogenic archaebacteria. A balanced coverage of halophiles, methanogens, and thermophiles can be found in Michael J. Danson et al., *Archaebacteria: Biochemistry and Biotechnology* (1992).

Artificial Selection

Field of study: Population genetics
Significance: *Artificial selection is the process through which humans have domesticated and improved plants and animals. It continues to be*

the primary means by which agriculturally important plants and animals are modified to improve their desirability. However, artificial selection is also a threat to genetic diversity of agricultural organisms as uniform and productive strains replace the many diverse, locally produced varieties that once existed around the globe.

Key terms

HERITABILITY: a proportional measure of the extent to which differences among organisms within a population for a particular character result from genetic rather than environmental causes (a measure of nature versus nurture)

GENETIC VARIATION: a measure of the availability of genetic differences within a population upon which artificial selection has potential to act

GENETIC MERIT: a measure of the ability of the parent to contribute favorable characteristics to its progeny

The Process of Selection

Selection is a process through which organisms with particular genetic characteristics leave more offspring than those organisms with alternative genetic forms. The process may occur because the genetic characteristics confer upon the organism a better ability to survive or reproduce (natural selection), or it may be caused by an act of human will (artificial selection). Natural and artificial selection may act in concert, as when a genetic characteristic confers a disadvantage directly to the organism. Dwarfism in cattle, for example, not only directly reduces the survival of the affected individuals but also reduces the value of the animal to the breeder. Conversely, natural selection may act in opposition to artificial selection. For example, a genetic characteristic that results in the seed being held tightly in the head of a grass such as wheat is an advantage to the farmer, as it makes harvesting easier. That same characteristic would be a disadvantage to wild wheat because it would impair the process of seed dissemination.

Initial Applications of Selection

The first intentional applications of artificial selection were likely conducted by early farmers who identified forms of crop plants that had characteristics that favored cultivation. Seeds from favored plants were preferentially kept for replanting. Any characteristics that were to some degree heritable would have had the tendency to be passed on to the progeny through the selected seeds. Some of the favored characteristics may have been controlled by a single gene and easily made permanent in the populations. Other favored characteristics may have been controlled by a large number of genes with individually small effects. Nevertheless, seeds selected from the best plants would tend to produce offspring that were better than average, resulting in gradual improvement in the population. It would not have been necessary to have knowledge of the mechanisms of genetics to realize the favorable effects of selection. Likewise, individuals that domesticated the first animals for their own use would have made use of selection to capture desirable characteristics within their herds and flocks. The first of those characteristics was probably docile behavior, a trait known to be heritable in present livestock populations.

Enhancements from Technology

Technology to improve organisms through selective breeding preceded the understanding of the genetic basis for the effectiveness of selection. Recording of pedigrees and performance records began with the formal development of livestock breeds in the 1700's. Some breeders, notably Robert Bakewell, began recording pedigrees and using progeny testing to determine which sires had superior genetic merit. Understanding of the principles of genetics through the work of Gregor Mendel enhanced but did not revolutionize applications to agricultural plant and animal improvement.

Genetic testing schemes to improve the efficiency of artificial selection dominated advances in both the fields of plant and animal genetics during the first two-thirds of the twentieth century. Genetic merit of progeny is expected to be equal to the average genetic merit of the parents. Increasing the rate of genetic change is dependent upon enhancing the ability to recognize potential parents with out-

Many varieties of livestock, such as this powerful "super bull," are the products of artificial selection. (Dan McCoy/Rainbow)

standing genetic merit. Use of computers and large-scale databases has had great impact on improving selection programs for crops and livestock. Genetic change on the order of 2 percent per year is possible. However, selection to improve horticultural species and companion animals has continued to rely largely on the subjective judgment of the breeder for identification of superior stock. Plant and animal genome mapping programs may facilitate the next leap forward in genetic improvement of agricultural organisms. Selection among organisms based directly on their gene sequences may permit bypassing much of the time-consuming data-recording programs upon which genetic progress of the 1990's relied.

Impact and Applications

Exhaustion of genetic variation forms the ultimate limit to what can be achieved by selection. One example of the extremes that can be accomplished by selection is dogs in which the heaviest breeds weigh nearly one hundred times as much as the lightest breeds. Experimental selection for body weight in insects and oil content in corn have resulted in changes of similar magnitude. However, modern breeding programs for crops and livestock seek to decrease variability while increasing productivity. Uniformity of the products enhances the efficiency with which they can be handled mechanically. A concern resulting from the success of modern breeding programs is that as indigenous crop and livestock varieties are replaced by high-producing varieties, the genetic variation that provides the source of potential future improvements is lost. Widespread use of uniform varieties may also increase the susceptibility to catastrophic losses from an outbreak of disease.

—William R. Lamberson

See Also: Genetic Engineering: Agricultural Applications; High-Yield Crops; Natural Selection; Population Genetics; Quantitative Inheritance.

Further Reading: Colin Tudge, *The Engineer in the Garden* (1993), provides a historical overview of genetics and explores the potential ramifications of past, present, and future genetic advances. I. Michael Lerner, *Heredity, Evolution, and Society* (1968), provides a clear discussion of polygenic inheritance, particularly with respect to humans. Selection from the perspective of the gene is explored in Richard Dawkins, *The Selfish Gene* (1990).

Autoimmune Disorders

Field of study: Immunogenetics

Significance: *Autoimmune disorders are chronic diseases that arise from a breakdown of the body's autotolerance, or the ability to distinguish between its own cells and invading substances. Autoimmune disorders can be caused by both genetic and environmental factors, and they can cause an individual's immune system to react against the organs or tissues of the individual's own body.*

Key terms

ANTIBODIES: molecules in blood plasma responsible for recognizing and binding to antigens

ANTIGENS: foreign substances recognized by the immune system that result in the production of antibodies and lymphocytes directed specifically against them

IMMUNE SYSTEM: the system that normally responds to foreign agents by producing antibodies and stimulating antigen-specific lymphocytes, leading to destruction of these agents

LYMPHOCYTES: sensitized cells of the immune system that recognize and destroy harmful agents via antibody and cell-mediated responses that include B lymphocytes from the bone marrow and T lymphocytes from the thymus

MAJOR HISTOCOMPATIBILITY COMPLEX (MHC): a system of protein markers on a cell's outer membrane following infection with a virus, malignant cell, or foreign cell that signals the immune system to destroy the cell

Autoimmune Disorders and Immune System Dysfunction

Autoimmune disorders involve a large group of chronic and potentially life-threatening diseases that are initiated by an attack of an individual's own immune system directed against the organs or tissues of the individual's own body. The main function of the immune system is to defend against invading microorganisms such as bacteria, fungi, viruses, protozoa, and parasites by producing antibodies or lymphocytes that recognize and destroy the harmful agent. The ability to distinguish normal body constituents (self) from foreign substances (nonself) and attack and destroy only nonself agents is crucial to appropriate immune functioning. Alterations in immune system functioning may result in the production of antibodies against the body's own cells, causing malfunction and destruction as the body becomes unable to control immune reactions against itself.

One important function of the immune system is the accurate surveillance of body cells to ensure that they are not inappropriately utilizing the major histocompatibility complex (MHC) protein marker system. Appropriate surveillance results in autotolerance, the specific lack of an immune response against self because of the accurate discrimination of self and nonself. Autoimmunity—the inaccurate recognition of a normal body component as foreign, followed by the mounting of an autoimmune response—results when autoantigens present within the internal cells stimulate the development of autoantibodies. These autoantibodies most often develop as a genetic defect during viral and bacterial infections or from environmental or chemical influences; they may also, according to some researchers, result as a natural consequence of the aging process.

History and Classification of Autoimmune Disorders

The concept of autoimmune disorders was first proposed in 1901, but it was not until the 1950's that autoimmunity was experimentally created in animals using immunization. By the 1960's, it was recognized that autoimmunity was a direct or indirect contributor to numerous human ailments. Many diseases were formerly classified as collagen-vascular diseases (collagenoses) until it was determined incorrect to classify them primarily as connective tissue lesions when the primary defect was initiated by the immune system.

Autoimmune disorders are generally categorized as organ-specific diseases and non-organ-specific (also called systemic) diseases. Organ-specific autoimmune diseases involve an attack directed against one main organ and have been documented for essentially every organ in the body. Common examples include multiple sclerosis (brain), insulin-dependent diabetes melletis (pancreas), Graves' disease (thyroid), Addison's disease (adrenal glands), pernicious anemia (stomach), myasthenia gravis (muscle), autoimmune hemolytic anemia (blood), primary biliary cirrhosis (liver), pemphigus vulgaris (skin), and glomerulonephritis (kidneys). Non-organ-specific autoimmune diseases involve an attack by the immune system on several body areas, potentially causing diseases such as systemic lupus erythematosus, rheumatoid arthritis, polyarteritis nodosa, scleroderma, ankylosing spondylitis, and rheumatic fever.

Some evidence exists to indicate that other diseases that affect, for example, the eye (uveitis) or reproductive systems (male and female infertility) may be autoimmune related. Allergies involve hypersensitivity reactions that result when immune reactions damage tissue, potentially leading to anaphylactic shock and death. Environmental antigens such as pollen, dust mites, food proteins, and bee venom may cause allergic reactions such as hay fever, asthma, and food intolerance in sensitive individuals via the antibody class known as IgE. Medications such as antibiotics may also be recognized as chemical antigens, causing adverse allergic reactions such as penicillin hypersensitivity. Autoimmunity may also result during surgical transplantation and blood transfusions as evidenced by all individuals (except those of blood group AB) possessing natural antibodies against red blood cell antigens other than their own type. Antibodies to the rhesus (Rh) factor antigen on red blood cells

may cause hemolytic disease of a newborn. This is evident when an Rh positive fetus is carried by an Rh negative mother and the mother produces antibodies against the Rh antigen on fetal red cells, resulting in life-threatening anemia and jaundice in the fetus when these antibodies cross the placenta.

Immunologists (medical researchers who investigate the body's response to antigens) do not know the precise origin of most autoimmune diseases. What researchers have shown is that most autoimmune diseases occur more frequently in females than in males and that the development of autoimmune disorders often requires both a genetic susceptibility and additional stimuli such as exposure to a toxin. Of the numerous theories proposed for the cause of autoimmunity development, three models have received the most consideration by clinical researchers. The clonal deletion theory suggests that autoimmunity develops if autoreactive T or B cell clones are not eliminated during the fetal period or very soon after birth. The body normally does not react to its own fetal or neonatal antigens, which are recognized because the corresponding T and B cell clones are eliminated from the immune system. In the unfortunate event that "forbidden clones" of autoreactive cells remain active, antibodies are produced that are directed against its own antigens, and autoimmunity develops, frequently involving the loss of the helper T cells' ability to regulate B-cell function. A second proposal suggests that some antigens that are normally nonimmunogenic (hidden antigens) somehow become autoimmunogenic and stimulate the immune system to react against itself. A third proposal suggests that autoimmunity can be initiated by an exogenous antigen, assuming that the antibodies produced to fight it cross-react with a similar determinant on the body's own cells.

Diagnosis and Treatment of Autoimmune Disorders

Diagnosis of autoimmune disorders generally begins with the often difficult task of documenting autoantibodies and autoreactive T cells. Other direct or indirect markers are also used to reveal if the immune mechanism is pathogenetically important to development of the lesion, such as a favorable response to immunosuppressive, corticosteroid, or anti-inflammatory drug treatment along with several other immunologic techniques.

Autoimmune treatment strategies lag behind the ability to diagnose these disorders. Initial management involves the control and reduction of both pain and loss of function. Correction of deficiencies in hormones such as insulin or thyroxin that are not being adequately produced by an affected gland is often performed first by a physician. Replacing blood components by transfusion is also considered, but treatment effectiveness is often limited by the lack of knowledge of the precise disease mechanisms. Diminishing the activity of the immune system is also often attempted, but achieving a delicate balance between controlling the autoimmune disorder and maintaining the body's ability to fight disease in general is critical.

Medication therapy commonly includes corticosteroid drugs, with more powerful immunosuppressant drugs such as cyclophosphamide, methotrexate, azathioprine, chloroquine derivatives, and small doses of antimetabolic or anticancer drugs often required. A majority of these medications can rapidly damage dividing tissues such as the bone marrow and thus must be used with caution. Plasmapheresis (removal of toxic antibodies) is often helpful in diseases such as myasthenia gravis, while other treatments involve drugs that target immune system cells such as the cyclosporines. Fish oil and antioxidant supplementation has been shown to be an effective anti-inflammatory intervention and may help suppress autoimmune diseases such as rheumatoid arthritis and systemic lupus erythematosus.

—*Daniel G. Graetzer*

See Also: Allergies; Developmental Genetics; Diabetes; Heredity and Environment.

Further Reading: Timothy J. Vyse and John A. Todd, "Genetic Analysis of Autoimmune Disease," *Cell* 85 (May, 1996), describes how ongoing study of the entire human genetic code will assist in the isolation and correction of aberrant genes that cause immunological disease and presents evidence that environmental factors have only minor effects on immune system abnormalities. *Clinical Aspects of Immunology* (1993), edited by P. J. Lachman et al., provides an excellent overview of immunology for the clinician. Gabriel Fernandes and Christopher A. Jolly, "Nutrition and Autoimmune Diseases," *Nutrition Reviews* 56 (January, 1998), summarizes several topics from a conference on nutrition and immunity and relates striking benefits of fish oil and antioxidant supplementation on gene and T-cell subsets.

Bacterial Genetics and Bacteria Structure

Field of study: Bacterial genetics
Significance: *The study of bacterial structure and genetics has made tremendous contributions to the fields of medicine and genetics. Knowledge of the structure of bacteria has led to the development of drugs for the treatment of disease, while studies of bacterial genetics have led to the discovery of DNA as the master chemical of heredity and to general knowledge about the regulation of gene expression in other organisms, including humans.*

Key terms

PROKARYOTIC: lacking a membrane-bound nucleus

EUKARYOTIC: made up of cells having a membrane-bound nucleus that contains chromosomes

COLONY: an accumulation of microorganisms on the surface of a host or within the body of a host

RECOMBINANT DNA: a deoxyribonucleic acid (DNA) sequence that has been constructed or engineered in a test tube from two or more distinct DNA sequences

CLONING: the generation of many copies of DNA by replication in a suitable host

MUTATION: the process by which a DNA base-pair change or a change in a chromosome is produced; the term "mutation" is also used to describe the change itself

Bacteria and Their Structure

Bacteria are contained in the kingdom *Prokaryotae* (*Monera*). Organisms in this kingdom are unicellular (one-celled) and prokaryotic (lacking a membrane-bound nucleus). Bacteria are among the simplest, smallest, and most ancient of organisms. They are found in nearly every environment on earth. While some bacteria are autotrophic (capable of making their own food), most are heterotrophic (forced to draw nutrients from their environment or from other organisms). For most of human history, the existence of bacteria was unknown. It was not until the late 1800's that bacteria were first identified. Their role in nature is that of decomposers: They break down organic molecules into their component parts. Along with fungi, they are the major recyclers in nature. They are also capable of changing atmospheric nitrogen to a form that is usable by plants and animals.

That some bacteria are pathogens, or causers of disease, has long been known. Scientists have expended tremendous effort in describing bacteria's role in disease and in creating agents that could kill them. Other bacteria, such as *Escherichia coli* (*E. coli*), may be part of a mutualistic relationship with another organism, such as humans. Bacteria have been used extensively in genetics research because of their small size and because they reproduce rapidly; some bacteria produce a new generation every forty minutes. Since they have been so thoroughly studied, a great deal is known about their structure and genetics.

Bacteria are microscopic unicells. Most are less than one micron (one millionth of a meter) in length. They do not contain mitochondria (organelles that produce the energy molecule adenosine triphosphate, or ATP), chloroplasts (plant organelles in which the reactions of photosynthesis take place), lysosomes (organelles that contain digestive enzymes), or interior membrane systems such as the endoplasmic reticulum or Golgi bodies. They do, however, contain ribonucleic acid (RNA), ribosomes (organelles that serve as the sites of protein synthesis), and deoxyribonucleic acid (DNA), which is organized as part of a single, circular chromosome. The circular chromosome is centrally located within the cell in a region called the nucleoid region and is capable of supercoiling. Bacteria often have genes in addition to those found on the main chromosome. These genes are carried on small rings called plasmids, which have been used extensively in genetic research. Some plasmids carry genes that impart antibiotic resistance to the cells that contain them.

There are three basic bacterial morphologies, or body shapes. Bacteria that are spherical are called cocci. Some coccus bacteria form

Types of Bacteria

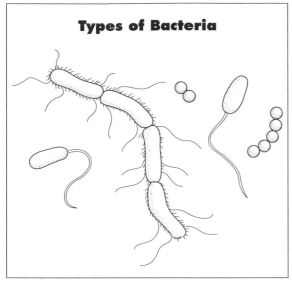

Bacteria occur in a variety of shapes and sizes.

clusters (staphylococcus), while others may form chains (streptococcus). Bacteria that have a rodlike appearance are called bacilli. Spiral or helical bacteria are called spirilla (sometimes called spirochetes).

Differentiating Bacteria

The kingdom *Prokaryotae* contains two subkingdoms that are defined by the nature of the bacterial cell wall. The subkingdom *Archaebacteria* contains bacteria that have cell walls that do not contain peptidoglycan, a complex organic molecule made of two unusual sugars held together by short peptides (short chains of amino acids). Archaebacters represent a small percentage of bacterial strains and are usually found in extreme environments, such as hot springs and sea vents. Methanogens are the most common archaebacters. They are strict anaerobes, which means that they are killed by oxygen. They live in oxygen-free environments, such as sewers and swamps, and produce methane gas as a waste product of their metabolism. Halobacteria live in only those environments that have a high concentration of salt, such as salt ponds. Thermoacidophiles are archaebacters that grow in very hot or very acidic environments.

The subkingdom *Eubacteria*, or true bacteria, is a very large and diverse group of organisms.

Bacteria in this subkingdom fall into three basic types: those that lack cell walls, those with thin cell walls, and those with thick cell walls. Mycoplasmas lack cell walls entirely (the bacterium that causes tuberculosis, *Mycobacterium tuberculosis*, is one example). However, most bacteria in this subkingdom do have cell walls, and, unlike the archaebacters, eubacterial cell walls are composed of peptidoglycan. In 1884, Hans Christian Gram, a Danish physician, found that certain bacterial cells absorbed a stain called "crystal violet," while others did not. Those cells that absorb the stain are called "gram-positive," and those that do not are called "gram-negative." It has since been found that gram-positive bacteria have thick walls of peptidoglycan, while gram-negative bacteria have thin peptidoglycan walls covered by a thick outer membrane. It is this thick outer membrane that prevents crystal violet from entering the bacterial cell. Distinguishing between gram-positive and gram-negative bacteria is an important step in the treatment of disease since some antibiotics are more effective against one class than the other.

Bacteria can be further differentiated by the presence or absence of certain surface structures. Some strains produce an outer slime layer called a "capsule." The capsule permits the bacterium to adhere to surfaces (such as human teeth, for example, where the build-up of such bacteria causes dental plaque) and provides some protection against other microorganisms. Some strains display pili, which are fine, hairlike appendages that also allow the bacterium to adhere to surfaces. Some pili, such as the F pili in *E. coli*, are involved in the exchange of genetic material from one bacterium to another in a process called "conjugation." Some bacterial strains have one or more flagella, which allow them to be motile, or capable of movement. Any bacterium may have one or more of these surface structures.

Bacterial Reproduction

Bacteria reproduce in nature by means of binary fission. This process is essentially the same as mitosis, wherein one cell divides to produce two daughter cells that are genetically identical. As bacteria reproduce, they form clustered associations of cells called colonies. All members of a colony are genetically identical to each other, unless a mutagen (any substance that can cause a mutation) has changed the DNA sequence in one of the bacteria. Changes in the DNA sequence of the chromosome often lead to changes in the physical appearance or nutritional requirements of the colony. While a bacterium is microscopic, bacterial colonies can be seen with the naked eye; changes in the colonies are relatively easy to perceive. This is one of the reasons bacteria have been favored organisms for genetic research.

For the most part, there is very little genetic variation between one bacterial generation and the next unless a mutation arises. Unlike higher organisms, bacteria do not engage in sexual reproduction, which is the major source of genetic variation within a population. In laboratory settings, however, bacteria can be induced to engage in a unidirectional (one-way) exchange of genetic material. This transfer of genes from one bacterium to another is called "conjugation" and was first observed in 1946 by biochemists Joshua Lederberg and Edward Tatum. The unidirectional nature of the gene transfer was discovered by William Hayes in 1953. He found that one bacterial cell was a donor cell while the other was the recipient. In the 1950's, molecular biologists François Jacob and Elie Wollman used conjugation and a technique called "interrupted mating" to map genes onto the bacterial chromosome. By breaking apart the conjugation pairs at intervals and analyzing the times at which donor genes entered the recipient cells, they were able to determine a correlation between time and the distance between genes on a chromosome. The use of this technique led to a complete picture of the sequence of genes contained in the chromosome. It also led to a surprise: It was use of interrupted mating with *E. coli* that first demonstrated the circularity of the bacterial chromosome. The circular structure of the chromosome was a striking contrast to eukaryotic chromosomes, which are linear.

Transformation and Transduction

The bacterium *Streptococcus pneumoniae* (*S. pneumoniae*) was used in one of the early studies

that eventually led to the identification of DNA as the master chemical of heredity. Two strains of *S. pneumoniae* were used in a study conducted by microbiologist Frederick Griffith in 1928. One strain (S) produces a smooth colony that is virulent (infectious) and causes pneumonia. The other strain (R) produces a rough colony that is avirulent (noninfectious). When Griffith injected mice with living type R bacteria, the mice survived and no bacteria were recovered from their blood. When he injected mice with living type S, the mice died, and type S bacteria were recovered from their blood. However, if type S was heat-killed before the mice were injected, the mice did not die, and no bacteria were recovered from their blood. This confirmed what Griffith already knew: Only living type S *S. pneumoniae* caused lethal infections. Something interesting happened when Griffith mixed living type R with heat-killed type S, however: Mice injected with this mixture died, and virulent type S bacteria were recovered from their blood. An unknown agent apparently transformed avirulent type R into virulent type S. Griffith called the agent the "transforming principle." It was his belief that the transforming principle was a protein.

Sixteen years later, in 1944, bacteriologists Oswald Avery, Colin MacLeod, and Maclyn McCarty designed an experiment that showed conclusively that the transforming principle was DNA rather than protein. They showed that R bacteria could be transformed to S bacteria in a test tube. They then progressively purified their extract until only proteins and the two nucleic acids, RNA and DNA, remained. They placed some of the mixture onto agar plates (glass dishes containing a gelatin growth medium). At this point, transformation still occurred; therefore, it was clear that one of these three molecules was the transforming agent. They treated their extract with protein-degrading enzymes, which denatured (destroyed) all the proteins in the extract. Despite the denaturing of the proteins, transformation still occurred when some of the extract was plated; had protein been the transforming agent, no transformation could have occurred. Protein was eliminated as the transforming agent. The next step was to determine which of two nucleic acids was responsible for the transformation of the R strain into the S strain. They introduced RNase, an enzyme that degrades RNA, to the extract. The RNA was destroyed, yet transformation took place. RNA was thus eliminated. At this point, it was fairly obvious that DNA was the transforming agent. To conclusively confirm this, they introduced DNase to the extract. When the DNA was degraded by the enzyme, transformation did not take place, proving conclusively that DNA was the transforming agent.

Another way that genetic material can be exchanged between bacteria is transduction. Transduction requires the presence of a bacteriophage (a virus that infects bacteria). A virus is a simple structure consisting of a protein coat called a "cap" or "capsid" that contains either RNA or DNA. Viruses are acellular, nonliving, and extremely small. To reproduce, they must infect living cells and use the host cell's internal structures to replicate their genetic material and manufacture viral proteins. Bacteriophages, or phages, infect bacteria by attaching themselves to a bacterium and injecting their genetic material into the cell. Sometimes, during the assembly of new viral particles, a piece of the host cell's DNA may be enclosed in the viral cap. When the virus leaves the host cell and infects a second cell, that piece of bacterial DNA enters the second cell, thus changing its genetic makeup. Generalized transduction (the transfer of a gene from one bacterium to another) was discovered by Joshua and Esther Lederberg and Norton Zinder in 1952. Using *E. coli* and a bacteriophage called P1, the Lederbergs and Zinder were able to

show that transduction could be used to map genes to the bacterial chromosome.

Hershey-Chase Bacteriophage Experiments

The use of bacteriophages has been instrumental in confirming DNA as the genetic material of living cells. Alfred Hershey and Martha Chase devised a series of experiments using *E. coli* and the bacteriophage *T2* that conclusively established DNA as genetic material in 1953. Bacteria are capable of manufacturing all essential macromolecules by uptaking material from their environment. Hershey and Chase grew cultures of *E. coli* in a growth medium enriched with a radioactive isotope of phospho-

Martha Chase, whose work with Alfred Hershey helped to confirm that DNA is the genetic material. (California Institute of Technology)

rous (^{32}P). DNA contains phosphorous; as the succeeding generations of bacteria pulled phosphorous from the growth medium to manufacture DNA, each DNA strand also carried a radioactive label. *T2* phages were used to infect the cultures of *E. coli*. When the new *T2* viruses were assembled in the bacterial cells, they too carried the radioactive label ^{32}P on their DNA. A second culture of *E. coli* was grown in a medium enriched with radioactive sulfur (^{35}S). Proteins contain sulfur (but no phosphorous). *T2* viruses were used to infect this culture. New viruses contained the ^{35}S label on their protein coats.

Since the *T2* phage consists of only protein and DNA, one of these two molecules had to be the genetic material. Hershey and Chase infected unlabeled *E. coli* with both types of radioactive *T2* phages. Analysis showed that the ^{32}P label passed into the bacterial cells, while the ^{35}S label was found only in the protein coats that did not enter the cells. Since the protein coat did not enter the bacterial cell, it could not influence protein synthesis. Therefore, protein could not be the genetic material. The Hershey-Chase experiment confirmed DNA as the genetic material.

Restriction Enzymes and Gene Expression

Using the aforementioned methods, it has been possible to construct a complete genetic map showing the order in which genes occur on the chromosome of *E. coli* and other bacteria. Certain genes are common to all bacteria. There are also several genes that are shared by bacteria and higher life-forms, including humans. Further research showed that genes could be either inserted into or deleted from the bacterial DNA. In nature, only bacteria contain specialized enzymes called restriction enzymes. Restriction enzymes are capable of splicing DNA at specific sites called restriction sites. The function of restriction enzymes in bacteria is to protect against invading viruses. Bacterial restriction enzymes are designed to destroy viral DNA without harming the host DNA. Over 150 such enzymes have been isolated from bacteria, and each is named for the bacterium from which it comes. The discovery and isolation of restriction enzymes has led to

a new field of biological endeavor: genetic engineering.

Use of these enzymes has made gene cloning possible. Cloning is important to researchers because it permits the detailed study of individual genes. Restriction enzymes have also been used in the formation of genomic libraries (a collection of clones that contains at least one copy of every DNA sequence in the genome). Genomic libraries are valuable because they can be searched to identify a single DNA recombinant molecule that contains a particular gene or DNA sequence.

Bacterial studies have been instrumental in understanding the regulation of gene expression, or the translation of a DNA sequence first to a molecule of messenger RNA (mRNA) and then to a protein. Bacteria live in environments that change rapidly. To survive, they have evolved systems of gene regulation that can either "turn on" or "turn off" a gene in response to environmental conditions. François Jacob and Jacques Monod discovered the lac operon, a regulatory system that permits *E. coli* to respond rapidly to changes in the availability of lactose, a simple sugar. Other operons, such as the tryptophan operon, were soon discovered as well. An operon is a cluster of genes whose expression is regulated together and involves the interaction of regions of DNA with regulatory proteins. The discovery of operons in bacteria led to searches for them in eukaryotic cells. While none has been found, several other methods of regulating the expression of genes in eukaryotes have been described.

Impact and Applications

Diabetes mellitus is a disease caused by the inability of the pancreas to produce insulin, a protein hormone that is part of the critical system that controls the body's metabolism of sugar. Prior to 1982, people who suffered from diabetes controlled their disease with injections of insulin that had been isolated from other animals, such as cows. In 1982, human insulin became the first human gene product to be manufactured using recombinant DNA. The technique is based on the knowledge that genes can be inserted into the bacterial chromosome; that once inserted, the gene product, or protein, will be produced; and that once produced, the protein can be purified from bacterial extracts. Human proteins are usually produced by inserting a human gene into a plasmid vector, which is then inserted into a bacterial cell. The bacterial cell is cloned until large quantities of transformed bacteria are produced. From these populations, human proteins, such as insulin, can be recovered.

Many proteins used against disease are manufactured in this manner. Some examples of recombinant DNA pharmaceutical products that are already available or in clinical testing include atrial natriuretic factor, which is used to combat heart failure and high blood pressure; epidermal growth factor, which is used in burns and skin transplantation; factor VIII, which is used to treat hemophilia; human growth hormone, which is used to treat dwarfism; and several types of interferons and interleukins, which are proteins that have anticancer properties.

Bacterial hosts produce what are called the "first generation" of recombinant DNA products. There are limits to what can be produced in and recovered from bacterial cells. Since bacterial cells are different from eukaryotic cells in a number of ways, they cannot process or modify most eukaryotic proteins, nor can they add sugar groups or phosphate groups, additions that are often required if the protein is to be biologically active. In some cases, human proteins produced in prokaryotic cells do not fold into the proper three-dimensional shape; since shape determines function in proteins, these proteins are nonfunctional. For this reason, it may never be possible to use bacteria to manufacture all human proteins. Other organisms are used to produce what are called the second generation of recombinant DNA products.

The impact of the study of bacterial structures and genetics, and the use of bacteria in biotechnology, cannot be underestimated. Bacterial research has led to the development of an entirely new branch of science, that of molecular biology. Much of what is currently known about molecular genetics, the expression of genes, and recombination comes from research involving the use of bacteria. Moreover, bacteria have had and will continue to have applications in the production of pharmaceuticals and the treatment of disease. The recombinant DNA technologies developed with bacteria are now being used with other organisms to produce medicines and vaccines.

—*Kate Lapczynski*

See Also: Archaebacteria; Biopharmaceuticals; Biotechnology; Cloning; *Escherichia coli*; Gene Regulation: Bacteria; Gene Regulation: Lac Operon; Genetic Code, Cracking of; Molecular Genetics.

Further Reading: *Understanding DNA and Gene Cloning: A Guide for the Curious* (1992), by Karl Drlica, is an overview with good illustrations and graphics. *A Dictionary of Genetics* (1997), by Robert C. King and William D. Stansfield, is helpful for understanding the terminology of genetics. *Recombinant DNA: A Short Course* (1983), by James D. Watson et al., is an interesting book written by one of three men shared a Nobel Prize in Physiology or Medicine for describing the molecular structure of DNA. *Fundamentals of Genetics* (1994), a textbook written by Peter J. Russell, is enriched with many illustrations and graphics that make complex subjects understandable.

Bacterial Resistance and Super Bacteria

Field of study: Bacterial genetics
Significance: *Antibiotic-resistant bacteria have become a significant worldwide health concern. Some strains of bacteria (such as super bacteria) are now resistant to most, if not all, of the available antibiotics and threaten to return health care to a preantibiotic era. Understanding how and why bacteria become resistant to antibiotics may aid treatment, the design of future drugs, and efforts to prevent other bacterial strains from becoming resistant to antibiotics.*

Key terms

ANTIMICROBIAL DRUGS: chemicals that destroy disease-causing organisms without damaging body tissues; chemicals made naturally by bacteria and fungi are also known as antibiotics

RESISTANCE FACTOR (R FACTOR): a piece of deoxyribonucleic acid (DNA) that carries a gene encoding for resistance to an antibiotic

PLASMIDS: small, circular pieces of DNA that can exist separately from the bacterial chromosome; plasmids can be transferred among bacteria, and they may carry more than one R factor

TRANSPOSONS: also known as "jumping genes," transposons are larger pieces of DNA that carry R factors and can integrate into a bacterial chromosome; they are also responsible for the spread of drug resistance in bacteria and fungi, and, like plasmids, each transposon may carry more than one R factor

History of Antibiotics

Throughout history, illnesses such as cholera, pneumonia, and sexually transmitted diseases have plagued humans. However, it was not until the early twentieth century that antibiotics were discovered. Until then, childhood diseases such as diphtheria, cholera, and influenza were considered serious and sometimes deadly. With the advent of the antibiotic era, it appeared that common childhood diseases would no longer be a serious health concern. As the twentieth century progressed, however, it became apparent that this would not be the case. Emerging infectious diseases such as multidrug-resistant tuberculosis, vancomycin-resistant enterococci, and penicillin-resistant gonorrhea have become serious global health care concerns.

A laboratory accident led to the discovery of the first mass-produced antibiotic. In 1928, Scottish bacteriologist Alexander Fleming grew *Staphylococcus aureus* in petri dishes, and the plates became contaminated with a mold. Before Fleming threw out the plates, he noticed that there was no bacterial growth around the mold. The mold, *Penicillium notatum*, produced a substance that was later called penicillin, which was instrumental in saving the lives of countless soldiers during World War II. From the 1950's until the 1980's, antibi-

Sir Alexander Fleming's discovery of penicillin opened a new era in medicine but also led to the development of resistant strains of bacteria. (The Nobel Foundation)

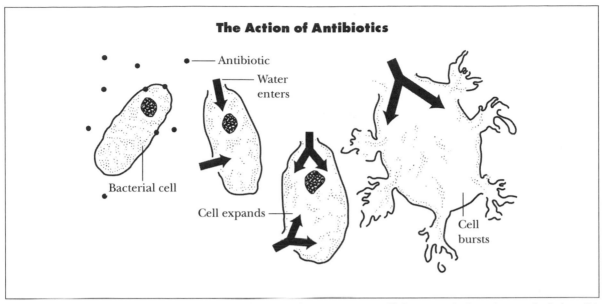

The Action of Antibiotics

Antibiotic

Water enters

Bacterial cell

Cell expands

Cell bursts

An antibiotic destroys a bacterium by causing its cell walls to deteriorate; water will then enter the bacterium unchecked until it bursts.

otics were dispensed with great regularity for most bacterial infections, for earaches, for colds, and as a preventive measure. The seemingly haphazard dispensing of antibiotics for viral infection (antibiotics are ineffective against viral infection such as the common cold or influenza) is unnecessary and can lead to antibiotic-resistant bacterial infections.

Bacterial Resistance

On average, bacteria can replicate every twenty minutes. Several generations of bacteria can reproduce in a twenty-four-hour period. This quick generation time leads to a rapid adaptation to changes in the environment. English naturalist Charles Darwin's *On the Origin of Species* (1859) first explained the theory of natural selection. If an organism has an advantage over other organisms (such as the ability to grow in the presence of a potentially harmful substance), that organism will survive to pass that characteristic on to its offspring while the other organisms die. The emergence of antibiotic-resistant bacteria is an excellent example of Darwin's theory of natural selection.

In the early twentieth century, German microbiologist Paul Ehrlich coined the term

"magic bullet" in regard to chemotherapy (the treatment of a disease with a chemical compound). For a drug such as an antibiotic to be a "magic bullet," it must have a specific target that is unique to the disease-causing agent and cannot harm the host in the process of curing the disease. In 1910, Ehrlich discovered that salvarsan, a derivative of arsenic, could be used to treat syphilis (a very common sexually transmitted disease in the early twentieth century). Until that time syphilis had no known cure. The use of salvarsan did cure some patients of syphilis, but, since it was a rat poison, it killed other patients. Generally speaking, antimicrobials have specific targets (or modes of action) within bacteria. They target the following structures or processes: synthesis of the bacterial cell wall, injury to the plasma membrane, and inhibition of synthesis of proteins, deoxyribonucleic acid (DNA), ribonucleic acid (RNA), and other essential metabolites (all of these substances are building blocks for the bacteria). A good antibiotic will have a target that is unique to the bacteria so the host (the patient) will not be harmed by the drug.

Bacteria and fungi are, of course, resistant to the antibiotics they naturally produce. Other bacteria have the ability to acquire resistances to antimicrobials, and this drug resistance occurs either through a mutation in the DNA or by acquiring resistance genes on plasmids or transposons. Plasmids are small, circular pieces of DNA that can exist within or independently of the bacterial chromosome. Transposons, or "jumping genes," are pieces of DNA that can jump from one bacterial species to another and be integrated into the bacterial chromosome. The spread of plasmids and transposons that carry resistance genes to antibiotics has led to bacteria becoming resistant to many, if not all, currently available antibiotics.

There are several antimicrobial resistance mechanisms that allow bacteria to become drug resistant. The first mechanism does not allow the drug to enter the bacterial cell. A decrease in the permeability of the cell wall will inhibit the antimicrobial drug from reaching its target. An alteration in a penicillin-binding protein (pbp), a protein found in the bacterial cell wall, will allow the cell to "tie up" the penicillin. Also, the pores in the cell wall can be altered so the drug cannot pass through. A second strategy is to pump the drug out of the cell after it has entered. Such systems are found in pathogenic (disease-causing) *Escherichia coli*, *Pseudomonas aeruginosa*, and *Staphylococcus aureus*. These pumps are usually nonspecific and can cause bacteria to become resistant to more than one antibiotic at a time. Another method of resistance is through chemical modification of the drug. Penicillin is inactivated by breaking a chemical bond found in its ring structure. Other drugs are inactivated by the addition of other chemical groups. Finally, the target of the drug can be altered in such a manner that it is no longer affected by the drug. For example, *Mycobacterium tuberculosis*, which causes tuberculosis, can become resistant to the drug rifampin by altering the three-dimensional structure of a specific protein.

Antibiotic Misuse and Drug Resistance

The misuse of antibiotics over several decades has caused many strains of bacteria to become resistant. For some bacterial infections, there may be only one or no effective drug treatment. Many different factors of misuse, overuse, and abuse of antibiotics have led

to drug-resistant diseases. Perhaps one of the most important factors in the emergence of drug-resistant bacteria is the overprescription or inappropriate use of antibiotics. Another major factor is misuse by the patient. After several days of taking an antibiotic, a patient may begin to feel better and decide not to finish all of the prescription. By not completing the full course of treatment, the patient merely kills the bacteria that are sensitive to the antibiotic. The resistant bacteria can then grow, multiply, and cause the same infection. This time, another antibiotic (if there is one available that is effective) must be used.

Another contributing factor is the ease with which the newest and best antibiotics may be obtained in many countries. In several countries in Central America, for example, one can walk into the local pharmacy and receive any antibiotic that is available without a prescription. Another factor in the worldwide spread of drug-resistant infectious diseases is the ease of travel. Infected people can carry bacteria from one continent to another in a matter of hours and infect anyone with whom they come in contact.

The use of antibiotics is not limited to humans. They also play an important role in agriculture. Antibiotics are added to animal feed on farms to help keep herds healthy, and they are also used on fish farms for the production of fish for market. Antibiotics are used to treat domestic animals such as cats, dogs, birds, and fish and are readily available in pet stores to clear up fish aquariums. This widespread use of antibiotics allows bacteria in all environmental niches the possibility of becoming resistant to potentially useful drugs.

Emerging Resistant Infections and Super Bacteria

The misuse of antibiotics over the decades has led to more infectious diseases becoming resistant to the current arsenal of drugs. Some diseases that could be treated effectively in the 1970's and 1980's can no longer be controlled with the same drugs. Two very serious problems have emerged: vancomycin-resistant enterococci and multidrug-resistant tuberculosis. The enterococcus is naturally resistant to many types of antibiotics, and the only effective treatment has been vancomycin. With the appearance of vancomycin-resistant enterococci, however, there are no reliable alternative treatments. The fear that vancomycin resistance will spread to other bacteria such as staphylococci seems well-founded: A report from Japan in 1997 indicated the existence of a strain of staphylococcus that had become partially resistant to vancomycin. If a strain of methicillin-resistant *Staphylococcus aureus* (MRSA) also becomes resistant to vancomycin, there will be no effective treatment available against this super bacteria.

A second problem is the appearance of multidrug-resistant tuberculosis. *Mycobacterium tuberculosis* is a slow-growing bacterium that requires a relatively long course of antibiotic therapy. Tuberculosis is spread easily, and it is a deadly disease. In the United States in 1900, tuberculosis was the number one cause of death. In the 1990's, it was still a leading cause of death worldwide. Treatment of multidrug-resistant tuberculosis requires several antibiotics taken over a period of at least six months, with a success rate of approximately 50 percent; on the other hand, susceptible strains of TB have a cure rate of nearly 100 percent.

Another contributing factor to the emergence of drug-resistant infectious diseases is the lack of basic knowledge about some bacteria. Funding for basic genetic research on tuberculosis was cut dramatically in the mid-twentieth century when it appeared that TB would be eradicated just as smallpox had been. The appearance of multidrug-resistant tuberculosis caught scientists and physicians unprepared. Little was known about the genetics of tuberculosis or how drug resistance occurred.

The next obvious concern about drug-resistant infections is how to control them. Hospitals are vigilant, and, in some cases, very proactive in screening for drug-resistant infections. People can be asymptomatic carriers (that is, they carry the disease-causing organism but are still healthy) of a disease such as methicillin-resistant *Staphylococcus aureus* and could infect other people without knowing it. The role of the infection-control personnel is to find the source of the infection and remove it.

Impact and Applications

There is little encouraging news about the availability of new antibiotics. The crisis of super bacteria has altered the view that few new antibiotics would be needed. Pharmaceutical companies are scrambling to discover new antimicrobial compounds and modify existing antibiotics. Policy decisions of the 1970's and 1980's requiring more and larger clinical trials for antibiotics before they are approved for use by the Food and Drug Administration has increased the price of antibiotics and the amount of time it takes to market them. It may take up to ten years from the time of "discovery" for an antibiotic to be approved for use. The scientific community has therefore had to meet the increase of drug-resistant bacterial strains with fewer and fewer new antibiotics.

The emergence of antibiotic-resistant bacteria and super bacteria is a serious global health concern that will lead to a more prudent use of available antibiotics. It has also prompted pharmaceutical companies to search for potentially new and novel antibiotics in the ocean depths, outer space, and other niches. "Rationale" drug design (based on knowledge of how bacteria become drug resistant) will also be important. Exactly how scientists and physicians will be able to combat super bacteria is a question that remains to be answered. Until a more viable solution is found, prudent use of antibiotics and surveillance of drug-resistant infections appear to be the answers.

—*Mary Beth Ridenhour*

See Also: Bacterial Genetics and Bacteria Structure; DNA Replication; Mutation and Mutagenesis; Natural Selection; Transposable Elements.

Further Reading: Stuart Levy, *The Antibiotic Paradox* (1992), provides an overview of antibiotic resistance in bacteria. Levy also discusses mechanisms of resistance, reasons for the spread of antimicrobial resistance, and ways to combat this spread. The *Manual of Clinical Microbiology* (1995), edited by Patrick Murray, provides an in-depth discussion of testing for antimicrobial resistances. Gerard Tortora, *Microbiology: An Introduction* (1998), gives a general overview of antibiotics and how bacterial resistances to antibiotics occurred. T. J. Franklin and G. A. Snow, *Biochemistry of Antimicrobial Action* (1981), provides an explanation of the chemistry of antimicrobials and how bacteria may become resistant to their effects.

Behavior

Field of study: Population genetics
Significance: *The last great frontier that faces biologists in general and geneticists specifically may be the exploration of genetic and environmental influences on the development and control of the mammalian brain. By the mid-1990's, researchers had identified human genes that had been linked to such behavioral characteristics as depression, homosexuality, schizophrenia, and alcoholism; however, such findings were complicated by methodological questions and by the problem of distinguishing between the effects of genetic and environmental factors.*

Key terms

VERTEBRATE: an organism with a backbone; a member of the phylum Chordata

NEUROTRANSMITTER: a chemical messenger that transmits a nerve impulse between neurons

GENE: a unit of heredity; a segment of DNA that contains the instructions to build a protein

DEOXYRIBONUCLEIC ACID (DNA): a macromolecule containing the genetic code or hereditary blueprint

PROTEIN: a macromolecule made of amino acids; used for function and structure of cells

ENZYME: a special protein that controls metabolism by regulating chemical reactions in cells

GENOME: the entire set of genes required by an organism; a set of chromosomes

Basic Biology of the Brain

As the first organ system to begin development and the last to be completed, the nervous system, with the brain at the control, remains something of an enigma to biologists. The vertebrate nervous system (brain, spinal cord, and nerves) is based on neurons, special cells that generate and transmit bioelectrical impulses. The vertebrate brain consists of as many as three major areas: the brain stem, the cerebellum, and the cerebrum. A reptilian brain consists of only the brain stem, while the mammalian brain has all three, including a well-developed cerebrum (the two large hemispheres on top). The brain stem controls basic body functions such as breathing and heart rate, while the cerebrum is the ultimate control center. Consisting of billions of neurons (commonly called "brain cells"), the cerebrum controls functions such as memory, speech, hearing, vision, and analytical skills.

The brain is an exceedingly complex network of billions of neurons. As messages enter the brain stem from the spinal cord, groups of neurons either respond directly or transfer information to higher levels. A good analogy for the functioning of the brain is a small business with receptionists, a switchboard operator, general workers, supervisors, managers, and a board of trustees. The brain stem would consist of receptionists, general workers, and supervisors who handle much of the incoming action or work. At the top of the brain stem sits a central operator who collects all incoming calls or work orders and sends them to the appropriate manager in the cerebrum for action. Finally, the board of trustees would be in charge of ruling on the most important actions and making final decisions.

In order to communicate with one another, each individual neuron generates an impulse much like the impulse that carries a voice over a telephone line, and this message travels from the beginning to the end of each neuron. At the end of one neuron and the beginning of the next in line, a small open space occurs. This space is filled with fluids, and the message is carried across to the next neuron by a chemical known as a neurotransmitter. Neurotransmitters may be of several biochemical classifications, including acetylcholines, amines, amino acids, and peptides. An individual neuron and an entire neuronal circuit may fire or not fire an impulse based on the messages carried by these neurotransmitters. For example, the signal for pain is transmitted from neuron to neuron by a peptide-based neurotransmitter known as substance P, while another peptide transmitter (endorphin) acts as a natural painkiller. Thought, memory, and behavior, then, are produced by the activity along neuronal circuits. A genetic link occurs here, since neurotransmitters are either directly or indirectly based on information in genes.

By birth, the collection of approximately 100,000 genes in humans has directed the development of the nervous system. At birth, the brain consists of approximately 100 billion neurons and trillions of supporting glial cells to protect and nourish neurons. However, the intricate wiring between these neurons is yet to be determined. Studies from the 1980's and 1990's suggest that the critical networking and circuit formation between these billions of neurons that control later brain function is determined not from genes but from environmental input and experiences from birth until the brain is fully developed around age seven.

Fundamentals of Genetics

A gene is a sequence of approximately one thousand nitrogenous bases along a stretch of a deoxyribonucleic acid (DNA) molecule. A single gene or specific segment of a DNA molecule does nothing more or nothing less than provide a cell with the instructions on how to make a specific protein. In other words, genes contain a code for building proteins. Each human cell uses the approximately 100,000 genes with which it is provided to make the approximately 50,000 proteins it needs to function. The amount of DNA that a cell typically inherits depends on the complexity and needs of the

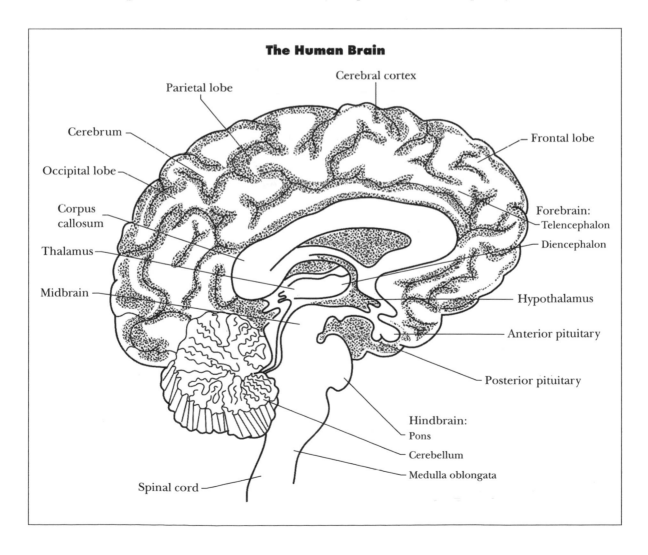

The Human Brain

Parietal lobe
Cerebral cortex
Cerebrum
Frontal lobe
Occipital lobe
Corpus callosum
Forebrain:
Telencephalon
Diencephalon
Thalamus
Midbrain
Hypothalamus
Anterior pituitary
Posterior pituitary
Hindbrain:
Pons
Cerebellum
Medulla oblongata
Spinal cord

organism involved. A common bacteria living in the human gut, for example, has approximately 1.5 million bases making up about 1,400 genes. A human genome, by comparison, has approximately 3 billion bases and about 100,000 genes, meaning 95 percent or so of the human genome does not code for proteins. Much of this 95 percent of the DNA is thought to be "junk" or not involved in a cell's daily activity, but much of this noncoding DNA is involved in the direct control and functioning of the 100,000 genes themselves.

One of the most common misconceptions about genes is that they have absolute control over a cell. A gene is ultimately under the control of other noncoding regions of DNA known as regulatory sequences. Likewise, these regulatory sequences often answer to cellular or environmental factors. One of the first well-understood examples of environmental influence on genes came from bacteria. Certain bacteria take up lactose (a two-unit sugar), then break lactose down into glucose and galactose for energy. The enzyme used for this breakdown is made by a specific gene. Bacteria are very efficient in their biochemical workings, so if there is no lactose to be broken down, the cell does not make the enzyme, and the gene will be "turned off." When lactose is present, however, the gene is "turned on" and directs the construction of the enzyme needed.

It is useful to think of genes as specific messages or blueprints to be used to build specific proteins. If the message is altered, the protein will be altered. A mutation of a gene is a change in the message. For example, a reader who saw the letters P-R-O-T-E-I-N would interpret this to mean the word "protein." If the reader saw the letters P-R-O-T-I-E-N, it would be easy enough to see that the letters were probably meant to make the word "protein," though the I and E are reversed. However, the letters T-E-N-P-R-O-I would not likely be mistaken for "protein" and would have no apparent meaning. A mutation is simply a change in the original base sequence of a gene that may or may not alter the final protein to be built. Mutations occur in nature either by random chance or in response to radiation or chemical influence. Theoretically, no two organisms other than identical twins are

genetically the same. These slight variations in genes occur from individual to individual, and there is thought to be considerable variation among humans in the 95 percent of DNA that is noncoding. It makes sense that most of the variation among humans would be in the "junk" DNA, since it takes the same basic set of genes (with modest variations) to build a functional human being.

A cell is ultimately built by and with protein molecules. Special proteins called enzymes direct all the chemistry that keeps a cell alive and working. In addition, proteins are part of a cell's membrane and intracellular machinery. Special receptor and identification proteins lie on the outer surfaces of cells. A cell depends on its proteins for daily functioning. In addition, signaling molecules called hormones (some of which are proteins) attach to specific protein receptors on the cell surface for communication purposes. Cells are known to be very efficient and do not typically waste time and energy building proteins unless they are needed. The environment is commonly the deciding factor that prompts a cell to make certain proteins when the need arises.

Genes and Behavior

The stage is now set for a discussion of what links, if any, are known between an organic's genetic makeup and its behavior. A very basic sequence of events to remember is that a gene makes a protein, and a protein may cause a specific or general biochemical response in a cell. The behavior of an animal takes place under the combined influences of its genes and its environment. A good example is the phenomenon of mating seasons in many animals. As the number of daylight hours gradually increases toward spring and summer, a critical day length is reached that signals the release of hormones that result in increased sexual activity, with the ultimate goal of seasonal mating. The production and activity of hormones involve genes or gene products. If the critical number of daylight hours is not reached, the genes will not be activated, and sexual behavior will not increase.

Each neuron making up the intricate networks and circuits throughout the cerebrum

(80 percent of the human brain) has protein receptors that respond to specific signaling molecules (chemoreceptors). The production of the receptors and signaling molecules used for any type of brain activity is directly tied to genes. A slightly different gene may lead to a slightly different signaling molecule or receptor and thus a slightly different cell (neuron) response. A larger difference between genes may lead to a larger difference between signaling molecules or receptors and thus a larger variation in cell response. Since human behavior involves the response of neurons and neuron networks in the brain to specific signals, and since the response of neurons occurs because of the interaction between a signaler and a receptor built by specific genes, the genetic link seems straightforward: input, signal, response, behavior. However, when the slight variations between genes are added to the considerable variation among noncoding or regulatory sequences of DNA, the genetic connection to behavior becomes much less direct. Since a gene is under the control of one or several regulatory sequences that in turn may be under the control of various environmental inputs, the amount of genetic variation between organisms is compounded with two other critical factors: the environmental variations under which each brain develops and the daily environmental variations to which each brain is exposed. A convenient way to think of genetics and behavior is to consider that genes simply allow humans to respond to a specific stimulus by building the pathway required for a response, while behavior is defined by the degree and the manner of human response.

The concept of eugenics was born during the evolution and study of basic genetics in the early twentieth century. Eugenics is the categorization of a specific human behavior to an underlying genetic cause. Human characteristics such as alcoholism and laziness were thought to be caused entirely by inherited genes. Since then, research has provided a much clearer picture of a genetic-behavior link. People inherit specific genes to build specific pathways that allow them to respond in certain ways to environmental input. With vari-

ations possible from the gene to gene regulators to the final cellular response, it is virtually impossible to disconnect the "nature versus nurture" tie that ultimately controls human behavior. Genes are simply the tools by which the environment shapes and reshapes human behavior. There is a direct correlation between gene and protein: Change the gene, change the protein. However, there is no direct correlation between gene and behavior: Changing the gene does not necessarily change the behavior. Behavior is a multifaceted, complex response to environmental influences that is only partially related to genetic makeup. Most studies conducted on humans based on twin and other relative data suggest that most behavioral characteristics have between a 30 and 70 percent genetic basis, leaving considerable room for environmental influence. For example, studies of twins indicate that homosexuality may be as much as 50 percent genetic, leaving 50 percent under environmental control.

Several genes were identified during the late 1980's and early to mid-1990's with possible direct behavioral links. A gene has been identified that seems to be involved in neurotic behavior such as anxiety, depression, hostility, and impulsiveness. This gene appears to be involved in the production of a chemical called serotonin, which is the chemical that is affected by the antidepressant drug Prozac. Another gene has been mapped to a region of the X chromosome that is thought to be involved in homosexuality, possibly by its influence on the development of a specific area of the brain stem thought to be connected to sexual preference. Scientists have also identified a gene that may be related to schizophrenia and a gene that may determine how well alcohol is cleared from the brain after overindulgence. Researchers are actively seeking additional and stronger links between behavior and genetics, but even when such links are found, the degree to which a particular gene is involved and the amount of variation among humans may be hard to uncover. One team has a project under way to determine the degree to which human personality traits are controlled by genes; the lead researcher speculates from early results that the ability to attain long-term happiness, for example, may be 80 percent genetic. If and when concrete evidence reveals a strong genetic basis for a particular type of behavior, humankind will be forced to face entirely new types of problems dealing with public reaction to and use of such information, as well as many additional ethical dilemmas.

—*W. W. Gearheart*

See Also: Classical Transmission Genetics; Eugenics; Genetic Code; Heredity and Environment.

Further Reading: Both *Nature and Nurture: An Introduction to Human Behavioral Genetics* (1990), by Robert Plomin, and *Born That Way: Genes, Behavior, Personality* (1998), by William Wright, provide thorough overviews of the connection between genes and behavior. R. Grant Steen's *DNA and Destiny: Nurture and Nature in Human Behavior* (1996), summarizes research regarding the relative contributions of genetics and environment to shaping the human personality. A view that genes have little to do with human behavior is presented by Robert Sapolsky in "A Gene for Nothing," *Discover* (October, 1997). An overview of how the human mind most likely develops is provided in "Special Report: Fertile Minds," *Time* (February, 1997).

Biochemical Mutations

Field of study: Molecular genetics
Significance: *The study of the biochemistry behind a particular phenotype is often necessary to understand why certain mutation genes are dominant, recessive, incompletely dominant, epistatic, or any other type. Knowledge of this biochemistry is especially useful in determining treatments for genetic diseases.*

Key terms

ALLELE: a form of a gene; most genes have at least two naturally occurring alleles
BIOCHEMICAL PATHWAY: the steps in the production or breakdown of biological chemicals in cells; each step usually requires a particular enzyme
ENZYME: a molecule that can increase the rate of chemical reactions in living organisms

GENOTYPE: the genetic characteristics of a cell or organism, expressed as a set of symbols representing the alleles present
HETEROZYGOUS: a genotype composed of two alleles that are different; also called hybrid
HOMOZYGOUS: a genotype composed of two alleles that are the same; "pure-bred"
PHENOTYPE: the physical characteristics of an organism, such as "tall" versus "short"

Proteins and Simple Dominant/Recessive Alleles

In order to understand how certain genotypes are expressed as phenotypes, knowledge of the biochemistry behind gene expression is essential. It is known that the various sequences of nitrogenous bases in the deoxyribonucleic acid (DNA) of genes code for the amino acid sequences of proteins. How the proteins thus coded act and interact in an organism determines that organism's phenotype.

Simple dominant and recessive alleles are easily described using the genetic disease phenylketonuria (PKU) as an example. Two alleles of the PKU gene exist: p^+, which codes for phenylalanine hydroxylase, an enzyme that converts phenylalanine (a common amino acid in proteins) to tyrosine (another common amino acid); and p, which is unable to code for this enzyme. Individuals with two normal alleles, p^+p^+, have the enzyme and are able to perform this conversion. However, individuals with two abnormal alleles, pp, do not have any of this enzyme and are unable to make this conversion. Since phenylalanine is not converted to tyrosine, the phenylalanine accumulates in the organism and eventually forms phenylketones, which are toxic to the nervous system and lead to mental retardation. The heterozygote, p^+p, has one normal and one abnormal allele. These individuals have phenylalanine and tyrosine levels within the normal range, since the enzyme can be used over and over again in the conversion. In other words, even when there is only one normal allele present, there is enough enzyme produced for the conversion to proceed at the maximum rate.

Many other inborn errors of metabolism follow this same scheme. In the case of albinism, afflicted people are missing the enzyme

The distinctive pigmentation pattern of Himalayan rabbits results from a biochemical mutation in a single gene. (Dan McCoy/Rainbow)

necessary to produce the brown-black melanin pigments. Galactosemics are missing an essential enzyme for the breakdown of galactose. The amino acid sequence of hemoglobin is altered in individuals with sickle-cell disease, and these individuals suffer many health problems on account of their altered hemoglobin. In the sex-linked disease hemophilia, individuals who have only the hemophilia allele are missing a factor that is necessary for the normal clotting of blood. In all these examples, heterozygous individuals who have one normal and one abnormal allele appear absolutely normal, since enough normal protein is produced for normal functioning.

Other Single-Gene Phenomena

Many other genetic phenomena can be explained by looking at the biochemistry behind them. For example, the "chinchilla coat" mutation in rabbits causes a gray appearance in the homozygous state, $c^{ch}c^{ch}$. This occurs because

the c^{ch} allele codes for a pigment enzyme that is partially defective. The partially defective enzyme works much more slowly than the normal enzyme, and the smaller amount of pigment produced leads to the gray phenotype. When this allele is heterozygous with the fully defective c allele, $c^{ch}c$, there is only half as much of an enzyme that works very slowly. As one might expect, there is less pigment produced, and the phenotype is an even lighter shade of gray called "light chinchilla." The enzyme concentration does affect the rate of the reaction and, ultimately, the amount of product made. This phenomenon is known as incomplete, or partial, dominance. Genes for the red pigments in such flowers as four-o'clocks and snapdragons show incomplete dominance, as do the hair, skin, and eye pigment genes of humans and the purple pigment genes of corn kernels.

Sometimes a mutation occurs that creates an enzyme with a different function instead of creating a defective enzyme. The B allele in the ABO blood-group gene codes for an enzyme that adds galactose to a short sugar chain that exists on the blood cell surface forming the B antigen. The A allele codes for an enzyme that adds N-acetylgalactosamine to the same previously existing sugar chain, forming the A antigen. Anyone with two B alleles, $I^B I^B$, makes only the B antigen and is type B. Those with two A alleles, $I^A I^A$, make only the A antigen and are type A. Heterozygotes, $I^A I^B$, have the enzymes to make both antigens, and they do. Since they have both antigens on their blood cells, they are classified as type AB. This phenomenon is known as codominance and is also seen in other blood-type genes.

Biochemistry can also explain other single-gene phenomena such as the pigmentation pattern seen in Siamese cats and Himalayan rabbits. The Siamese-Himalayan allele codes for an enzyme that is so unstable that it falls apart and is completely nonfunctional at the normal body temperature of most mammals. Only at cooler temperatures can the enzyme retain its stability and function. Since mammals have lower temperatures at their extremities, it is there that the enzyme produces pigment; at more centrally located body areas, it cannot function. This leaves a pattern of dark pigmentation on the tail, ears, nose, feet, and scrotum, with no pigmentation at other areas.

Multiple-Gene Phenomena

Few genes act completely independently, and biochemistry can be used to explain gene interactions. One simple interaction can be seen in fruit-fly eye pigmentation. There are two separate biochemical pathways to make pigment. One produces the red pteridines, and the other produces the brown omochromes. If b is an allele that cannot code for an enzyme necessary to make red pigments, a $b b r^+ r^+$ fly would have brown eyes. If r is an allele that cannot code for an enzyme necessary to make brown pigment, a $b^+ b^+ r r$ fly would have red eyes. When mated, the resulting progeny would be $b^+ b r^+ r$. They would make both brown and red pigments and have the normal brick-colored eyes. Interbreeding these flies would produce some offspring that were $bbrr$. Since these off-

spring make neither brown nor red pigments, they would be white-eyed.

Another multigene phenomenon that is seen when looking at the genes of enzymes that are in the same biochemical pathway is epistasis. Consider the following pathway in dogs:

$$colorless \rightarrow brown \rightarrow black$$

The a^+ allele codes for the enzyme that converts colorless to brown, but the a allele cannot, and the b^+ allele codes for the enzyme that converts brown to black, but the b allele cannot. The phenotype of an organism that is aab^+b^+ depends only on the aa genotype, since an aa individual produces no brown and the b^+b^+ enzyme can make black only by converting brown to black. The cross $a^+ab^+b \times a^+ab^+b$ produces the normal $9a^+\_b^+\_$ (black) : $3a^+\_bb$ (brown) : $3aab^+\_$ (white) : $1aabb$ (white) phenotypic ratio, but this collapses to a 9 black : 3 brown : 4 white ratio. (The symbol "_" is used to indicate that the second gene can be either dominant or recessive; for example, $A\_$ means that both AA and Aa will result in the same phenotype.) Other pathways give different epistatic ratios such as the following pathway in peas:

$$white \rightarrow white \rightarrow purple$$

If A codes for the first enzyme, B codes for the second enzyme, and a and b are the non-functional alleles, both $AAbb$ and $aabb$ are white. Their progeny when they are crossed, $AaBb$, is purple because it has both of the enzymes in the pathway. Interbreeding the $AaBb$ progeny gives a ratio of 9 purple : 3 white : 3 white : 1 white or 9 purple : 7 white.

Human pigmentation is another case in which many genes are involved. In this case, the various genes determine how much pigment is produced by nonalbino individuals. Several genes are involved, and the contributions of each allele of these genes is additive. In other words, the more functional alleles one has, the darker the pigmentation; the fewer one has, the lighter. Since many of the genes involved for skin, eye, and hair color are independent, ranges of color in all three areas are seen that may or may not be the same. In addition, there are genes that code for enzymes that produce chemicals that modify the expression of the pigment genes (for example, to change blue eyes to gray, convert hazel eyes to green, or change brown hair to auburn). This gives rise to the great diversity of pigmentation seen in humans today.

—*Richard W. Cheney, Jr.*

See Also: Epistasis; Incomplete Dominance; Mutation and Mutagenesis; One Gene-One Enzyme Hypothesis.

Further Reading: *The Metabolic Basis of Inherited Disease* (1989) has extensive coverage of the biochemistry behind many genetic diseases. Tom Strachan and Andrew Read, *Human Molecular Genetics*, also offers good coverage of the topic.

Bioethics

Field of study: Human genetics

Significance: *Bioethics is the practice of helping families, patients, and medical teams make tough health-care decisions. This branch of philosophy focuses on helping patients decide what is right for them while addressing the needs of families, health-care providers, and society.*

Key terms

INFORMED CONSENT: the right of patients to know the risks of medical treatment and to determine what is done to their bodies

GENETIC TESTING: the use of the techniques of genetics research to determine a person's risk of developing, or status as a carrier of, a disease or other disorder

The Emergence of Bioethics

As early as the mid-1960's, scholars began to turn their attention to ethical issues in medicine and scientific research. Advances in genetics and reproduction, life support, and transplantation technologies spurred the increase in attention. From the late 1960's through the mid-1970's, bioethicists were preoccupied with the moral difficulties of obtaining voluntary, informed consent from human subjects in scientific research. They concentrated on the development of ethical guidelines in research that would ensure the protection of individuals vulnerable to exploitation, including mentally or physically handicapped individuals, prisoners, and children. Beginning in the mid-1970's and continuing through the mid-1980's, bioethicists became increasingly involved in discussions of the definitions of life, death, and what it means to be human. In the mid-1980's, practitioners began to focus on cost containment in health care and the allocation of scarce medical resources.

Bioethicists worry about such matters as the guarantee of privacy, especially when compulsory testing for genetic disorders is involved, and about the limits of a person's right to threaten the health of others versus their personal right to freedom of choice. For example, the dissemination of information about genetic predispositions to chronic, costly, or incapacitating conditions can result in the denial of insurance coverage, job opportunities, and admittance to educational programs. Bioethicists also debate such matters as the use of people's reproductive materials—their eggs or sperm—to create embryos or fetuses without their explicit consent.

Beginning in 1992, the Joint Commission on Accreditation of Health Care Organizations, the U.S. agency that accredits hospitals and health-care institutions, required these organizations to establish committees to formulate ethics policies and address ethical conflicts and issues. Centers for the study of biomedical ethics such as the Society for Health and Human Values and the Park Ridge Center for the Study of Health, Faith, and Ethics have become important forums for public debate and research.

The overriding principle of bioethics and U.S. law is to respect each person's right to decide, free of coercion, what treatments or procedures he or she will undergo, except when the person making the decision is not competent because of youth, mental retardation, or medical deterioration. Other important rights discussed by bioethicists include a patient's right to know that medical practitioners are telling the truth and the right to know the risks of proposed medical treatment.

Impact and Applications

Advances in genetics and genetic testing have created a host of dilemmas for bioethicists, patients, and the health-care establishment. For example, as the ability to forecast and understand the genetic code progresses, people will have to decide whether knowing the future, even if it cannot be altered or changed, is a good thing for them or their children.

Bioethicists help people to decide whether genetic testing can be valuable for them. Factors typically considered before a person undergoes genetic testing include the nature of the test, the timing of the test, and the options that having the test information will bring. Testing can be done prenatally to detect disorders in fetuses; it can also be done before conception to determine whether a prospective parent is a carrier of a gene for a particular disorder or disease. Tests can also provide information about whether an adult is susceptible to or even in a presymptomatic state for a genetic disorder.

Practicing bioethicists help patients to focus on whether genetic testing will help them with the nature and severity of any disorders they or their children may have, the degree of disability or discomfort they may face, the costs and rigors of treatment, and the options that might be opened or closed as a result of testing. The key for consumers of genetic testing is whether the information obtained can be provided in time and at a time when it can help to guide treatments or family planning. Some affected persons need only to make lifestyle changes or take medications to help prevent or manage a disease; others learn that they or their offspring are at risk for, or even likely to develop, serious and often untreatable disorders. Knowing one's genetic fate may be more of a burden than a person wants, particularly if there is nothing that can be done to change or alter the risks the person faces. Bioethicists act as guides through the complicated and often wrenching decision process.

—*Fred Buchstein*

See Also: Gene Therapy: Ethical and Economic Issues; Genetic Counseling; Genetic Screening; Genetic Testing; Genetic Testing: Ethical and Economic Issues.

Further Reading: *Does It Run in the Family? A Consumer's Guide to DNA Testing for Genetic Disorders* (1997), by Doris Teichler Zallen, is an excellent introduction to genetic testing. Other helpful resources are *Due Consideration: Controversy in the Age of Medical Miracles* (1998), by Arthur Caplan, and *Bioethics: A Committee Approach* (1995), by Brendan Minogue.

Biofertilizers

Field of study: Genetic engineering and biotechnology

Significance: *Biofertilizers were used in agriculture long before chemical fertilizers became prevalent during and after the Industrial Revolution. The depleted soil fertility and contamination of ecosystems caused by the extensive use of chemical fertilizers, however, has prompted the redevelopment of biofertilizers, which are designed to work according to basic principles at work in nature, taking advantage of plants and other organisms to maintain healthy soil.*

Key terms

ALGAE: minute plants that live in fresh water; they are used as biofertilizers because of their high productivity and ability to fix atmospheric nitrogen

SYMBIOTIC RELATIONSHIP: a mutually beneficial association between two living organisms

NODULE: a symbiotic relationship between bacteria and plant roots that causes the conversion of nitrogen gas into a form readily accessible by plants

Overview

Biofertilizers are living microorganisms that either work alone or in association with other ingredients to enhance the fertility of soil. For many centuries, biofertilizers were used in organic farming in countries such as China, India, and Egypt until modernization resulted in a move toward the use of environmentally destructive chemical fertilizers. Organic agriculture integrates livestock, aquatic organisms, plants, and the scientific enhancement of natural processes to maintain ecological equilibrium, thus maximizing the production of foods

Species of algae have been used for centuries as naturally occurring biofertilizers. (Dan McCoy/Rainbow)

and goods through complete recycling of all resources. Biofertilizers may include microorganisms, nitrogen-fixing algae, green manure, plant residues, and sewer sludge. Biofertilizers not only provide an alternative method of farming but may also provide the only way to reduce environmental contamination and soil fertility depletion caused by chemical fertilizers.

Microorganisms and Algae

Microorganisms must be applied to the soil or mixed with seeds and other ingredients before they can establish "mycorrhiza," a symbiotic relationship with a plant's root system. They have been shown to stabilize manure, increase the amount of nitrogen in the soil, increase root surface area for absorption, and control the leaking of nitrogen into the groundwater. Another type of association between plant roots (particularly legumes) and nitrogen-fixing bacteria is a symbiotic relationship called a "nodule." Nodules are natural "factories" that do not consume fossil fuel yet produce ample fertilizers by converting nitrogen into ammonium, a form of nitrogen that may be used directly by plants or deposited slowly into the soil, thus enhancing its fertility. While continuous application of chemical fertilizers depletes the soil's natural fertility and destroys beneficial microorganisms, natural fertilizers produced by nodules alleviate the contamination to the ecosystem. Unfortunately, nodules are formed only between legumes and bacteria, thus limiting their use. Genetically engineered bacteria are being used to improve the efficiency of nitrogen fixation, and some researchers have attempted to transfer genes responsible for forming such symbiotic relationships from legumes to other plant species.

Algae are minute plants that are almost entirely aquatic. They grow and reproduce very rapidly during the growing season but die off during the nongrowing season. Such a "boom and bust" life cycle provides the soil with substantial amounts of nutrients through the degradation of dead algae and the deposit of the nitrogen fixed by algae during the growing season. Two types of algae, azospirillum and azotobacter, have been used as biofertilizers in rice fields for centuries in Southeast Asia. Algae are the only plants that are able to fix nitrogen by themselves. The cost of raising algae is minimal, but if they are grown too densely, they may become weeds and suffocate aquatic animals.

Green Manure, Plant Residues, and Treated Sewer Sludge

Some crops, particularly legumes, are grown and harvested to be used as green manure to restore the soil's fertility. They may be used directly or mixed with microorganisms. Green manure releases nutrients slowly and provides long-term fertility for soil. After grains are harvested for food, plant residues are processed into fertilizer in one of two ways. They may be burned to extract energy, after which the ashes are applied to the soil. Alternatively, they may be fermented in a sealed underground tank to make methane to be used as "natural" gas or ethanol, which may become an increasingly important fuel source. Genetically engineered microbes convert fiber-rich crop residues such as wheat straw and corn stalks into ethanol. Solid residue left over after fermentation is used as a biofertilizer. The fermentation approach is promising because it is environmentally friendly and also creates alternative energy sources, additional income, and market opportunities for farmers.

Even though sewer sludge has been used as a fertilizer in developing countries for some time, people are more resistant to it because of the animal and human wastes present in the sludge. Proper treatments must be in place to get rid of possible heavy metals or pathogens. More research is needed on sewer sludge before it is considered for wide use as a fertilizer.

—*Ming Y. Zheng*

See Also: Biopesticides; Biotechnology; Genetic Engineering: Agricultural Applications; Genetically Engineered Foods; High-Yield Crops.

Further Reading: *Organic Farming: Current Technology and Its Role in a Sustainable Agriculture* (1984), edited by D. F. Bezdicek et al., contains easy-to-understand discussions on many aspects of organic farming, including nutrient needs from organic sources, reducing soil erosion and nutrient loss, and decreasing energy inputs. *Mycorrhizae in Sustainable Agriculture*

(1992), by G. J. Bethlenfalvay and R. G. Linderman, provides excellent discussions on the knowledge and importance of mycorrhizae to plants and the soil. P. A. Matson et al., "Agricultural Intensification and Ecosystem Properties," *Science* (July 25, 1997), provides useful information on agriculture intensification and its sustainability, a strategy that can help reduce negative consequences.

Biological Determinism

Field of study: Human genetics

Significance: *Biological determinists argue that there is a direct causal relationship between the biological properties of human beings and their behavior. From this perspective, social and economic differences between human groups can be seen as a reflection of inherited and immutable genetic differences. This contention has been used by groups in power to claim that stratification in human society is based on innate biological differences. In particular, biological determinism has been used to assert that certain ethnic groups are biologically defective and thus intellectually, socially, and morally inferior to others.*

Key terms

DETERMINISM: the doctrine that everything, including one's choice of action, is determined by a sequence of causes rather than by free will

INTELLIGENCE QUOTIENT (IQ): performance on a standardized test, often assumed to be indicative of an individual's level of intelligence

REIFICATION: the oversimplification of an abstract concept such that it is treated as a concrete entity

REDUCTIONISM: the explanation of a complex system or phenomenon as merely the sum of its parts

The Use of Inheritance to Promote Social Order

The principle of biological determinism lies at the interface between biology and society. A philosophical extension of the use of determinism in other sciences such as physics, biological determinism views human beings as a reflection of their biological makeup and hence simple extensions of the genes that code for these biological processes. Long before scientists had any knowledge of genetics and the mechanisms of inheritance, human societies considered certain groups to be innately superior by virtue of their family or bloodlines (nobility) while others were viewed as innately inferior (peasantry). Such views served to preserve the social order. According to evolutionary biologist Stephen Jay Gould, Plato himself circulated a myth that certain citizens were "framed differently" by God, with the ranking of groups in society based on their inborn worth.

As science began to take a more prominent role in society, scientists began to look for evidence that would justify the social order. Since mental ability is often considered to be the most distinctive feature of the human species, the quantification of intelligence was one of the main tactics used to demonstrate the inferiority of certain groups. In the mid-1800's, measurements of the size, shape, and anatomy of the skull, brain, and other body features were compiled by physician Samuel George Morton and surgeon Paul Broca among others. These measurements were used to depict races as separate species, to rank them by their mental and moral worth, and to document the subordinate status of various groups, including women. In the first decades of the twentieth century, such measurements were replaced by the intelligence quotient (IQ) test. Although its inventor, Alfred Binet, never intended it to be used in this way, psychologists such as Lewis M. Terman and Robert M. Yerkes promoted IQ as a single number that captured the complex, multifaceted, inborn intelligence of a person. IQ was soon used to restrict immigration, determine occupation, and limit access to higher education. Arthur Jensen, in 1979, and Richard Herrnstein and Charles Murray, in 1994, have reasserted the claim that IQ is an inherited trait that differs among races and classes.

Problems with the Principle of Biological Determinism

Geneticists and sociobiologists (who study the biological basis of social behavior) have uncovered a variety of animal behaviors that are influenced by biology. However, the genetic makeup of an organism ("nature") is expressed only within the specific context of its environment ("nurture"). Thus genes that are correlated with behavior usually code for predispositions rather than inevitabilities. For such traits, the variation that occurs within a group is usually greater than the differences that occur between groups. In addition, the correlation between two entities (such as genes and behavior) does not necessarily imply a causal relationship (for example, the incidence of ice cream consumption and drowning are correlated only because both increase during the summer). Complex, multifaceted behaviors such as intelligence and violence are often reified, or treated as discrete concrete entities (as IQ and impulse control, respectively), in order to make claims about their genetic basis. Combined with the cultural and social bias of scientific researchers, reification has led to many misleading claims regarding the biological basis of social structure.

Biological and cultural evolution are governed by different mechanisms. Biological evolution occurs only between parents and offspring (vertically), while cultural evolution occurs through communication without regard to relationship (horizontally) and thus can occur quickly and without underlying genetic change. Moreover, the socially fit (those who are inclined to reproduce wealth) are not necessarily biologically fit (inclined to reproduce themselves). The reductionist attempt to gain an understanding of human culture through its biological components does not work well in a system (society) shaped by properties that emerge only when the parts (humans) are put together. Cultures cannot be understood as biological behaviors any more than biological behaviors can be understood as atomic interactions.

Impact and Applications

Throughout history, biological determinism has been used to justify or reinforce racism, genocide, and oppression, often in the name of achieving the genetic improvement of the human species (for example, the "racial

health" of Nazi Germany). Gould has noted that claims of biological determinism tend to be revived during periods when it is politically expedient. In times of economic hardship, many find it is useful to adopt an "us against them" attitude to find a group to blame for social and economic woes or to free themselves from the responsibility of caring for the "biologically inferior" underprivileged. As advances in molecular genetics lead to the identification of additional genes that influence behavior, society must guard against using this information as justification for the mistreatment or elimination of groups that are perceived as "inferior" or "undesirable" by the majority.

—*Lee Anne Martínez*

See Also: Behavior; Eugenics; Eugenics: Nazi Germany; Heredity and Environment; Intelligence; Race.

Further Reading: In *The Scientific Attitude* (1992), Frederick Grinnell discusses the social and cultural nature of scientific research. *The Bell Curve: Intelligence and Class Structure in American Life* (1994), by Richard Herrnstein and Charles Murray, reintroduces the claim of innate group differences in IQ. Their claim is soundly refuted by Stephen Jay Gould in *The Mismeasure of Man* (1996), which presents an engaging historical overview of how pseudoscience has been used to support racism and bigotry. In "Gray Matters," *Newsweek* (March 27, 1995), Sharon Begley demonstrates the popular appeal of differences between the brains of males and females. In "The Rise of Neurogenetic Determinism," *Nature* 373 (February, 1995), Steven Rose discusses how such advances in neuroscience have led to a resurgence of the interpretation that genes are responsible for deviant human behavior.

Biopesticides

Field of study: Genetic engineering and biotechnology

Significance: *As an alternative to chemical pesticides, agricultural scientists have begun using ecologically safer methods such as biopesticides to protect plants from insects.*

Key terms

TRANSGENIC SPECIES: an organism synthesizing a foreign protein the gene of which was obtained from a different species of organism

TRANSFORMATION: the process of transferring a foreign gene into an organism

BACULOVIRUS: a strain of virus that is capable of causing disease to a variety of insects

BACILLUS THURINGIENSIS: a species of bacteria that produces a toxin deadly to caterpillars, moths, beetles, and certain flies

AGROBACTERIUM TUMEFACIENS: a species of bacteria that is able to transfer genetic information into plant cells

Bacillus Thuringiensis

Hungry insects are the bane of gardeners. This problem is worsened for farmers, whose livelihoods depend on keeping fields free of destructive insects. Although effective, chemical pesticides have a variety of drawbacks. The increasing popularity of organically grown produce that is untreated by chemicals suggests that consumers are wary that human-made pesticides may hold hidden dangers. In response to consumers' worries over chemical-pesticide safety, agricultural biologists have turned to nature to solve pest problems. Biopesticides are insecticides taken from nature. They are designed not in a laboratory but through evolution, making them very specific and effective. A biopesticide may be sprayed directly on crops or may be genetically engineered to be produced by a crop itself.

Since the 1950's, the bacterial pesticide *Bacillus thuringiensis* (*Bt*) has been used on crops susceptible to destruction by insect larvae. Upon sporulation, *Bt* produces a crystallized protein that is toxic to many forms of larvae. The protein is synthesized by the bacteria as an inactive proenzyme. Upon digestion, enzymes in the insect's gut cleave the protein into an active, toxic fragment. The active toxin binds to receptors in the insect's midgut cells and

The drawbacks associated with chemical pesticides have prompted agricultural scientists to develop biopesticides and other genetically engineered treatments for crops. (Ben Klaffke)

blocks those cells from functioning. Only caterpillars (tobacco hornworms and cotton bollworms), beetles, and certain flies have the gut biochemistry to activate the toxin. The toxin does not kill insects that are not susceptible, nor does it harm vertebrates in any way.

The drawbacks of *Bt* are its expense and short-lived effect. In the early 1990's, scientists overcame these two drawbacks through applied genetic engineering. They produced transgenic cotton plants that generated their own *Bt* toxin. The toxin gene was first isolated from *Bt* cells and ligated (enzymatically attached) into a Ti-plasmid. A Ti-plasmid is a circular string of double-stranded deoxyribonucleic acid (DNA) that originates in the *Agrobacterium tumefaciens* bacteria. The *A. tumefaciens* has the ability to take a portion of that Ti-plasmid, called the T-DNA, and transfer it and whatever foreign gene is attached to it into a plant cell. Cotton plants were exposed to the *A. tumefaciens* carrying the toxin gene and were transformed. The transgenic plants synthesized the *Bt* toxin and became resistant to many forms of larvae.

Many crystal toxins have been isolated from various strains of *Bt*. These toxins make up a large collection of proteins active against pests from nematodes to aphids. Researchers are in the process of reengineering the toxin genes to improve upon their characteristics and to design better methods of transporting genes from one *Bt* strain to another.

Other Biopesticides

Several species of fungi have been found to be toxic to insects, including *Verticillium lecanii* and *Metarhizium anisopliae*. Scientists have just begun to study these fungi, which are not yet commercially available to farmers. In the mid-1990's, a viral biopesticide called baculovirus became widely popular. Baculoviruses are sprayed onto high-density pest populations just like chemical pesticides. Baculoviruses have several advantages over conventional pesticides. The most important advantage is their strong specificity against moths, sawflies, and beetles but not against beneficial insects. Also, viruses, unlike bacteria, tend to persist in the environment for a longer period. Finally, baculoviruses are ideal for use in developing countries because they can be produced cheaply and in great quantity with no health risks to workers. One limitation of baculovirus is that it must be administered in a precise temporal and spatial framework to be effective. Knowledge of insect behavior after hatching, the insect population's distribution within the crop canopy, and the volume of foliage ingested by each larva is essential. For example, moths usually do the most damage at the late larval stage. To minimize crop damage from moths, one must spray baculovirus as early as possible before the insects reach that late stage.

One final biopesticide approach has been to make transgenic plants that manufacture proteins isolated from other insect-resistant plant species. Tomatoes naturally make an enzyme inhibitor that deters insects by keeping their digestive enzymes (trypsin and chymotrypsin) from functioning. These inhibitors were isolated by Clarence Ryan at the University of Washington. Ryan transformed tobacco plants with two different forms of inhibitor (inhibitors I and II from tomato). The tomato proteins were effectively produced in tobacco and made the transgenic plants resistant to tobacco hornworm larvae.

Biopesticide Resistance

As with chemical pesticides, over time insect populations grow resistant to biopesticides. *Bt*-resistant moths can now be found around the world. Resistance arises when pesticides are too effective and destroy more than 90 percent of a pest population. The few insects left are often very resistant to the pesticide, breed among themselves, and create large, resistant populations.

Entomologists have suggested strategies for avoiding pesticide-resistant insect populations. One strategy suggests mixing biopesticide-producing and nonproducing plants in the same field, thereby giving the pesticide-susceptible part of the insect population places of refuge. These refuges would allow resistant and nonresistant insects to interbreed, making the overall species less resistant. Other strategies include synthesizing multiple types of *Bt* toxin in a single plant to increase the toxicity range and reduce resistance, making other biological toxins besides *Bt* in a single plant, and reducing the overall exposure time of insects to the biopesticides.

—*James J. Campanella*

See Also: Genetic Engineering; Genetic Engineering: Agricultural Applications; Genetically Engineered Foods; Transgenic Organisms.

Further Reading: James Watson et al., *Recombinant DNA* (1992), furnishes background on the agricultural applications of biotechnology. Joseph Sambrook et al., *Molecular Cloning: A Laboratory Manual* (1989), provides the best source for methods and background in cloning and transformation. Jenny Cory and David Bishop, "Use of Baculoviruses as Biological Insecticides," *Molecular Biotechnology* (March, 1997), presents a complete discussion on baculoviruses. Anne Gibbons, "Moths Take the Field Against Biopesticides," *Science* (November 1, 1991), discusses resistance to biopesticides and its implications.

Biopharmaceuticals

Field of study: Genetic engineering and biotechnology

Significance: *Biopharmaceuticals are medicines that are mass-produced through the use of biotechnology. Hormones, interferon, blood-clotting factors, antibiotics, and enzymes are examples of biopharmaceuticals that have been produced.*

Key terms

CLONE: a population of cells or organisms descended from one cell or one organism and hence genetically identical

DEOXYRIBONUCLEIC ACID (DNA): the genetic material for most organisms, a long-chain macromolecule made of units called "nucleotides"

RECOMBINANT DNA TECHNOLOGY: methods used to splice a DNA fragment from one organism into DNA from another organism and then clone the new recombinant DNA molecule

TRANSGENIC ORGANISM: an organism that contains foreign DNA

Transgenic Animals

In 1976, the methods of recombinant deoxyribonucleic acid (DNA) technology, such as DNA cloning, nucleotide synthesis, and gene expression, were used to express a human protein from recombinant DNA for the first time. The protein was somatostatin, a fourteen-amino-acid neurotransmitter. The first commercial product of biotechnology was human insulin produced in bacteria. Insulin for the treatment of diabetes had previously been extracted from the pancreases of pigs and cows, but some people developed allergic reactions to such insulin. The cloning and expression of human insulin made it possible to treat diabetics with insulin that was identical to that normally produced by humans.

Transgenic livestock can be used to produce biopharmaceuticals. Through recombinant DNA methods, genes for useful pharmaceutical proteins can be produced in the milk of transgenic livestock. To do this, a cloned gene of a useful pharmaceutical protein is fused to the transcriptional promoter of a gene that is expressed in milk, such as b-lactoglobin. The recombinant DNA is then introduced into the animal. Typically, the gene is introduced into a fertilized egg by microinjection, and the egg is implanted into a female animal. Female offspring are checked for the presence of the transgene and the expression of the recombinant protein in their milk. If the recombinant protein is detected, it is relatively simple to isolate it from the milk. Although sheep and goats have been used, cows may be the preferred animals to use to create biopharmaceuticals because they produce large quantities of milk (10,000 liters per year). An example of a transgenic biopharmaceutical is tissue plasminogen activator (tpa), which is administered to heart-attack victims to dissolve clots.

Hemophilia A is a genetic disease in which affected individuals lack clotting factor VIII, which is needed for efficient clotting of blood. Hemophiliacs are treated with factor VIII concentrates prepared from human plasma or from a recombinant cloned factor VIII gene expressed in mammalian cell culture. Both of these procedures are costly. The use of human plasma also includes the risk of infecting the patient with blood-borne viruses. A less expensive way to obtain factor VIII may be to produce it in the milk of transgenic animals. In 1997, Rekha Paleyanda and coworkers reported the successful production of functional human factor VIII in the milk of transgenic pigs.

The production of transgenic animals is not very efficient. Many fertilized eggs must be microinjected with the recombinant DNA, and many injected eggs must be implanted before a transgenic animal that produces the desired protein product in its milk at high levels is obtained. In the future, once such a transgenic female animal is bred, it may be more efficient to clone that entire animal to create genetically identical animals than to try to duplicate the results by producing different transgenic animals.

Basic Research

Basic research using recombinant DNA methods has been essential in gaining a better understanding of biological systems, which allows better drugs to be designed to affect the systems. For example, many microbes are used to make antibiotics. Biotechnology has facilitated an increase in the knowledge of the biosynthetic pathways that microbes use to make these compounds. This increased understanding of the pathways has made it easier to modify the pathways to produce modified compounds that may serve as better antibiotics. Another example is the development of more specific and less toxic drugs for cancer treatment. Many researchers have worked for decades to study the biology of cancer. Genes such as oncogenes and tumor suppressors have been identified, and many of the steps that lead toward the production of cancers have been determined. This knowledge is being used to design drugs that are aimed at counteracting the genetic changes that cause cancer or to design drugs that block some of the signaling pathways that are overactive in cancer.

DNA Vaccines

The use of recombinant DNA technology has led to the development of better vaccines. A conventional vaccine uses a killed or disarmed pathogen that is injected into an organism. The organism develops an immune response to the material injected and becomes resistant to the pathogen. Relatively rare side effects of conventional vaccines include an allergic reaction in some individuals and the possibility that a disarmed pathogen could become active and cause the disease. Recombinant DNA technology is used to produce vaccines that never contained the live pathogen. In such vaccines, only a protein of the pathogen (for example, the coat protein of a pathogenic virus) is used to stimulate the organism to develop immunity. DNA vaccines, in which only a small piece of the pathogen DNA is used to immunize an organism, are being developed and appear to be promising in the development of a vaccine for malaria. In the early 1990's, Steve Hoffman and coworkers at the Naval Medical Research Institute tried to make vaccines against malaria by using the whole malaria-causing protozoan or genetically engineered proteins from the protozoan. Their attempts with conventional vaccines failed to produce a strong immune response in the mice on which they tested the materials. In a DNA vaccine, a gene from the pathogen is placed in a circular bacterial DNA called a "plasmid," which is injected into the mice. The malaria DNA vaccine produced a strong immune response in mice and protected them from the disease. In late 1997, about six DNA vaccines for diseases such as acquired immunodeficiency syndrome (AIDS), influenza, and cancer were in clinical trials. Although DNA vaccines sound promising, scientists do not completely understand how they work, and no DNA vaccine has been effective in a human study.

Creating edible vaccines by introducing genes from pathogens into food plants was suggested by Charles J. Arntzen of Cornell University. In 1998, William H. R. Langridge and coworkers at Loma Linda University reported that they cloned a gene into potatoes that caused the potato to produce a nontoxic component of the cholera toxin. When mice ate potatoes with the gene, they produced antibodies against cholera. The researchers found that if they cooked the potatoes, at least half of the vaccine survived in the biologically active form.

In a 1996 editorial in *Science* magazine, Philip

H. Abelson estimated that one thousand U.S. biotechnology companies were involved in the development of new medicines. In early 1996, sixteen pharmaceuticals based on biotechnology had been approved by the Food and Drug Administration (FDA), and an additional 150 biopharmaceuticals were in clinical trials. Abelson predicted that the pharmaceutical companies would become involved in the determination of the DNA sequences of genomes of humans and other organisms. This genome information will lead to a better understanding of disease processes that will aid in the creation of better drugs to treat those diseases.

—*Susan J. Karcher*

See Also: Biotechnology; Cloning; Genetic Engineering; Knockout Genetics and Knockout Mice; Transgenic Organisms.

Further Reading: Eric S. Grace gives a basic overview of biotechnology in *Biotechnology Unzipped: Promises and Realities* (1997). William H. Velander et al. provide an overview of biopharmaceutical applications of transgenic animals in their article "Transgenic Livestock as Drug Factories," *Scientific American* 276 (January, 1997). James D. Watson et al., *Recombinant DNA* (1992), provides an overview of biotechnology methods and applications. Biopharmaceutical drugs used to fight cancer are summarized in *Science* 278 (November 7, 1997), in "From Benchtop to Bedside," by Marcia Baringa, and "On the Biotech Pharm: A Race to Harvest New Cancer Cures," by Wade Roush. Gary Taubes discusses DNA vaccines in the article "Salvation in a Snippet of DNA?" *Science* 278 (December 5,1997).

Biotechnology

Field of study: Genetic engineering and biotechnology

Significance: *Biotechnology is a technique used to reveal the complex processes of how genes are inherited and expressed, provide better understanding and effective treatment of various diseases (particularly genetic disorders), and generate economic benefits that include improved plants and animals for agriculture and efficient production of valuable biopharmaceuticals. Biotechnology provides both vast promise and a potential threat to humankind: It will not only shape medicine and agriculture in the future but also challenge human ethical systems and religious beliefs.*

Key terms

DEOXYRIBONUCLEIC ACID (DNA): the genetic material for most organisms

GENE CLONING: the development of a line of genetically identical organisms

GENE THERAPY: the insertion of a working gene or genes into a cell to correct a genetic abnormality

POLYMERASE CHAIN REACTION (PCR): a process by which specific parts of a gene or DNA can be made into millions or billions of copies within a short time

RECOMBINANT DNA: a hybrid DNA molecule created in a test tube by joining a DNA fragment of interest with a carrier DNA

RESTRICTION ENDONUCLEASES: enzymes that recognize and cut DNA at specific sites

SOUTHERN BLOT: a procedure used to transfer DNA from a gel to a nylon membrane, which in turn allows researchers to locate genes that are complementary to particular DNA sequences called "probes"

A Brief History of Biotechnology

After the discovery of the double-helical structure of deoxyribonucleic acid (DNA) by English physicist Francis Crick and American microbiologist James Watson in 1953, human curiosity regarding this magical molecule propelled the advancement of biological sciences in unprecedented fashion. The real story of biotechnology, or recombinant DNA, began with the discovery of the first restriction endonuclease by Hamilton Smith in 1970. These enzymes, which are produced naturally by bacteria, are able to recognize specific amino acid sequences within DNA molecules and to cut the DNA chain at these sites. The specificity exhibited by each enzyme ensures that DNA fragments of reproducible size can be generated and that a DNA fragment bearing a particular gene can be cut out from the rest of the DNA molecule and moved from one organism to another. The discovery of another enzyme, reverse transcriptase, by American virologists David Baltimore and Howard Temin in 1970 proved to be another important step toward developing biotechnology. The genetic information stored in DNA is usually transferred from DNA to messenger RNA (mRNA) through a process called transcription. The message on mRNA is then translated into protein or enzyme. The reverse transcriptase, produced by only some viruses, uses mRNA to make copies of DNA. In other words, this enzyme provides an important method of making a complementary copy of a gene in the laboratory by exposing the relevant mRNA to the enzyme. For instance, by using mRNA extracted from a mouse's pituitary gland, it is possible to make a copy of a growth hormone gene in DNA form.

The next important development occurred in 1972, when DNA fragments from two different organisms were joined together to produce a biologically functional hybrid DNA molecule. The first group of "vectors" used to carry DNA were plasmids, small, circular, naturally occurring DNAs in bacteria that replicate independently of the host bacterial DNA. Plasmids provide vehicles for the movement of DNA fragments or genes; when taken up by a bacterium, the latter can then be grown in vessels to produce clones with multiple copies of the incorporated DNA fragment.

The next milestone came in 1975 when Edward Southern introduced a technique with many applications that proved invaluable for subsequent development of biotechnology. This technique, Southern blotting, is used to identify a particular gene or DNA fragment from a mixture of thousands of different genes or DNA fragments. It consists of several steps. First, the DNA fragments generated by restriction endonuclease are separated according to their size through a process called gel electrophoresis. Second, the double-helix DNA in the gel is made into a single strand with an alkali solution (a process of denaturation). Third, the denatured DNA fragments are transferred to a nylon membrane by laying the membrane over the gel.

The 1970 discovery of the reverse transcriptase enzyme by American virologists David Baltimore (above) and Howard Temin represented a milestone in the development of biotechnology. (The Nobel Foundation)

At the end of blotting, DNA becomes firmly bound to the membrane. This makes it possible to fix DNA fragments or particular genes on the membrane at the exact positions they occupied on the original gel. Finally, a particular gene or DNA fragment may be detected by a probe DNA that is complementary to the gene or particular DNA fragment. The probe DNA is usually made radioactive, which allows the DNA to be seen when the membrane is exposed to an X-ray film in the dark (a technique referred to as autoradiography).

Using such techniques, researchers made the first prenatal diagnosis of a genetic disease in 1976 for α-thalassemia, a genetic disorder caused by the absence of globin genes. This represented a monumental step forward in the use of biotechnology in prenatal diagnosis. It paved the way for later developments in which mutations in many genes could be detected in early pregnancy. Three years later, insulin was first synthesized using recombinant DNA. In 1982, the commercial production of genetically engineered human insulin became a reality. Other developments included the invention of polymerase chain reaction (PCR) in 1987, publication of the first complete human genetic map in 1993, and experimentation with various new techniques in DNA fingerprinting and isolation of specific genes. Trials of gene therapy began with cystic fibrosis in 1994, after which an increasing number of genetic disorders were treated in preclinical or clinical trials with gene therapy.

Prevention of Genetic Disorders

Many human diseases such as cystic fibrosis, Down syndrome, Huntington's chorea, muscular dystrophy, sickle-cell anemia, and Tay-Sachs disease are inherited. There are usually no conventional treatments for these disorders. However, commercial production of pharmaceuticals using biotechnology has emerged as a rapidly developing field. Though tremendous progress has been made, it is far too early to conclude that a cure for genetic disorders is in place. In fact, because of the technical difficulty in permanently correcting defective genes, the best technique for curing genetic disorders seems to be prevention, which can be achieved

through various means. The best prevention, if methods are available, is to reduce or eliminate the frequency of mutations. Prevention may also be achieved by avoiding the environmental factors that cause the abnormality. However, genetic diseases are most commonly prevented by ascertaining those individuals in the population who are at risk of passing a serious genetic disorder to their offspring and offering them genetic counseling and prenatal diagnosis with selective abortion of affected fetuses.

Genetic counseling is the process of communicating information gained through classic genetic studies and biotechnology research to those individuals who are themselves at risk or have a high likelihood of passing a defect to their offspring. During counseling, information about the disease itself—its severity and prognosis, the availability of effective therapy, and the risks of recurrence—is generally presented. For those couples who find the risks unacceptably high, this process may also include discussions about contraceptive methods, adoption, prenatal diagnosis, abortion, and artificial insemination by a donor. Even though the final decision must still rest with the couple, the great improvement in the accuracy of risk assessment made possible with biotechnology makes it easier for parents to make well-informed decisions.

Prenatal diagnosis can be performed for a variety of genetic disorders. It requires samples of fetal cells or chemicals produced by the fetus through either amniocentesis or chorionic villus sampling. After sampling, several analyses can be performed. Biochemical analysis is used to determine the concentration of chemicals in the sample and therefore diagnose whether a particular fetus is deficient or low in enzymes that facilitate specific biological reactions. Analysis of the chromosomes of the fetal cells can show if all the chromosomes are present and whether there are structural abnormalities in any of them. The most effective means of analysis is to detect the defective genes through recombinant DNA techniques. This has become possible with the rapid increase of DNA copies through a technique called polymerase chain reaction (PCR), which can produce virtually unlimited copies of a specific gene or

DNA fragment, starting with as little as a single copy. Routine prenatal diagnosis can be performed to screen a fetus for Down syndrome, Huntington's chorea, sickle-cell anemia, and Tay-Sachs disease; meanwhile, biotechnological procedures are being developed for prenatal diagnosis of an increasing number of severe genetic disorders.

Treatment of Diseases and Genetic Disorders

Biotechnology may be used for direct treatments of diseases or genetic disorders through various means, including the production of a possible vaccine for acquired immunodeficiency syndrome (AIDS), treatment for various cancers, and synthesis of biopharmaceuticals for various metabolic, growth, and developmental diseases. These potentials may be reached through the use of biosynthesis and gene therapy.

In general, biosynthesis is a process in which genetic coding for a particular product is isolated, cloned into another organism (mostly bacteria), and later expressed in that organism. By cultivating bacteria, scientists can harvest and purify large quantities of the gene products. For example, insulin is essential for the treatment of insulin-dependent diabetes. Historically, insulin has been obtained from beef or pig pancreases. However, many pancreases are needed to extract enough insulin for continuous treatment of one patient, and insulin obtained in this manner is not chemically identical to human insulin; therefore, some patients may produce antibodies that can seriously interfere with the treatment. Human insulin produced through biotechnology is quite effective and is without any side effects. It was first produced commercially and made available to patients in 1982.

Another successful story in biosynthesis is the production of human growth hormone (HGH), which is used in the treatment of children with a form of growth retardation called pituitary dwarfism. The successful biosynthesis of HGH is important for several reasons. The conventional source of HGH has been human pituitary glands removed at autopsy, which only exist in the brain and the liver. Each child afflicted with pituitary dwarfism needs two injections per week until the age of twenty. Such a treatment regime requires over one thousand pituitaries. Autopsy supply could hardly keep up with the demand. Furthermore, because of a small amount of viral contamination in extracted HGH, many children receiving treatment developed virus-related diseases. Other biopharmaceuticals under development or in clinical trials include anticancer drugs, antiaging agents, and a possible vaccine for AIDS.

Though the idea of correcting defective genes once and for all sounds attractive, gene therapy has been directed toward temporary correction of the symptoms. A cloned human gene is transferred into a viral vector, which is used to infect white blood cells removed from the patient. The transferred (normal) gene is then inserted into a chromosome and becomes active. After they are allowed to multiply in sterile conditions, the cells are reimplanted into the patient, where they produce a gene product that is missing in the untreated patient, allowing the individual to function normally. Several disorders have been treated with this technique, including severe combined immunodeficiency (SCID). Individuals with SCID have no functional immune system and usually die from infections that would be minor in normal people. Gene therapy has also been used or contemplated as a treatment for cystic fibrosis, skin cancer, breast cancer, brain cancer, and AIDS. By 1997, nearly twenty biotechnology companies in the United States alone had been created specifically to develop products for gene therapy. A number of such products have reached the marketplace. However, most of these treatments are only partially successful and are prohibitively expensive. They can only be performed at major medical centers with well-equipped facilities.

Agriculture and Food Production

There are numerous ways that biotechnology may be used to benefit agriculture and food production. The production of vaccines and the application of methods for transferring genes for commercially important traits such as milk yield, butter fat, and higher proportion of lean meat is likely to benefit animal husbandry.

For example, a bovine growth hormone produced through biotechnology was developed in the late 1980's to boost milk production by cows. A mutant form of the myostatin gene nicknamed the "Schwarzenegger gene" has been identified and found to cause heavy muscling of mice and, more important, Belgian blue bulls. This marks the first step toward breeding cows and meat animals with lower fat and a higher proportion of lean meat. Other examples of using biotechnology in animal husbandry include hormones for faster growth rate in poultry and the production of recombinant human proteins in the milk of livestock.

Biotechnology is expected to dramatically alter the conventional approaches to developing new strains of crops through breeding. The technology allows transferral of genes for nitrogen fixation; improvement of photosynthesis (and therefore yield); resistance to pests, pathogens, and herbicides; tolerance to frost, drought, and increased salinity; and improvement of nutritional value and consumer acceptability. The first biotechnologically produced potato was approved for human consumption by the U.S. government in 1995 and by Canada in 1996. The NewLeaf potato, developed by corporate giant Monsanto in St. Louis, Missouri, carries a gene from the bacterium *Bacillus thuringiensis*. This gene produces a protein toxic to the Colorado potato beetle, an insect that causes substantial loss of the crop if left uncontrolled. The production of this protein by potato plants equips them with resistance to beetles and thus alleviates crop loss, saves costs on pesticides, and reduces the risk of environmental contamination. Antiviral genes have been successfully transferred and expressed into cotton, and the release of a new cotton strain with resistance to multiple viruses is just a matter of time. At least five transgenic corn strains with resistance to herbicides or pathogens have been developed. Some genes coding for tolerance to drought and subfreezing temperatures have been cloned and transferred into or among crop plants, some of which have already made a great impact on agriculture in developing countries. Initial efforts have been made to replace chemical fertilizers with more environ-

mentally friendly biofertilizers. Secondary metabolites produced naturally by plants have also been purified and used as biopesticides.

Impact and Applications

As biotechnology expands, it will produce an endless list of possible biomedical applications. Two areas in particular are worth noting: microbial biotechnology and the use of biotechnology in forensic science and DNA profiling. It is necessary to develop new antimicrobial therapies because of the progressive rise of antibiotic-resistant infectious diseases. The need for new antimicrobial drugs is so prominent that some have placed it as the highest priority of biotechnology. Also, accurate DNA fingerprinting can help solve crimes, identify parentage, settle disputes over who owns the right to certain property, and advance the study of the genealogies of various species.

Biotechnology has opened new avenues for possible correction or treatment of genetic disorders and renovation of agriculture and food production, and it also presents a mixed blessing of invaluable benefits and dilemmas, which science and technology have always offered to humankind. There are those who would like to restrict the uses of biotechnology and who might prefer that such technology had never been developed. Others feel that the benefits far outweigh the possible risks and that any potential threat can easily be overcome by simple legal means. Others do not take either side on biotechnology in general but are greatly concerned with some specific applications. In short, the new powers of biotechnology demand a new set of decisions, both ethical and economic, by individuals, governments, and all of society.

Considerable concern has been expressed about possible biohazards by both scientists and the general public. What happens if engineered organisms prove resistant to all known antibiotics or carry cancer genes that might spread throughout the community? What if a genetically engineered plant becomes an uncontrollable super weed? Would these kinds of risks outweigh the potential benefits? Others argue that the risk has been exaggerated and therefore do not want to impose limits on research. Biotechnology has also generated legal problems concerning intellectual properties and patents for different aspects of the technology. Many controversies are unsettled because of the differences in perception of patents and the imperfection of patent laws.

Biotechnology and its applications have also given rise to social, medical, and economic debates. For example, some people are vigorously opposed to the use of bovine growth hormone produced through biotechnology to boost milk production for two main reasons: First, the recombinant hormone may change the composition of the milk (this view was dismissed by experts from the National Institutes of Health and the Food and Drug Administration after a thorough study); second, many dairy farmers fear that greater milk production per cow will drive prices down and put some small farmers out of business.

Numerous aspects of the application of biotechnology to humans also present ethical challenges. Should couples who carry a defective gene and have an appreciable chance of having an affected child refrain from reproduction? For genetic disorders caused by chromosomal abnormalities, such as Tay-Sachs disease, prenatal diagnosis can detect the defect in a fetus with great precision. If such a defect is detected, should the fetus be aborted? Should screening tests of infants for genetic disorders be required? If so, would such a requirement infringe on the rights of the individual? Perhaps the greatest concern is the possibility of designing or cloning a human through recombinant DNA techniques. These concerns are real. However, biotechnology itself cannot be blamed. It is the manner in which the technology is used rather than the technology itself that is at the heart of the matter.

—*Ming Y. Zheng*

See Also: Biopharmaceuticals; Biotechnology, Risks of; Cloning; DNA Fingerprinting; Gene Therapy; Genetic Engineering; Polymerase Chain Reaction; Transgenic Organisms.

Further Reading: *An Introduction to Recombinant DNA in Medicine* (1995), by Alan Emery and Sue Malcolm, provides a straightforward outline of the general principles and medical applications of biotechnology in a jargon-free style. "Innovation and the Patenting of Genes," by Jan Leschly, *Genetic Engineering News* 16 (May 1, 1996), is a good summary of points by two opposing sides of the gene patenting debate. "Biotechnology Medicine and Vaccine Approved and Under Development," by the Pharmaceutical Research and Manufacturers of America, *Genetic Engineering News* 15 (August, 1995), includes a complete list of biotechnology drugs and vaccines up to 1995. "Altered Plant Gene Concerns," by John Morrow, Jr., *Genetic Engineering News* 16 (September 1, 1996), provides an overview of the debate on the environmental transfer of engineered plant genes. *Biotechnology* (1998), by Susan Barnum, gives an overview of biotechnology and its applications.

Biotechnology, Risks of

Field of study: Genetic engineering and biotechnology

Significance: *The application of biotechnology, specifically genetic engineering, creates real and foreseeable risks to humans and to the environment. Furthermore, like any new technology, biotechnology may cause unforeseen problems. Predicting the occurrence and severity of both anticipated and unexpected problems resulting from biotechnology is a subject of much debate in the scientific community.*

Key terms

FITNESS: the probability of offspring of a genotype surviving to maturity; a measure of how well specific individuals can adapt to their environment

GENOME: the genetic content of a single set of chromosomes

GENOTYPE: the genetic makeup of an individual, referring to some or all of its specific genetic traits

GERMPLASM: the existing genotypes that constitute a species

SELECTION: a natural or artificial process that removes genotypes of lower fitness from the population and results in the inheritance of traits from surviving individuals

TRANSGENIC ORGANISM: an organism that has had its genome deliberately modified using genetic engineering techniques and that is usually capable of transmitting those changes to offspring

The Nature of Biotechnological Risks

Most of the potential risks of biotechnology center on the use of transgenic organisms. Potential hazards can result from the specific protein products of newly inserted or modified genes; interactions between existing, altered, and new protein products; the movement of transgenes into unintended organisms; or changes in the behavior, ecology, or fitness of transgenic organisms. It is not the process of removing, recombining, or inserting deoxyribonucleic acid (DNA) that usually causes problems. Genetically modifying an organism using laboratory techniques creates a plant, animal, or microbe that has DNA and RNA that is fundamentally the same as that found in nature.

Risks to Human Health and Safety

The problem most likely to result from ingesting transgenic food is unexpected allergenicity. Certain foods such as milk or Brazil nuts contain allergenic proteins that, if placed into other foods using recombinant DNA technology, would cause the same allergic reactions as the food from which the allergenic protein originally came. Scientists and policymakers will, no doubt, guard against or severely restrict the movement of known allergens into the food supply. New or unknown allergens, however, will necessitate extensive safety testing of each transgenic food product prior to general public consumption. Safety testing will be especially important for proteins that have no known history of human consumption.

Unknown, nonfunctional genes that produce compounds harmful or toxic to humans and animals could become functional as a result of the random insertion of transgenes into an organism. Unlike traditional breeding methods, recombinant DNA technology provides scientists with the ability to introduce specific genes without extra genetic material. These methods, however, usually cannot control where the gene is inserted within the target genome. As a result, transgenes are randomly placed among all the genes that an organism possesses, and sometimes "insertional mutagenesis" occurs. This is the disruption of a previously functional gene by the newly inserted gene. This same process may also activate previously inactive genes residing in the target genome. Early testing of transgenic organisms would easily reveal those with acute toxicity problems; however, testing for problems caused by the long-term intake of new proteins is difficult.

Many human and animal disease organisms are becoming resistant to antibiotics. Some scientists worry that biotechnology may accelerate that process. Recombinant DNA technologies usually require the use of antibiotic resistance genes in order to find cells that have been genetically transformed. Consequently, most transgenic plants contain antibiotic resistance genes that are actively expressed. Although unlikely, it is possible that resistance genes could be transferred from plants to bacteria or that the existence of plants carrying active antibiotic resistance genes could encourage the selection of antibiotic-resistant bacteria. As long as scientists continue using naturally occurring antibiotic resistance genes that are already commonly found in native bacterial populations in surprisingly high numbers, there is little reason to believe that plants with these genes will affect the rate of bacteria becoming resistant to antibiotics.

Another possible problem associated with antibiotic resistance genes is the reduction or loss of antibiotic activity in individuals who are taking antibiotic medication while eating foods containing antibiotic resistance proteins. Would the antibiotic be rendered useless if transgenic foods were consumed? Scientists have found that this in not the case for the most commonly used resistance gene, NPTII (neomycin phosphotransferase II), which inactivates and provides resistance to kanamycin and neomycin. Studies have shown this protein to be completely safe to humans, to be broken down in the human gut, and to be present in the current food supply. Each person consumes, on average, more than one million kanamycin-resistant bacteria daily through the ingestion of fresh fruit and vegetables. These results are probably similar for other naturally occurring resistance genes of bacterial origin.

Risks to the Environment

If environmentally advantageous genes are added to transgenic crops, then those crops, or crop-weed hybrids, may become weeds, or their weediness may increase. For example, tolerance to high-salt environments is a useful and highly desirable trait for many food crops. The addition of transgenes for salt tolerance may allow crop-weed hybrids to displace naturally occurring salt-tolerant species in high-salt environments. Most crop plants are poor competitors in natural ecosystems and probably would not become weeds even with the addition of one or a few genes conferring some competitive advantage. Hybrids between crops and related weed species, however, can show increased weediness, and certain transgenes may also contribute to increased weediness.

Biotechnology may accelerate the development of difficult-to-control pests. Crops and domesticated animals are usually protected from important diseases and insect pests by specific host resistance genes. Genetic resistance is the most efficient, effective, and environmentally friendly means for controlling and preventing agricultural losses caused by pests. Such genes are bred into plants and animals by mating desirable genotypes to those that carry genes for resistance. This method is limited to those species that can intermate. Biotechnology provides breeders with methods for moving resistance genes across species barriers, which scientists could not do prior to the 1980's. Bacteria and viruses, however, have been moving bits of DNA in a horizontal fashion (that is, across species and kingdom barriers) since the beginning of life. The widespread use of an effective, specific host resistance gene in any domesticated species historically has led to the rapid development of high levels of virulence to that gene in the pest population. Once pathogens or pests develop virulence, the host gene is useless, and breeders must seek other genes. Recombinant methods will likely accelerate the loss of resistance genes as compared with traditional methods because one

resistance gene can be expressed simultaneously in many species, is often continuously expressed at high levels within the host, and will more likely be used over large areas because of the immediate economic benefits such a gene will bring to a grower or producer.

Hybrid plants carrying genes that increase fitness (through, for example, disease resistance or drought tolerance) may decrease the native genetic diversity of a wild population through competitive or selection effects. As new genes or genes from unrelated species are developed and put into domesticated species, engineered genes may move, by sexual outcrossing, into related wild populations. Gene flow from nontransgenic species into wild species has been taking place ever since crops were first domesticated, and there is little evidence that such gene flow has decreased genetic diversity. In most situations, transgene flow will likewise have little or no detrimental effects on the genetic diversity of wild populations; however, frequent migration of transgenes for greatly increased fitness could have a significant impact on rare native genes in the world's centers of diversity. A center of diversity harbors most of the natural genetic resources for a given crop and is a region in which wild relatives of a crop exist in nature. These centers are vital resources for plant breeders seeking to improve crop plants. The impact of new transgenes on such centers should be fully investigated before transgenic crops are grown near their own center of diversity.

Impact and Applications

The risks associated with genetically modified organisms have been both overstated and understated. Proponents of biotechnology have downplayed likely problems while opponents have exaggerated the risks of the unknown. As with any new technology, there will be unforeseen problems; however, as long as transgenic organisms are scientifically and objectively evaluated on a case-by-case basis prior to release or use, society will avoid the obvious or most likely problems associated with biotechnology and benefit from its application.

—Paul C. St. Amand

See Also: Genetic Engineering: Agricultural Applications; Genetically Engineered Foods; Synthetic Genes; Transgenic Organisms.

Further Reading: *Biotechnology and Safety Assessment* (1993), edited by John Thomas et al., covers a wide range of topics related to safety in biotechnology. Jack Heinemann, *Trends in Genetics* 7 (June, 1991), reviews the remarkable range of species known to participate in naturally occurring gene transfer between species. *Agricultural Biotechnology and the Environment: Science, Policy, and Social Issues* (1996), edited by Sheldon Krimsky et al., thoroughly covers biotechnology risks related to agriculture. The policy and safety issues regarding food biotechnology are detailed in *Genetically Modified Foods: Safety Issues* (1995) by Karl-Heinz Engel et al.

Breast Cancer

Field of study: Human genetics

Significance: *Inherited breast cancer is one of the most common genetic diseases in the industrialized world. While the great majority of breast cancers are caused by acquired mutations, about 5 percent of all breast cancers are caused by inherited mutations that greatly increase the chances of developing the disease. Germ-line mutations in the* BRCA1 *and* BRCA2 *genes are associated with the majority of inherited breast cancers.*

Key terms

CELL CYCLE: the sequence of events of a dividing cell

EXON: the coding sequence (part of a messenger ribonucleic acid, or mRNA) that specifies the amino acid sequence of the protein produced during translation

GERM-LINE MUTATION: a heritable change in the chromosomes of an individual's reproductive cells; often linked to hereditary diseases such as breast cancer

TUMOR SUPPRESSOR GENE: a gene that produces a protein product that limits cell division and therefore acts to inhibit the uncontrolled cell growth of cancers

p53: a tumor suppressor gene that encodes a protein transcription factor that stops the cell cycle until deoxyribonucleic acid (DNA)

repair has occurred; a defective *p53* gene no longer stops cell division, and unrepaired DNA can be replicated, resulting in accumulated mutations in the cell

Genes Associated with Breast Cancer

Approximately one in eight women develops breast cancer over the course of her lifetime. In the United States alone, there are approximately 180,000 new cases of breast cancer yearly. By 1997, at least forty different genes had been found to be altered in breast cancers. Those breast cancers that are not familial are termed "sporadic." It is estimated that about 5 to 10 percent of all breast cancers are familial (inherited). Approximately 80 to 85 percent of inherited breast cancer can be attributed to mutations in the *BRCA1* or *BRCA2* genes.

The first gene identified in an inherited breast cancer was *p53*, which is mutated in Li-Fraumeni syndrome. About 1 percent of women who develop breast cancer before the age of thirty have germ-line mutations in *p53*. Families with this syndrome have extremely high rates of brain tumors and other cancers in both children and adults.

Some gene mutations may predispose an individual to develop breast cancer. For example, there is an increased incidence of breast cancer associated with the ataxia telangiectasia gene (*AT*) and the HRAS1 gene. In 1995, the *AT* gene was identified on chromosome 11q22-23. A mutated form of the *AT* gene is found in the rare recessive hereditary disorder ataxia telangiectasia, which has a very wide range of symptoms, including cerebellar degeneration, immunodeficiency, balance disorder, high risk of blood cancers, extreme sensitivity to ionizing radiation, and an increased risk of breast cancer. Individuals with one mutated copy of the *AT* gene have an increased risk of cancer. The *AT* gene was identified as a phosphatidylinositol-3 kinase (an enzyme that adds a phosphate group to a lipid molecule) that transmits growth signals and other signals from the cell membrane to the cell interior. The *AT* gene was found to be similar in sequence to other genes that are known to have a role in blocking the cell cycle in cells whose deoxyribonucleic acid (DNA) is damaged by ultraviolet light or X rays.

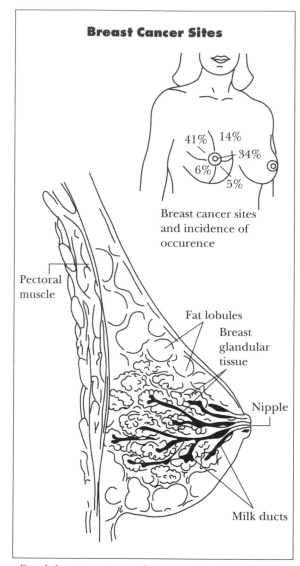

Breast Cancer Sites

41% 14%
34%
6%
5%

Breast cancer sites
and incidence of
occurence

Pectoral
muscle

Fat lobules

Breast
glandular
tissue

Nipple

Milk ducts

Female breast anatomy and common sites of breast cancer.

It is possible that the mutated *AT* gene does not stop the cell from dividing, and the damaged DNA may lead to cancers. It is disturbing to note that individuals with a mutated *AT* gene may be more sensitive to ionizing radiation. It must be determined if these individuals should then avoid low X-ray doses, such as those received from a mammogram used to detect the early stages of breast cancer.

Possible Functions of Breast Cancer Genes

The *BRCA1* gene is on chromosome 17q21 and encodes a protein that is 1,863 amino acids

long. Germ-line mutations of *BRCA1* are associated with 50 percent of hereditary breast cancer and with an increased risk of ovarian cancer. The *BRCA2* gene is on chromosome 13q12-13 and encodes a protein of 3,418 amino acids. The gene is 10,248 nucleotides long encoded by twenty-six exons. Germ-line mutations of *BRCA2* are thought to account for approximately 35 percent of families with multiple-case, early onset female breast cancer. Mutations of BRCA2 are also associated with an increased risk of male breast cancer, ovarian cancer, prostate cancer, and pancreatic cancer.

Although *BRCA1* was cloned in 1994 and *BRCA2* in 1995, the search for a function of these genes has been difficult. Part of the difficulty has been that the proteins coded by these genes do not resemble any proteins in the data bases. In 1997, David Livingston and coworkers of the Dana-Farber Cancer Institute found that the *BRCA1* gene product associates with repair protein *RAD51.* A few months later, Allan Bradley of Baylor College of Medicine and Paul Hasty of Lexicon Genetics reported that the *BRCA2* protein binds to the *RAD51* repair protein. This work suggests that both genes may be in the same DNA-repair pathway. Bradley and Hasty also showed that embryonic mouse cells with inactivated mouse *BRCA2* genes are unable to survive radiation damage, again suggesting that the *BRCA* genes are DNA-repair genes. Initially, it had been thought that the breast cancer genes were typical tumor suppressor genes that normally function to control cell growth. The 1997 work suggests that the breast cancer gene mutations act indirectly to disrupt DNA repair and allow cells to accumulate mutations, including mutations that allow cancer development. Other studies show that the *BRCA2* protein can activate gene transcription. The understanding of the function of *BRCA1* and *BRCA2* is incomplete, but what is known will encourage additional studies.

Social Implications of Screening for *BRCA1* and *BRCA2*

With the cloning of the *BRCA1* and *BRCA2* genes, it became possible to test them for mutations. The testing for mutations of these genes has been controversial, raising a number

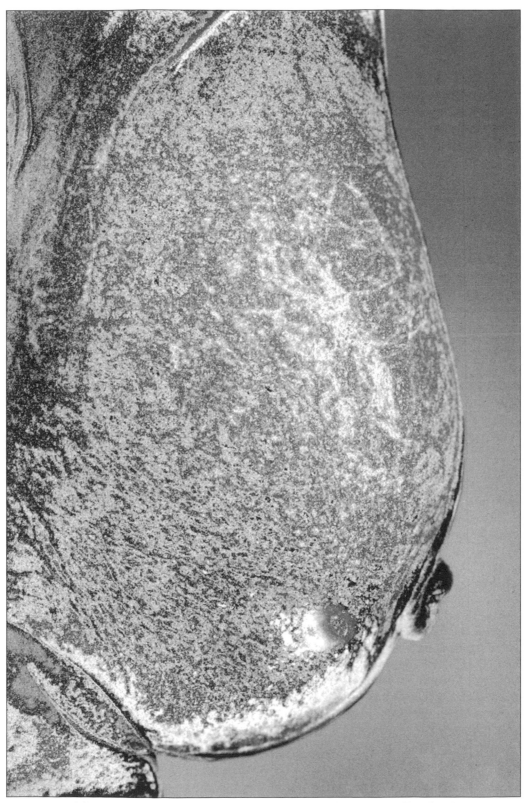

A breast tumor shows on a mammogram as a dark spot. (Dan McCoy/Rainbow)

of social and psychological issues. There is a concern that the technical ability to test for genetic conditions is ahead of the ability to predict outcomes or risks, prescribe the most effective treatment, or counsel individuals. Part of the dilemma about testing is the uncertainty about the meaning of the test results. If a test confirms the presence of a mutation in a breast cancer gene in a woman with a family history of breast cancer, there is a high risk, but not a certainty, that the woman will develop breast cancer. Even if a test is negative, it does not mean the woman is not at risk for breast cancer, because the large majority of breast cancers are not inherited. If a test is positive, it is not clear what the best course for the woman would be. Increased monitoring with mammography and even removal of both breasts as a preventative measure should reduce the chances of developing cancer but are not a guarantee that the cancer will not develop. Even if a woman does not yet have cancer, she may feel the additional psychological stress of knowing she has a high risk of developing cancer. There is also concern that test results may be misused by employers or insurers. A number of states have passed laws that prevent health insurance companies from using genetic test results to discriminate against patients. In 1996, the National Cancer Institute established the National Cancer Genetics Network as a means for individuals with a family history of cancer to enroll in research studies and learn of their genetic status while receiving counseling.

—*Susan J. Karcher*

See Also: Genetic Screening; Genetic Testing; Mutation and Mutagenesis; Oncogenes; Tumor-Suppressor Genes.

Further Reading: Ricki Lewis, *Human Genetics: Concepts and Applications* (1997), provides a general description of the genetics of cancer (chapter 16) as well as an overview of breast cancer genes. *Scientific American* 275 (September, 1996) contains many articles about cancer, including "How Cancer Arises," by Robert A. Weinberg. Mary-Claire King, "Breast Cancer Genes: How Many, Where and Who Are They?" *Nature Genetics* 2 (October, 1992), and Jean Marx, "Possible Function Found for Breast Cancer Genes," *Science* 276 (April 25, 1997),

describe the identification of breast cancer genes and possible functions of these genes. *Science* 278 (November 7, 1997) contains a section devoted to "Frontiers in Cancer Research," including the article "Human Cancer Syndromes: Clues to the Origin and Nature of Cancer," by Eric R. Fearon, which discusses inherited cancers, including the breast cancer genes. Perhaps the most comprehensive book on breast health, including breast cancer, is *Dr. Susan Love's Breast Book* (1991), by Susan M. Love, which includes discussion of the genetic risks for breast cancer.

Burkitt's Lymphoma

Field of study: Human genetics
Significance: *Burkitt's lymphoma, a cancer of B lymphocytes, is the most common tumor among children and young adults in Central Africa and New Guinea. It is one of the most aggressive malignancies known. However, early clinical and laboratory diagnosis usually leads to effective treatment and survival.*

Key terms

B LYMPHOCYTE: an antibody-producing lymphocyte
ONCOGENE: a gene that codes for an abnormal protein that makes cells become cancerous
RECIPROCAL TRANSLOCATION: a chromosomal abnormality in which there is an exchange of chromosome segments between nonhomologous chromosomes

The Discovery of Burkitt's Lymphoma

Burkitt's lymphoma was first described by Denis Burkitt, a surgeon working in Uganda in the 1950's, as a sarcoma (a cancer arising from cells of mesodermal origin) of the jaw in African children. Males are affected more commonly than females. The mean age for affected children in Africa is seven years, whereas the mean age in the United States is eleven years. Tumor infiltration usually occurs in abdominal sites such as bowels, kidneys, ovaries, or other organs. Rare cases occur as acute leukemia with circulating Burkitt tumor cells (L3-ALL). The acute lymphocytic leukemia (ALL) presentation is particularly common in cases associated with acquired immunodeficiency syndrome (AIDS). Burkitt's lymphoma grows very rapidly. A healthy child may become critically ill in about four to six weeks. These children often exhibit a head or neck mass or a large abdominal mass with fluid (ascites) accumulating in their abdomens. Other symptoms include vomiting, pain, anemia, and increased bleeding. Prompt diagnosis is essential because of the rapid growth of these tumors, which have been estimated to have a doubling time of approximately twenty-four hours.

Diagnosis

The diagnosis of Burkitt's lymphoma is usually made by a needle biopsy from a suspected disease site such as the bone marrow, ascites, or a lymph node. The pathologist uses the microscope for staging, or evaluating the degree of development of, the disease. Early clinical and laboratory diagnosis spares the child any life-threatening complications from the rapid tumor growth. Common tests that are performed include a complete blood count (CBC), a platelet count, a bone marrow aspiration, a biopsy, and a lumbar puncture. Further tests may include specialized radiographic exams such as a computer-assisted tomography (CAT) scan to look for hidden tumor masses. The National Cancer Institute (NCI) stages Burkitt's lymphoma according to the amount of the disease present. The less disease, the better the outlook for improvement after treatment. Patients who remain free of disease for over one year from the time of diagnosis are considered to be cured.

EBV: Culprit or Consort?

As with many other cancers, the exact cause of Burkitt's lymphoma is not known. In patients from equatorial Africa, however, there is a close correlation with the Epstein-Barr virus (EBV). Over 97 percent of lymphomas from equatorial Africa carry the EBV genome. By contrast, only 15 to 20 percent of sporadic cases of Burkitt's lymphoma in Europe and North America are positive for EBV. EBV has a single, linear, double-stranded deoxyribonucleic acid (DNA) genome and was the first herpes virus to be completely sequenced. EBV infection is not limited to areas where Burkitt's lymphoma is found. It infects people worldwide without producing symptoms. EBV is also the causative agent of infectious mononucleosis, a common disease in which B cells are infected.

At least two EBV subtypes have been identified in human populations: EBV-1 is detected more commonly in Western societies, whereas EBV-1 and EBV-2 subtypes seem to be equally distributed in Africa. Although EBV is identified as a possible causative agent of African Burkitt's lymphoma, it appears that non-African Burkitt's lymphoma is associated with more than one causative factor in a multistep process of development. Burkitt's lymphoma is a mono-

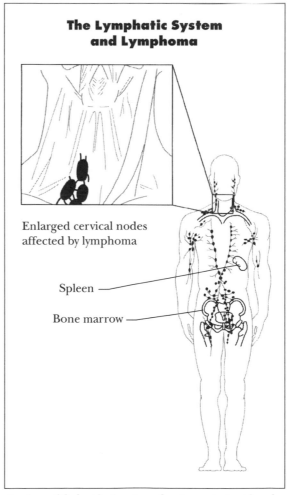

The Lymphatic System and Lymphoma

Enlarged cervical nodes affected by lymphoma

Spleen

Bone marrow

Anatomy of the lymphatic system, showing major lymph nodes; enlarged lymph nodes may occur for a wide variety of reasons, including but not limited to lymphoma (cancer).

clonal proliferation of B lymphocytes. The lymphocytes have membrane receptors for EBV and are its target. African children who develop Burkitt's lymphoma are thought to be unable to mount an appropriate immune response to primary EBV infection, possibly because of co-existent malaria, which is immunosuppressive. As time passes, excessive B-cell proliferation occurs. The precise role of EBV in the development of Burkitt's lymphoma remains unclear, but much research in this area continues to be done.

Good Genes and Bad Genes

The cells of Burkitt's lymphoma are characterized by a specific chromosomal defect known as a balanced reciprocal translocation. Observation of fresh Burkitt's lymphomas and cultured cells from a laboratory revealed an additional hand at the end of the long arm of chromosome 14, while the end of one chromosome 8 was consistently absent. Researchers suggested that the missing part of chromosome 8 was translocated to chromosome 14. This 8/14 translocation has also been observed in sporadic Burkitt's tumors from America, Japan, and Europe and has been observed in Burkitt's tumors with or without EBV markers.

The part of chromosome 8 involved in the translocation is known as the c-myc proto-oncogene. Proto-oncogenes are "good" genes that normally control the cell cycle by regulating the number of cell divisions. They are especially active when high rates of cell division are needed, as in embryonic development, wound healing, or regeneration. They can become "bad" oncogenes when the chromosomes break and reunite during chromosome translocations. The rearrangement of genes in a translocation causes oncogenic activity in the DNA that results in an abnormal fusion protein that triggers the onset of cancer. More than sixty human proto-oncogenes have now been localized to a specific chromosome or chromosome region. The new location of the c-myc gene results in deregulation and subsequent overexpression of c-myc, which becomes activated as a result of the reciprocal translocation event. An inactive proto-oncogene is transformed into an active oncogene.

Ninety percent of Burkitt's tumors are associated with a reciprocal translocation involving chromosomes 8 and 14. As additional tumors were examined, two other related translocations were observed. All three translocations involved chromosome 8. The variant translocations involved chromosome 2 or chromosome 22. However, no unified theory exists to explain the role of chromosome abnormalities in the activation of oncogenes. The Epstein-Barr virus has been implicated in Burkitt's lymphoma and is known to be a B-cell mitogen. As a mitogen, it stimulates inactive cells to transform into actively dividing cells. Perhaps EBV plays a role in the origin of 8/14 translocation abnormalities simply by increasing the number of B cells undergoing DNA replication. This could increase the chances for developing a chromosome abnormality with the potential to become cancerous.

—Phillip A. Farber

See Also: Cancer; Chromosome Structure; DNA Replication; Oncogenes; Protein Synthesis.

Further Reading: R. A. Weinberg, "How Cancer Arises," *Scientific American* (September, 1996), possesses excellent illustrations, tables, and discussions of important research. S. Heim et al., *Cancer Cytogenetics* (1995), provides excellent correlation of molecular and chromosomal findings in Burkitt's lymphoma. R. S. Cotran et al., *Robbins Pathologic Basis of Disease* (1994), provides a photograph of a child with Burkitt's lymphoma and microscopic observations by the pathologist.

Cancer

Field of study: Human genetics

Significance: *Scientists have discovered the genetic cause of several different kinds of cancer. Most cancers, however, are caused by both genetic and environmental factors. The role of genes in the development of cancer has given researchers several important clues to the mystery of how cells go wrong and begin to grow in uncontrollable ways.*

Key terms

CARCINOGEN: a substance that produces or encourages cancer

DEOXYRIBONUCLEIC ACID (DNA): the genetic material found in all cells

ONCOGENES: one of two types of genes involved in the development of tumors and cancer

TUMOR: a mass formed by the uncontrolled growth of cancer cells

TUMOR-SUPPRESSOR GENES: a gene found in normal cells that prevents the growth of cancer cells

The Problem of Cancer

Cancer is any disorder of cell growth that leads to the invasion and destruction of healthy tissue by abnormal cells. Cancer cells begin as normal cells but start to grow in an uncontrollable way for an unknown reason, rapidly multiplying and escaping from the normal rules regulating cell growth and behavior. Normal cells in the human body are continuously growing, but this growth is controlled by hormones and enzymes; cancer cells are able to avoid these regulators.

Living cells grow because it is necessary to replace older cells that become damaged because of daily wear and tear. Humans lose millions of cells every day from their skin and other parts of the body. These losses are replaced through the growth and multiplication of new cells. Cells also respond to injuries. Cut fingers or bruised knees lead to rapidly increased growth of new cells. Several billion cells can be grown in a few days to help wounds heal, after which cell growth returns to its normal rate. Normal cells stop growing when damage is repaired probably because the substances involved in stimulating growth return to normal levels. This involves the production of what are called "inhibitory growth factors." Cancer cells avoid these factors and ignore their signal to stop growing. They continue to multiply and refuse to die. Scientists do not completely understand the process of cell replacement, but they are working to find the substances that enable cells to grow and multiply. Understanding the processes involved in normal replacement rates would help them understand how and why cancer cells are able to escape this normal behavior.

Cancer cells have five basic properties that make them different from normal cells: They keep growing long after they should have stopped, they can invade surrounding areas and tissues of the body, they can travel through the body in the bloodstream or lymph vessels, they can produce substances that interfere with normal cell activity, and they can establish a mass of cells (tumors) in distant areas of the body, far from the site where the original cancer began.

The area where cancer begins to form a tumor is called the primary site. Most types of cancer begin in one place (the breast, lung, or bowel, for example) from which the cells invade neighboring areas and form secondary tumors. To make matters more complicated, some types of cancer, such as leukemia, lymphoma, and myeloma, begin in several places at the same time, usually in the bone marrow or lymph nodes. Primary tumors begin with one abnormal cell. This cell, as is true of all cells, is extremely small, no more than 0.002 or 0.003 millimeters across (about one-twentieth the width of a human hair). Therefore early cancer is very difficult to locate. Even if there are more than 100,000 cancer cells in a tumor, it is barely visible except under a microscope.

Cancer cells divide and reproduce about every two to six weeks. If they divide on the average of once per month, a single cell will multiply into approximately four thousand cells by the end of a year. After twenty months, there will be one million cells, which would form a tumor about the size of a pinhead and would still be undetectable. A tumor can be discovered only when a lump of approximately one billion cells is present. This would be about the size of a small grape. It would take about two and one-half years for a single cancer cell to reach this size. Within seven months, the one billion cells would grow to more than 100 billion cells, and the tumor would weigh about four ounces. By the fortieth month of growth, the lump of cancer cells would weigh about two pounds. The human body cannot tolerate a growth of this size, and the patient will usually die. Death normally occurs about three and one-half years after the first cancer cell begins to grow. It takes about forty-two cell doublings to reach the lethal stage. The problem is that, in most cases, tumors are detectable only after thirty doublings have taken place. By this time, cancer cells may have invaded many other areas of the body beyond the primary site.

How Cancer Cells Grow and Invade

Cancer cells are able to break down the barriers that normally keep cells from invading other cells. With the aid of a microscope, cancer cells can be observed breaking through the boundary between cells, called the "basement membrane." Researchers have discovered several of the methods used by cancer cells for this invasion. Cancers are able to make substances that can break down the intercellular matrix, the "glue" that holds cells together. This glue is made up of many chemicals, including collagen, the protein that makes tissues strong. A tissue is a collection of similar cells organized to carry out a particular function. For example, nervous tissue is specialized to receive and transmit sensations. A lung is an organ that contains many different tissues. Collagen gives these tissues strength, holds them together, and gives them flexibility. Cancer cells produce collagenase, an enzyme that breaks down collagen, thereby destroying its function of holding cells together. Enzymes are chemicals the body produces that are necessary for many normal body functions. Lactase, for example, is a enzyme that breaks down the lactose in dairy products, making it available for healthy bone marrow production. Enzymes produced by cancer cells destroy such substances rather than making them useful.

Cancer cells also produce hyaluronidase, an enzyme that breaks down the acid in cells that helps protect the cell wall from invasions. This allows cancer cells to push through normal boundaries and establish themselves in surrounding tissues. Cancer cells are thus able to move throughout the entire body. They can be observed under a microscope because they look much different from normal cells. Cancer cells have jagged edges, are irregular in shape, and have hard-to-detect borders. Normal cells, on the other hand, have a regular, smooth edge and shape. What makes cancer cells so dangerous is this ability to spread to distant parts of the body by violating the rules of cell conduct and establishing secondary tumors, called metastases.

There are many steps involved in the process of metastasizing, not all of which are understood by researchers. First is the entry into a blood vessel or lymph channel. Lymph channels, or lymphatics, comprise a network of vessels that carry lymph from the tissues to the bloodstream. Lymph is a colorless liquid that drains from spaces between cells. It consists mainly of water, salts, and proteins and eventually enters the bloodstream near the heart. The function of lymph is to filter out bacteria and other foreign particles that might enter the blood and cause infections. A mass of lymph vessels is called a lymph node. In the human body, lymph nodes are found in the neck, under the arms, and in several other places. Every body tissue has a network of lymph and blood vessels running through it.

If a cancer develops, cells may gain entry into a nearby lymph vessel by breaking down defensive enzymes. Once in the lymph system, cancer cells can travel to nodes and eventually into the bloodstream. Whatever route they take, groups of cancer cells can break away from the primary site of the tumor and float along whatever vessel they have invaded, forming numerous secondary tumors along the way. Because cancer cells are not considered foreign substances, such as bacteria or viruses, they are able to evade the body's normal reaction to invasion. These normal invaders are usually trapped by the body's defense mechanisms in a web of white cells or a clot of blood. Cancer cells, on the other hand, fool the body into thinking they are normal and therefore not dangerous.

Cancer cells eventually enter narrow blood vessels called capillaries and stay there for a brief period; from there, clumps of cancer cells enter tissues such as lungs, bones, skin, and muscle. The secondary tumors then capture their own territory. As the tumor establishes itself, other tissues build new blood vessels in order to bring the blood's supply of nutrients directly into the cancerous tumor. The body is thus fooled into feeding the very substance that will eventually kill it.

The Causes of Cancer

The causes of cancer can be divided into two categories. The first is through a tendency to break the normal rules of cell growth, a defect that is probably inherited through the generations in certain families. The second cause is a "trigger" that can induce certain cells to become cancerous. In most cases, cancer requires both a tendency and a trigger. For example, not all heavy smokers develop lung cancer because some people's lung cells do not have the tendency to follow that path. Smoking cigarettes, however, is certainly a trigger that causes cells to become cancerous in most circumstances. Most cancers require such a triggering device. Only a few are the direct result of genetic inheritance. One such genetic cancer is retinoblastoma, a cancer of the eye found in some children. Another example is the breast cancer gene (*BRCA1*), which is responsible for about 5 percent of breast cancer cases.

Cancer can be caused by damage to genetic material such as deoxyribonucleic acid (DNA), which is the major constituent of the chromosomes within the cell nucleus. It is possible to identify the specific parts of DNA that play a role in producing cancer. For example, sufficient exposure to sunlight increases the damage to the DNA of some skin cells. This damage can cause skin cancer. It is also true that certain chemicals in cigarette smoke cause a great deal of damage to the genetic material of lung cells. Sunlight and cigarette smoke can therefore be considered cancer triggers. Other substances

in the smoke can cause bronchitis and emphysema but do not cause cancer. Triggers can be found in many different forms. Researchers have discovered certain viruses that can trigger cancer of the cervix. It is also quite possible that a low-fiber, high-fat diet eaten by many Americans can trigger bowel cancer. Many cancer triggers have not been identified.

Two of the many genetic factors involved in the development of cancer have been found. The first of these factors is a group of genes called "oncogenes." An oncogene is a gene that can transform a normal cell into a cancerous one. They are involved in the formation of several different types of cancer. Another group of genes, called tumor suppressor genes, are also involved in the development of cancer.

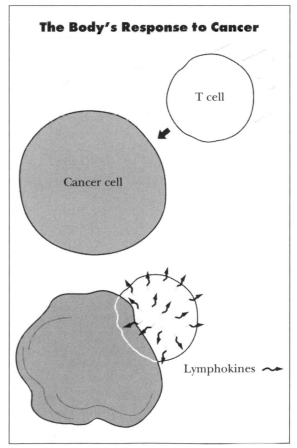

The Body's Response to Cancer

T cell

Cancer cell

Lymphokines

Cancer cells are fast-growing, irregular cells; normally, the body releases killer T cells that interact with antigens on the surface of the cancer cell, releasing lymphokines that are toxic to the cancer cell.

Their purpose is to prevent cells from becoming cancerous or malignant. When they are removed from a chromosome or suffer damage, cancer cells may develop.

Oncogenes are one kind of the more than 100,000 genes arranged on each of a human cell's forty-six chromosomes. Like all genes, they are short segments of DNA. About forty oncogenes have been identified. They are different from all other genes in the way they behave. Each normal cell has about three hundred to four hundred genes whose job it is to control important cellular functions, such as aspects of growth, cell-to-cell signaling, and movement across the cell wall. These are the proto-oncogenes. If these useful genes are damaged, they can turn into oncogenes that can produce cancer. Proto-oncogenes can be damaged by chemicals in food, drugs, or other environmental factors. They can also be changed and damaged by X rays, ultraviolet light, or errors in DNA reproduction. If such damage results, an oncogene can develop and cause cancer. Scientists understand some of the steps involved in the development of oncogenes, and they are able to tell whether a oncogene is involved in a particular cancer. With such knowledge available, it might eventually be possible to predict which individuals are at high risk to develop a particular cancer given the presence or absence of a particular oncogene. However, since some cancers develop without the presence of any oncogenes, this method of prediction is only part of the answer.

Tumor suppressor genes control abnormal growth in cells. If these genes are damaged or lost, cancer may develop. The role of tumor suppressor genes has been demonstrated in several different cancers, including retinoblastoma. This very rare form of eye cancer, found in a small number of children, seems to result from the absence of tumor suppressor genes in certain eye cells. Most people have two of these genes, labelled *RB*, which allow the retina to develop normally. In cases in which one *RB* is damaged or missing, there is a likelihood that cancer will develop; if both genes are damaged or missing, there is a 95 percent chance of a tumor developing. Retinoblastoma was the first case in which specific genes were directly re-lated to the occurrence of a disease.

Other types of genetic factors have also been shown to be responsible for certain kinds of cancer. A gene known as *BRCA1* seems to be involved in about 5 percent of occurrences of ovarian and breast cancer, and women with this gene have a much higher chance of developing some form of breast or ovarian cancer than women who do not. A number of other genes have been identified that cause a wide variety of cancers such as colon cancer, Wilms' tumor (a childhood kidney cancer), and pancreatic cancer. The most devastating gene associated with cancer is known as *p53*, a protein that attaches itself to other genes and causes them to reproduce at an alarming rate.

Cancer Research

Tumor biologists look at cells in a laboratory and attempt to find out what causes them to turn cancerous, how and why cancer cells behave the way they do, and how cancer cells invade and influence other cells. Human cells grown in the laboratory usually live for only a few days before they die. More than fifty years ago, scientists discovered ways to keep cancer cells alive and growing for an indefinite period of time. Research on these cells has shown that cancers grow faster and become much larger than normal cells.

There are many ways that tumor biologists can turn normal cells into cancer. Investigations into the environmental (carcinogenic) causes of cancer began more than two hundred years ago when a British doctor named Percivall Pott noticed that chimney sweeps had a high incidence of cancer of the scrotum. This was caused by the soot that covered their skin as they crawled into chimneys to clean them. Simple washing of the skin proved sufficient to reduce the incidence of cancer among chimney sweeps and their apprentices. Beginning in the 1930's, scientists began to investigate the effects of the soot found in the lungs of cigarette smokers. A key factor in the development of most cancers is the presence of a trigger. How carcinogens such as soot trigger cancer is not yet well understood. The "double hit" hypothesis provides one possible explanation. The first hit might be a genetic flaw such as a missing or damaged gene. The second hit, which is necessary for cancer to develop, is the result of an environmental factor. For example, a woman might possess a breast cancer gene (*BRCA1*), but this would remain undetected unless a second hit occurred. This second factor could be something in the diet or a carcinogenic chemical in cigarette smoke, a pesticide, or auto exhaust.

Epidemiologists take another approach in looking for the causes of cancer. They look at the rates of cancer in different populations in different areas of the world. For example, an epidemiologist might study rates of bowel cancer in different countries. Research shows that this type of cancer has been quite common among North Americans but very rare among some Bantu people of Africa and among elderly Japanese citizens. With this information in hand, investigators decided to look at cancer rates among Japanese immigrants to the United States. A study of these people indicated that Japanese Americans who had been in the United States for more than twenty years had the same rate of bowel cancer as did other Americans. It was concluded that genes probably played little role in this type of cancer. Instead, it was concluded that bowel cancer was probably caused by a factor in the lifestyle of people in the United States, such as high-sugar, high-starch diets.

In the United States, there are more than 1,250,000 new cases of cancer reported every year. For most of those cancers, the primary causes remain unknown, and scientists have hardly any knowledge about what triggers some cancers such as brain tumors and Hodgkin's disease. A few cancers, such as retinoblastoma, are known to be entirely inherited. Others, including breast and ovarian cancer, are partially caused by genes. Some forms of cancer seem to cluster in some families. These cancers seem to result from the complete loss of a gene or a tiny error in the chemistry of a DNA segment. This flaw is passed down through the generations.

An individual may inherit a faulty breast cancer gene from a parent, but the second copy inherited from the other parent should still be in good condition. Inheriting one defective gene would leave the person more vulnerable to breast cancer. However, if the other gene suffers damage, it still might not lead to cancer. Some genes do not have important functions in the human body, but if the lost or defective gene is a tumor suppressor or plays an important role in cell growth, the results can be devastating. If key tumor suppressors are missing, the result is almost always some kind of cancer.

—Leslie V. Tischauser

See Also: Breast Cancer; Burkitt's Lymphoma; Oncogenes; Tumor-Suppressor Genes.

Further Reading: Robert Buckman, *What You Really Need to Know About Cancer: A Comprehensive Guide for Patients and Their Families* (1997), provides detailed information on all types of cancers and describes their causes and treatment. Robert M. McAllister et al., *Cancer: What Cutting-Edge Science Can Tell You and Your Doctor About the Causes of Cancer and the Impact on Diagnosis and Treatment* (1993), describes the "battle against cancer" in a very detailed manner. Kevin Davies and Michael White, *Breakthrough: The Race to Find the Breast Cancer Gene* (1996), gives a history of the research into the genetic causes of breast cancer and other types of cancer. James T. Patterson, *The Dread Disease* (1987), gives a historical view of the search for genetic and other causes of cancer. L. A. Liotta, "Cancer Cell Invasion and Metastasis," *Scientific American* (1992), provides a basic description of cancer genetics.

Central Dogma of Molecular Biology

Field of study: Molecular genetics

Significance: *The central dogma of molecular biology describes the role of deoxyribonucleic acid (DNA) in heredity and its relationship to the traits of organisms. The elucidation of the detailed mechanisms of the central dogma has led to a revolution in genetic engineering, an understanding of the causes of many genetic diseases, and the diagnosis and treatment of some of these diseases.*

Key terms

DEOXYRIBONUCLEIC ACID (DNA): the genetic material for the vast majority of organisms, a long, double-stranded helical molecule composed of subunits called nucleotides

GENE: a discrete unit of hereditary information composed of DNA

RIBONUCLEIC ACID (RNA): a type of genetic material very similar to DNA but usually single-stranded and composed of subunits called ribonucleotides

PROTEIN: a type of large biological molecule composed of subunits called amino acids

TRANSCRIPTION: the process that converts the information in a sequence of DNA

Relationship of Genotype to Phenotype

General rules for the inheritance of traits were first proposed by Austrian botanist Gregor Mendel in the mid-nineteenth century but without knowledge of the cellular structures and biological molecules involved. Cell biologists observed that the behavior of chromosomes (long, threadlike structures) inside cells correlated with Mendelian inheritance patterns and, in 1902, proposed the chromosome theory of inheritance. In the early part of the twentieth century, many types of molecules were considered as the hereditary material, including nucleic acids (DNA and RNA) and proteins. It was not until 1944 that Oswald Avery, Maclyn McCarty, and Colin MacLeod proved that the hereditary material was DNA. James Watson and Francis Crick, with information from Rosalind Franklin and Maurice Wilkins, elucidated the structure of DNA in 1953. DNA is a major component of chromosomes.

The Watson-Crick model of DNA revealed that it is a long, double-stranded molecule called a double helix shaped like a ladder that has been twisted into a spiral. The subunits that makes up the double helix are nucleotides. Each nucleotide consists of a sugar and phosphate molecule, which make up the sides of the helix, and a nitrogen-

Oswald Avery, whose work with Maclyn McCarty and Colin MacLeod demonstrated that DNA was the material of heredity. (National Library of Medicine)

containing base, which makes up the rungs. There are four different bases in DNA: guanine (G), cytosine (C), adenine (A), and thymine (T). The two strands join together at the rungs so that the rungs are composed of two bases bonded together. A always binds with T, and G always binds with C. This complementary base pairing suggests a means for DNA to replicate and for the information contained in the DNA to be converted into traits. In 1958, Crick described the relationship of genotype (the DNA makeup of an individual) to phenotype (the traits of an individual) as the "central dogma." In addition, the central dogma describes the flow of genetic information between generations.

The central dogma combines two facts to explain the passage of genetic information from parent to offspring: DNA is contained in chromosomes in eukaryotic organisms (those with a membrane-bound nucleus), and chromosomes are known to follow Mendelian inheritance laws. DNA is a self-replicating molecule, and chromosomes are replicated and passed to offspring during reproduction. In DNA replication, the DNA molecule unwinds. Each half of the molecule acts as a template for the synthesis of a new, complementary strand of DNA that is synthesized precisely by a complex array of enzymes and proteins. Each new DNA molecule consists of one old and one new strand. DNA replication occurs in most types of cells to renew aging or injured cells, to enable growth and development, and to form eggs and sperm.

Biochemistry of the Information Transfer

Expression of the information contained in DNA into biologically active molecules requires several steps. In the first phase, DNA is copied precisely into a molecule called messenger RNA during a process called transcription. One strand of the DNA molecule acts as a template for the synthesis of a strand of messenger RNA. In eukaryotic cells, the messenger RNA is transported from the nucleus of the cell to the surrounding cytoplasm. Here, during a process called translation, the information in the messenger RNA is converted into a protein. Proteins (both enzymes, which catalyze chemi-

cal reactions, and structural proteins) mediate the development of physical traits and orchestrate the complex biochemical cycles needed for growth and the maintenance of life. The hereditary information flows in one direction, from DNA to protein, and not the reverse.

The Genetic Code

How is the information contained in DNA converted into protein? Each DNA nucleotide can be thought of as one of four letters in an alphabet. During transcription, each letter is converted into its complement in the messenger RNA molecule: G in the DNA strand is converted to C in the RNA strand and vice versa; T becomes A, but A becomes U (uracil) because RNA substitutes U for T. During translation, a structure called a ribosome holds the messenger RNA. Transfer RNA molecules "read" the messenger RNA molecule as if all the words were three letters. These three-letter "words" are called codons. Each possible three-letter combination from the four-letter alphabet (G, A, C, U; sixty-four possible combinations) specifies a "word." The words in this case are amino acids. There is some overlap between the codons; thus, there are only twenty amino acids plus three "stop codons" that specify the end of translation.

A molecule called a transfer RNA molecule recognizes a codon of the messenger RNA and brings in the amino acid specified by the codon. The next codon is read by another transfer RNA molecule, and the appropriate amino acid is brought in and attached to the first. This process continues until the entire messenger RNA molecule is read. At that time, the chain of amino acids is released and it folds into its three-dimensional protein structure and is transported to where it is needed in the organism. Each protein is specified by one gene (a segment of DNA on a very long DNA molecule). In eukaryotic cells, genes are transcribed individually from the DNA molecule.

Modifications of the Central Dogma

Understanding of the role of DNA in heredity is one of the great achievements of modern science. Research that has elucidated the mechanisms has also yielded a few surprises and modifications of the central dogma. In eukaryotic cells, the genes that are transcribed contain segments that do not code for any part of the final protein product. These regions, called introns, are transcribed into messenger RNA along with the coding sequences of the gene. After transcription is complete, enzymes cut out the introns and splice together the coding regions of the messenger RNA. Before the messenger RNA can leave the nucleus to proceed to translation in the cytoplasm, it is first biochemically modified at both ends, creating a molecule called a heterogeneous nuclear RNA molecule. It is then transported into the cytoplasm. The genes in prokaryotic cells such as bacteria do not contain introns and have no nucleus, so these modifications do not apply to them. Complex regulatory mechanisms control when a gene will be transcribed. In eukaryotes, for example, not all genes are expressed (converted into proteins) in all types of tissue and at all times of life; they are expressed only when needed.

Some viruses use RNA instead of DNA as their genetic material. Retroviruses such as human immunodeficiency virus (HIV), which causes acquired immunodeficiency syndrome (AIDS), are one example. When a retrovirus infects a host cell, it first converts its RNA to DNA with an enzyme called reverse transcriptase. The newly formed DNA integrates into the host cell, and the flow of information is the same as already described. While the majority of organisms have genes lined up successively one after another on a long strand of DNA, some viruses have overlapping genes. In this case, a single fragment of DNA may encode two or three different proteins. The "start" point of transcription would be at different sites so that a different series of codons is read during translation. Some genes do not occupy fixed locations on chromosomes. First studied in corn by Barbara McClintock, transposable elements or mobile genes have been found in most organisms and can change position within the chromosomes of one organism or, in some cases, can be transferred between organisms.

Research on the central dogma has shown that DNA meets all the requirements necessary for it to be the hereditary molecule. It has an immense capacity for encoding information and is available to every cell of the body. Its information is converted into a usable form through transcription and translation. Complementary base pairing provides an extremely accurate means of DNA replication and of passing this information to offspring. Furthermore, loss, addition, or change in a DNA base causes a mutation, or change, that can lead to both negative consequences to an organism and new proteins that ensure healthy genetic variability in a population. While some exceptions and variations have been discovered, the central dogma of molecular biology first proposed by Crick in 1958 stands as the primary scaffolding for understanding the relationship between genes and traits and the flow of hereditary information between generations.

Impact and Applications

Mendel's seminal paper on the inheritance of traits was published in 1865. In it he showed that the physical traits of an organism corresponded to *elemen* (his equivalent of genes) that were passed on to progeny during reproduction. In 1953, Watson and Crick determined the chemical components of the *elemen* to be DNA, and, in 1958, Crick first proposed the central dogma of molecular biology as the means of expression of hereditary traits encoded in DNA. Over the last half of the twentieth century, the detailed mechanisms of the dogma were elucidated with only a few modifications added over the years. This information directly led to the development of recombinant DNA technology and improved understanding, diagnosis, and treatment of genetic diseases.

In the late 1960's, scientists discovered that some bacteria made enzymes that would cut the DNA of other bacteria at specific sequences and that these "restriction enzymes" would also cut eukaryotic DNA. In the early 1970's, these restriction enzymes were used to create hybrid DNA molecules consisting of bacterial DNA and a eukaryotic gene of interest. When these "recombinant DNA molecules" were introduced into bacteria, they functioned normally, and the bacteria produced the protein specified by the eukaryotic gene. Several human proteins have been produced inexpensively in this manner, such as insulin since the early 1980's and human growth hormone since the mid-1980's. Many other organisms have been genetically engineered in a similar fashion to produce products they normally do not. Examples of these include cows modified to increase their milk production by being genetically engineered to produce excessive amounts of bovine growth hormone, crops designed to be resistant to selected herbicides, and tomatoes engineered to ripen slowly after being picked.

Once the connection between DNA and physical traits was established, techniques were developed to explore the basis of genetic diseases at the level of the gene. Sickle-cell anemia, a serious hereditary blood disease, was found to be caused by a change in a single DNA nucleotide out of the more than seventeen hundred nucleotides composing the hemoglobin gene, leading to a change in one amino acid of the hemoglobin protein. The protein molecules stick to each other in blood cells not carrying oxygen, distorting their shape and clogging small blood vessels. In some cases, once a defect in the DNA is identified as a cause of a hereditary disease, prenatal tests can be developed to screen fetuses for the disorder. Parents of an affected fetus can opt to terminate the pregnancy or prepare for a child with the particular genetic disease. For some genetic diseases, tests to detect carrier status of an adult have been developed. "Carriers" have one copy of a defective gene, but, since their other copy is normal, they do not express the disease. Knowing one's carrier status, especially if a particular genetic disease is present in their family, can help individuals make informed reproductive choices about their possibilities of passing the disease to a child. Carrier tests exist for sickle-cell anemia, cystic fibrosis (a disease of the respiratory and digestive tracts), Tay-Sachs disease (an incurable, always fatal metabolic disorder of early childhood), and a host of other genetic disorders.

The Human Genome Project was begun in 1990 to sequence the entire human genome and develop a physical map of all the human chromosomes. Scientists believe that sequencing the genome will help them identify genes involved with human development and health.

After the genes are identified, scientists will be able to connect genes to particular traits or diseases, which could lead to additional treatments for gene-based diseases. Gene therapy as a cure for genetic disorders was first attempted in 1992 by French Anderson on two young girls suffering from severe combined immunodeficiency syndrome (SCID). SCID is a genetic disorder caused by the lack of an enzyme called adenosine deaminase (ADA). Bone marrow cells were removed from the girls, genetically modified to contain ADA, and reintroduced into their bodies. Their bone marrow began producing the enzyme, and both girls were maintaining greatly improved health five years after the procedure. Gene therapy is being considered for a variety of immunodeficiency conditions and genetic diseases.

The many medical and technological advances that knowledge of the central dogma has facilitated have led to a host of ethical questions hotly debated in scientific, government, and public forums. The morality of the genetic engineering of human embryos, access to information on one's genetic status by employers and insurance companies, and the safety of genetically modified foods and crops are but a few of these issues. Education and discussion are the best vehicles to ensure the wise use of these technologies.

—*Karen E. Kalumuck*

See Also: Biochemical Mutations; DNA Replication; DNA Structure and Function; Genetic Code; Protein Synthesis.

Further Reading: Neil Campbell's *Biology* (1996) provides a detailed but accessible overview for the general reader. Another excellent overview geared to the nonspecialist is *Essential Cell Biology: An Introduction to the Molecular Biology of the Cell* (1997), edited by Bruce Alberts, which is lavishly illustrated, clear, and concise. A more advanced discussion can be found in *Molecular Biology of the Cell* (1995), by Bruce Alberts et al. Chapters titled "DNA as Information" and "From Gene to Protein" in Benjamin Lewin's *Genes VI* (1997) provide a thorough and understandable discussion of the topic. *Molecular Cell Biology* (1995), Harvey Lodish et al., is another clear, detailed resource on the central dogma.

Chemical Mutagens

Field of study: Molecular genetics

Significance: *Mutagens are naturally occurring or human-made chemicals in the environment that can interact with and damage DNA. These chemicals can directly or indirectly create mutations or changes in the information carried by the DNA. Mutations may cause birth defects or lead to the development of cancer.*

Key terms

DEOXYRIBONUCLEIC ACID (DNA): the genetic material for most organisms; a double-stranded substance composed of units called nucleotides

NUCLEOTIDE: the basic unit of DNA, consisting of a five-carbon sugar, a base containing nitrogens, and a phosphate group

BASE: the component of a nucleotide that gives it its identity and special properties

ENZYME: a molecule, usually a protein, that assists and accelerates cellular reactions without itself being altered by the reaction

The Discovery of Chemical Mutagens

The first report of mutagenic action of a chemical occurred in 1946, when Charlotte Auerbach showed that nitrogen mustard (a component of the poisonous "mustard" gas widely used in World War I) could cause mutations in fruit flies (*Drosophila melanogaster*). Since that time, it has been discovered that many other chemicals of a wide range of types are also able to induce mutations in a variety of experimental systems. This knowledge has led to the birth of the field of genetic toxicology, developed in the last half of the twentieth century to identify potentially mutagenic chemicals in food, water, air, and consumer products. The great interest in mutagens springs from the realization that many agents capable of inducing mutations are also able to cause cancer in experimental organisms. It is possible to distinguish chemical mutagens by their modes of action: Some cause mutations by interacting directly with DNA, while others have indirect effects and, in essence, trick the cell into mutating its own DNA.

Chemical Mutagens with Direct Action on DNA

Base analogues are chemicals that structurally resemble the organic bases purine and pyrimidine and may be incorporated into DNA in place of the normal bases during DNA replication. An example is bromouracil, an artificially created compound extensively used in research. It resembles the normal base thymine and differs only by having a bromine atom instead of a methyl (CH_3) group. Bromouracil is incorporated into DNA by DNA polymerase, which pairs it with an adenine base just as it would thymine. However, bromouracil is more unstable than thymine and is more likely to change its structure slightly in a process called tautomerization. After the tautomerization process, the new form of bromouracil prefers to pair with guanine rather than adenine. If this happens to a DNA molecule being replicated, DNA polymerase will insert guanine opposite bromouracil, thus changing an adenine-thymine pair to guanine-cytosine by way of the two intermediates involving bromouracil. This type of mutation is referred to as a transition, in which a purine is replaced by another purine and a pyrimidine is replaced by another pyrimidine. A mutation of this sort in a portion of DNA specifying a particular protein can lead to the synthesis of a slightly different protein.

Another class of chemical mutagens are those that alter the structure and the pairing properties of bases by reacting chemically with the bases. An example is nitrous acid, which is formed by digestion of nitrite preservatives found in some foods. Nitrous acid removes an amino (NH_3) group from the bases cytosine and adenine. When cytosine is deaminated, it becomes the base uracil, which is not a normal component of DNA but is found in ribonucleic acid (RNA). It is able to pair with adenine. Therefore, the action of nitrous acid on DNA will convert what was a cytosine-guanine base pair to uracil-guanine, which, if replicated, will give rise to a thymine-adenine pair. This is also a transition type of mutation.

Alkylating agents are a large class of chemical mutagens that act by causing an alkyl group (which may be methyl, ethyl, or a larger hydrocarbon group) to be added to the bases of

Automobile exhaust and other forms of air pollution are among the most widespread chemical mutagens. (Ben Klaffke)

DNA. Some types of alkylation cause the base to become unstable and lost from the DNA; this type of event can cause a mutation if the DNA is replicated with no base present or can lead to potentially fatal breaks in the DNA strand. Other alkylation products will change the pairing specificity of the base and create mutations when the DNA is replicated.

Intercalating agents such as acridine orange, proflavin, and ethidium bromide (which are used in labs as dyes and mutagens) have a unique mode of action. These are flat, multiple-ring molecules that interact with bases of DNA and insert themselves between them. This insertion causes a "stretching" of the DNA duplex, and the DNA polymerase is "fooled" into inserting an extra base opposite an intercalated molecule. The result is that intercalating agents cause frameshift mutations in which the "sense" of the DNA message is lost, just as if an extra letter were inserted into the phrase "the fat cat ate the hat" to make it "the ffa tca tat eth eha t." This occurs because genes are read in

groups of three bases during the process of translation. This type of mutation means that a very different protein will be made from the gene from that specified by the original sequence.

Chemical Mutagens with Indirect Action

Aromatic amines are large molecules that bind to bases in DNA and cause them to be nonreadable by DNA polymerase or RNA polymerase. An example is N-2-acetyl-2-aminofluorine (AAF), which was originally used as an insecticide. This compound and other aromatic amines are relatively inactive on DNA until they react with certain cellular enzymes, after which they react readily with guanine. Mutagens of this type and all others with indirect action work by triggering cells to induce mutagenic DNA repair pathways; the result is a loss of accuracy in DNA replication.

One of the oldest known environmental carcinogens is the chemical benzo(*a*)pyrene, a hydrocarbon found in coal tar, cigarette smoke, and automobile exhaust. An English surgeon, Percivall Pott, observed that chimney sweeps had a high incidence of cancer of the scrotum. The reason for this was later found to be their exposure to benzo(a)pyrene in the coal tar and soot of the chimneys. Like the aromatic amines, benzo(a)pyrene is activated by cellular enzymes and causes mutations indirectly.

Another important class of chemical mutagens with indirect action are agents causing cross-links between the strands of DNA. Such cross-links prevent DNA from being separated into individual strands as is needed during DNA replication and transcription. Examples of cross-linking agents are psoralens (compounds found in some vegetables and used in treatments of skin conditions such as psoriasis) and cis-platinum (a chemotherapeutic agent used to fight cancer).

Another important class of chemical mutagens are those that result in the formation of active species of oxygen (oxidizing agents). Some of these are actually created in the body by oxidative respiration (endogenous mutagens), while others are the result of the action of chemicals such as peroxides and radiation. Reactive oxygen species cause a wide variety of damage to the bases and the backbone of DNA and may have both direct and indirect effects.

Detection of Chemical Mutagens

The Ames test, developed by biochemistry professor Bruce Ames and his colleagues, is one of the most widely used screening methods for chemical mutagens. It employs particular strains of the bacterium *Salmonella typhimurium* that have a growth defect causing them to require the amino acid histidine because of mutations in genes controlling histidine production. The bacteria are exposed to the potential mutagen and then spread on an agar medium lacking histidine. The strains can grow only if they are mutated. The degree of growth indicates the strength of the mutagen; mutagens of different types are detected by using bacterial strains with different mutations. Mutagens requiring metabolic activation are detected by adding extracts of rat liver cells (capable of mutagen activation) to the tested substance

prior to exposure of the bacteria. The Ames test and others like it involving microorganisms are rapid, safe, and relatively inexpensive ways to detect mutagenic chemicals.

Impact and Applications

Mutations can have serious consequences for cells of all types. If they occur in gametes, they can cause genetic diseases or birth defects. If they occur in somatic (body) cells of multicellular organisms, they may alter a growth-controlling gene in such a way that the mutated cell begins to grow out of control and forms a cancer. DNA is subject to a variety of types of damage by interaction with a wide array of chemical agents, some of which are ubiquitous in the environment, while others are the result of human intervention. Methods of detection of chemicals with mutagenic ability have made it possible to reduce the exposure of humans to some of these mutagenic and potentially carcinogenic chemicals.

—*Beth A. Montelone*

See Also: DNA Repair; DNA Replication; DNA Structure and Function; Mutation and Mutagenesis.

Further Reading: Holly Ahern discusses oxidative damage to DNA in "How Bacteria Cope with Oxidatively Damaged DNA" in the American Society for Microbiology's *ASM News* (March, 1993). The Ames test and other bacterial tests are described by Raymond Devoret in "Bacterial Tests for Potential Carcinogens," *Scientific American* 241 (August, 1979). Errol C. Friedberg et al. provide extensive descriptions of the mechanisms of chemical mutagenesis in *DNA Repair and Mutagenesis* (1995).

Chloroplast Genes

Field of study: Molecular genetics
Significance: *Plants are unique among higher organisms in that they meet their energy needs through photosynthesis. The specific location for photosynthesis in plant cells is the chloroplast, which also contains a single, circular chromosome composed of deoxyribonucleic acid (DNA). Chloroplast DNA (cpDNA) contains many of the genes necessary for proper chloroplast functioning. A*

better understanding of the genes in cpDNA has improved the understanding of photosynthesis, and analysis of the DNA sequence of these genes has also been useful in studying the evolutionary history of plants.

Key terms

PHOTOSYNTHESIS: the process in which sunlight is used to take carbon dioxide from the air and convert it into sugar

CHLOROPLAST: the cell structure in plants responsible for photosynthesis

GENOME: all of the DNA in the nucleus or in one of the organelles such as a chloroplast

POLYPEPTIDE: a large, complex molecule composed of amino acids held together by chemical bonds

PROTEIN: a complex molecule composed of one or more polypeptides

The Discovery of Chloroplast Genes

The work of nineteenth century Austrian botanist Gregor Mendel showed that the inheritance of genetic traits follows a predictable pattern and that the traits of offspring are determined by the traits of the parents. For example, if the pollen from a tall pea plant is used to pollinate the flowers of a short pea plant, all the offspring are tall. If one of these tall offspring is allowed to self-pollinate, it produces a mixture of tall and short offspring, three-quarters of them tall and one-quarter of them short. Similar patterns are observed for large numbers of traits from pea plants to oak trees. Because of the widespread application of Mendel's work, the study of genetic traits by controlled mating is often referred to as Mendelian genetics.

In 1909, German botanist Carl Erich Correns discovered a trait in four-o'clock plants (*Mirabilis jalapa*) that appeared to be inconsistent with Mendelian inheritance patterns. He discovered four-o'clock plants had a mixture of leaf colors on the same plant: Some were all green, many were partly green and partly white (variegated), and some were all white. If he took pollen from a flower on a branch with all-green leaves and used it to pollinate a flower on a branch with all-white leaves, all the resulting seeds developed into plants with white leaves. Likewise, if he took pollen from a flower

on a branch with all-white leaves and used it to pollinate a flower on a branch with all-green leaves, all the resulting seeds developed into plants with green leaves. Repeated pollen transfers in any combination always resulted in offspring whose leaves resembled those on the branch containing the flower that received the pollen—that is, the maternal parent. These results could not be explained by Mendelian genetics.

Since Correns's discovery, many other such traits have been discovered. It is now known that the reason these traits do not follow Mendelian inheritance patterns is because their genes are not on the chromosomes in the nucleus of the cell where most genes are located. Instead, the gene for the four-o'clock leaf color trait is located on the single, circular chromosome found in chloroplasts. Because chloroplasts are specialized for photosynthesis, many of the genes on the single chromosome produce proteins or ribonucleic acid (RNA) that either directly or indirectly affect synthesis of chlorophyll, the pigment primarily responsible for trapping energy from light. Because chlorophyll is green and because mutations in many chloroplast genes cause chloroplasts to be unable to make chlorophyll, most mutations result in partially or completely white or yellow leaves.

Identity of Chloroplast Genes

Advances in molecular genetics have allowed scientists to take a much closer look at the chloroplast genome. The size of the genome has been determined for a number of plants and algae and ranges from 85 to 292 kilobase pairs (kb, or one thousand base pairs), with most being between 120 kb and 160 kb. The complete DNA sequence for several different chloroplast genomes of plants and algae have been determined. Although a simple sequence does not necessarily identify the role of each gene, it has allowed the identity of a number of genes to be determined, and it has allowed scientists to estimate the total number of genes. In terms of genome size, chloroplast genomes are relatively small and contain a little over one hundred genes.

Roughly half of the chloroplast genes produce either RNA molecules or polypeptides that are important for protein synthesis. Some of the RNA genes occur twice in the chloroplast genomes of almost all land plants and some groups of algae. The products of these genes represent all the ingredients needed for chloroplasts to carry out transcription and translation of their own genes. Half of the remaining genes produce polypeptides directly required for the biochemical reactions of photosynthesis. What is unusual about these genes is that their products only represent a portion of the polypeptides required for photosynthesis. For example, the very important enzyme ATPase—the enzyme that uses proton gradient energy to produce the important energy molecule adenosine triphosphate (ATP)—comprises nine different polypeptides. Six of these polypeptides are products of chloroplast genes, but the other three are products of nuclear genes that must be transported into the chloroplast to join with the other six polypeptides to make active ATPase. Another notable example is the enzyme ribulose biphosphate carboxylase (RuBP carboxylase), which is composed of two polypeptides. The larger polypeptide, called *rbcL,* is a product of a chloroplast gene, whereas the smaller polypeptide is the product of a nuclear gene.

The last thirty or so genes remain unidentified. Their presence is inferred because they have DNA sequences that contain all the components found in active genes. These kinds of genes are often called "open reading frames" (ORFs) until the function of their polypeptide products are identified.

Impact and Applications

The discovery that chloroplasts have their own DNA and the further elucidation of their genes have had some impact on horticulture and agriculture. Several unusual, variegated leaf patterns and certain mysterious genetic diseases of plants are now better understood. The discovery of some of the genes that code for polypeptides required for photosynthesis has helped increase understanding of the biochemistry of photosynthesis. The discovery that certain key chloroplast proteins such as ATPase and RuBP carboxylase are composed of

a combination of polypeptides coded by chloroplast and nuclear genes also raises some as yet unanswered questions. For example, why would an important plant structure like the chloroplast have only part of the genes it needs to function? Moreover, if chloroplasts, as evolutionary theory suggests, were once free-living bacteria-like cells, which must have had all the genes needed for photosynthesis, why and how did they transfer some of their genes into the nuclei of the cells in which they are now found?

Of greater importance has been the discovery that the DNA sequences of many chloroplast genes are highly conserved—that is, they have changed very little during their evolutionary history. This fact has led to the use of chloroplast gene DNA sequences for reconstructing the evolutionary history of various groups of plants. Traditionally, plant systematists (scientists who study the classification and evolutionary history of plants) have used structural traits of plants such as leaf shape and flower anatomy to try to trace the evolutionary history of plants. Unfortunately, there are a limited number of structural traits, and many of them are uninformative or even misleading when used in evolutionary studies. These limitations are overcome when gene DNA sequences are used.

A DNA sequence of a few hundred base pairs in length provides the equivalent of several hundred traits, many more than the limited number of structural traits available (typically much fewer than one hundred). One of the most widely used sequences is the *rbcL* gene. It is one of the most conserved genes in the chloroplast genome, which in evolutionary terms means that even distantly related plants will have a similar base sequence. Therefore, *rbcL* can be used to retrace the evolutionary history of groups of plants that are very divergent from one another. The *rbcL* gene, along with a few other very conservative chloroplast genes, has already been used in attempts to answer some basic plant evolution questions about the origins of some of the major flowering plant groups. Less conservative genes and ORFs show too much evolutionary change to be used at higher classification levels but are extremely useful in answering questions about the origins of closely related species, genera, or even families. As analytical techniques are improved, chloroplast genes show promise of providing even better insights into plant evolution.

—*Bryan Ness*

See Also: Classical Transmission Genetics; Evolutionary Biology; Extrachromosomal Inheritance; Mendel, Gregor, and Mendelism.

Further Reading: Jeff J. Doyle, "DNA, Phylogeny, and the Flowering of Plant Systematics," *Bioscience* 43 (June, 1993), introduces the reader to the basics of using DNA to construct plant phylogenies and discusses the future of using DNA in evolutionary studies in plants. John Svetlik, "The Power of Green," *Arizona State University Research Magazine* (Winter, 1997), provides a review of research at the Arizona State University Photosynthesis Center and provides good background for understanding the genetics of chloroplasts. Jeffry D. Palmer, "Comparative Organization of Chloroplast Genomes," *Annual Review of Genetics* 19 (1985), is one of the best overviews of chloroplast genome structure from algae to flowering plants.

Cholera

Field of study: Bacterial genetics
Significance: *Cholera is an extremely dangerous intestinal disease that has the potential to kill millions of people. Understanding of its genetic basis simplifies treatment and may lead to its eradication.*

Key terms

ENDEMIC: prevalent and recurring in a particular geographic region, or an organism that is specific to a particular region

PANDEMIC: occurring over a very wide geographic area, or an epidemic that occurs over a large area

ENZYME: a molecule, most often a protein, that accelerates a biochemical reaction without being altered by the reaction

HORMONE: a substance made in one body part and conveyed elsewhere by the blood to initiate a specific body function

PROPHYLAXIS: prevention or cure of a disease

PROTEIN: a complex, chainlike substance composed of many conjoined amino acids

Cholera, Its Symptoms, and Its Cure

Cholera arose centuries ago in India and was disseminated throughout Asia and Europe by trade and pilgrimage. It was devastating, causing epidemics that resulted in countless deaths. By the early twentieth century, cholera had been confined mostly to Asia. In 1961, however, a cholera pandemic beginning in Indonesia spread to Africa, the Mediterranean nations, and North America. In the poorer nations of the world, cholera is still widespread and occurs where sanitation is inadequate. In industrialized nations, where sanitation is generally good, only a few cases occur each year. These usually result from the return of afflicted travelers from regions where cholera is endemic. Because cholera has a 50 to 60 percent fatality rate when its symptoms are not treated quickly, occasional cases cannot be ignored; both the consequences to afflicted people and the potential for the outbreak of epidemics are great.

Cholera is an infection of the small intestine caused by the comma-shaped bacterium *Vibrio cholerae*. Infection is almost always caused by consumption of food or water contaminated with the bacterium. It is followed in one to five days by watery diarrhea that may be accompanied by vomiting. The diarrhea and vomiting may cause the loss of as much as a pint of body water per hour. This fluid loss depletes the blood water and other tissues so severely that if left unchecked it can cause death within a day. Treatment of cholera combines oral or intravenous rehydration of afflicted individuals with saline-nutrient solutions and chemotherapy with antibiotics, especially tetracycline. The two-pronged therapy replaces lost body water and destroys all *Vibrio cholerae* in infected individuals. Antibiotic prophylaxis, which destroys the bacteria, leads to the cessation of production of cholera toxin, the substance that causes diarrhea, vomiting, and death.

Genetics and Cholera

The disease occurs when cholera toxin binds to intestinal cells and stimulates the passage of water from the blood into the intestine. This water depletion and resultant cardiovascular

The poor sanitary conditions that exist in many Third World nations can contribute to cholera outbreaks. (Ben Klaffke)

collapse are major causes of cholera mortality. Study of the genetics and the biochemistry of cholera has shown that the toxin is a protein composed of portions called A and B subunits, each produced by a separate gene. When a bacterium secretes a molecule of cholera toxin, it binds to a cell of the intestinal lining (an intestinal mucosa cell) via B subunits. Then the A subunits cause the mucosal cell to stimulate the secretion of water and salts from the blood to produce diarrhea. Lesser amounts of the watery mix are vomited and add to dehydration.

The use of bacterial genetics to compare virulent *Vibrio cholerae* and strains that did not cause the disease helped in the discovery of the nature of the cholera toxin and enabled production of vaccines against the protein. These vaccines are useful to those individuals who visit areas where cholera is endemic, ensuring that they do not become infected with it during these travels. Unfortunately, the vaccines are only effective for about six months.

The basis for the operation of cholera toxin is production of a hormone substance called cyclic AMP. The presence of excess cyclic AMP in intestinal mucosa cells causes movement of water and other tissue components into the intestine and then out of the body. Cyclic AMP accumulation is caused by the ability of the cholera toxin to modify an enzyme protein, adenyl cyclase, to make it produce excess cyclic AMP via modification of a control substance called a G-protein. This modification, called adenine ribosylation, is a mechanism similar to that causing diphtheria, another dangerous disease that can be fatal, though in diphtheria other tissues and processes are affected.

Impact and Applications

Cholera has, for centuries, been a serious threat to humans throughout the world. During the twentieth century, its consequences to industrialized nations diminished significantly with the advent of sound sanitation practices that almost entirely prevented the entry of *Vibrio cholerae* into the food and water supply. In poorer nations with less adequate sanitation, the disease flourishes and is still a severe threat.

It must be remembered that handling cholera occurs at three levels. The isolation and the identification of cholera toxin, as well as development of current short-term cholera vaccines, was highly dependent on genetic methodology and protects most travelers from the disease. However, wherever the disease afflicts individuals, its treatment depends solely upon rehydration and use of antibiotics. Finally, modern cholera prevention is predicated solely on adequate sanitation. It is thus essential to produce a long-lasting vaccine for treatment of cholera to enable prolonged immunization at least at the ten-year level of tetanus shots. Efforts aimed at this goal are ongoing and utilize molecular genetics to define more clearly the basis for the intractability of cholera to long-term vaccination. Particularly useful will be fine genetic sequence analysis and the use of gene amplification followed by DNA fingerprinting.
—*Sanford S. Singer*

See Also: Diphtheria; DNA Fingerprinting; Genetic Engineering; Molecular Genetics.

Further Reading: *Conn's Current Therapy* (1995), edited by Robert E. Rakel, provides a succinct overview of cholera and its treatment for general readers. John Holmgren, "Action of Cholera Toxin and Prevention and Treatment of Cholera," *Nature* 292 (1981), clearly describes both the composition and bioaction of cholera toxin. *Vibrio Cholerae and Cholera: Molecular to Global Perspectives* (1994), edited by Kate Wachsmuth et al., contains everything one might want to know about the disease and its genetics.

Chromatin Packaging

Field of study: Molecular genetics
Significance: *The huge quantity of DNA present in each cell must be organized and highly condensed in order to fit into the discrete units of genetic material known as chromosomes. Gene expression can be regulated by the nature and extent of this DNA packaging in the chromosome, and errors in the packaging process can lead to genetic disease.*

Key terms

CHROMATIN: the material that makes up chromosomes; a complex of fibers composed of DNA, histone proteins, and nonhistone proteins

HISTONE PROTEINS: small, basic proteins that are complexed with DNA in chromosomes and that are essential for chromosomal structure and chromatin packaging

NONHISTONE PROTEINS: a heterogeneous group of acidic or neutral proteins found in chromatin that may be involved with chromosome structure, chromatin packaging, or the control of gene expression

NUCLEOSOME: the basic structural unit of chromosomes, consisting of 146 base pairs of DNA wrapped around a core of eight histone proteins

SOLENOID: a chromatin fiber, about 30 nanometers in diameter, composed of stacked and coiled nucleosomes complexed with histone *H1*

Chromosomes and Chromatin

Scientists have known for many years that an organism's hereditary information is encrypted in molecules of deoxyribonucleic acid (DNA) that are themselves organized into discrete hereditary units called genes and that these genes are organized into larger subcellular structures called chromosomes. James Watson and Francis Crick elucidated the basic chemical structure of the DNA molecule in 1952, and much has been learned since that time concerning its replication and expression. At the molecular level, DNA is composed of two parallel chains of building blocks called "nucleotides," and these chains are coiled around a central axis to form the well-known "double helix." Each nucleotide on each chain attracts and pairs with a complementary nucleotide on the opposite chain, so a DNA molecule can be described as consisting of a certain number of these nucleotide base pairs. The entire human genome consists of over six billion base pairs of DNA, which, if completely unraveled, would extend for over two meters. It is a remarkable feat of engineering that in each human cell this much DNA is condensed, compacted, and tightly packaged into chromosomes within a nucleus that is less than 10^{-5} meters in diameter. What is even more astounding is the frequency and fidelity with which this DNA must be condensed and relaxed, packaged and unpackaged, for replication and expression in each individual cell at the appropriate time and place during both development and adult life. The essential processes of DNA replication or gene expression (transcription) cannot occur unless the DNA is in a more open or relaxed configuration.

Chemical analysis of mammalian chromosomes reveals that they consist of DNA and two distinct classes of proteins, known as "histone" and "nonhistone" proteins. This nucleoprotein complex is called "chromatin," and each chromosome consists of one linear, unbroken, double-stranded DNA molecule that is surrounded in predictable ways by these histone and nonhistone proteins. The histones are relatively small, basic proteins (having a net positive charge), and their function is to bind directly to the negatively charged DNA molecule in the chromosome. Five major varieties of histone proteins are found in chromosomes, and these are known as *H1*, *H2A*, *H2B*, *H3*, and *H4*. Chromatin contains about equal amounts of histones and DNA, and the amount and proportion of histone proteins are constant from cell to cell in all higher organisms. In fact, the histones as a class are among the most highly conserved of all known proteins. For example, for histone *H3*, which is a protein consisting of 135 amino acid "building blocks," there is only a single amino acid difference in the protein found in sea urchins as compared with the one found in cattle. This is compelling evidence that histones play the same essential role in chromatin packaging in all higher organisms and that evolution has been quite intolerant of even minor sequence variations between vastly different species.

Nonhistones as a class of proteins are much more heterogeneous than the histones. They are usually acidic (carrying a net negative charge), so they will most readily attract and bind with the positively charged histones rather than the negatively charged DNA. Each cell has many different kinds of nonhistone proteins, some of which play a structural role in chromosome organization and some of which are more directly involved with the regulation of gene expression. Weight for weight, there is often as much nonhistone protein present in chromatin as histone protein and DNA combined.

Nucleosomes and Solenoids

The fundamental structural subunit of chromatin is an association of DNA and histone proteins called a "nucleosome." First discovered in the 1970's by Ada and Donald Olins and Chris Woodcock, each nucleosome consists of a core of eight histone proteins: two each of the histones *H2A*, *H2B*, *H3*, and *H4*. Around this histone octamer is wound 146 base pairs of DNA in one and three-quarter turns (approximately eighty base pairs per turn). The overall shape of each nucleosome is similar to that of a lemon or a football. Each nucleosome is separated from its adjacent neighbor by about fifty-five base pairs of "linker DNA," so that in its most unraveled state they appear under the electron microscope to be like tiny beads on a string. Portions of each core histone protein protrude outside of the wound DNA and interact with the DNA that links adjacent nucleosomes.

The next level of chromatin packaging involves a coiling and stacking of nucleosomes into a ribbonlike arrangement, which is twisted to form a chromatin fiber about 30 nanometers (nm) in diameter commonly called a "solenoid." Formation of solenoid fibers requires the interaction of histone *H1*, which binds to the linker DNA between nucleosomes. Each turn of the chromatin fiber contains about 1,200 base pairs (six nucleosomes), and the DNA has now been compacted by about a factor of fifty. The coiled solenoid fiber is organized into large domains of 40,000 to 100,000 base pairs, and these domains are separated by attached nonhistone proteins that serve to both organize and control their packaging and unpackaging.

Long DNA Loops and the Chromosome Scaffold

Physical studies using the techniques of X-ray crystallography and neutron diffraction have suggested that solenoid fibers may be further organized into giant supercoiled loops. The extent of this additional looping, coiling, and stacking of solenoid fibers varies, depending on the cell cycle. The most relaxed and extended chromosomes are found at interphase, the period of time between cell divisions. Interphase chromosomes typically have a diameter of about 300 nm. Chromosomes that are getting ready to divide (metaphase chromosomes) have the most highly condensed chromatin, and these structures may have a diameter of up to 700 nm. One major study on the structure of metaphase chromosomes has shown that a skeleton of nonhistone proteins in the shape of the metaphase chromosome remains even after all of the histone proteins and the DNA have been removed by enzymatic digestion. If the DNA is not digested, it remains in long loops (10 to 90 kilobase pairs) anchored to this nonhistone protein scaffolding.

In the purest preparations of metaphase chromosomes, only two scaffold proteins are found. One of these forms the latticework of the scaffold, while the other has been identified as toposiomerase II, an enzyme that is critical in DNA replication. This enzyme cleaves double-stranded DNA and then rapidly reseals the cut after some of the supercoiling has been relaxed, thus relieving torsional stress and preventing tangles in the DNA. Apparently this same enzyme activity is necessary for the coiling and looping of solenoid fibers along the chromosome scaffold that occurs during the transition between interphase and metaphase chromosome structure. In the most highly condensed metaphase chromosomes, the DNA has been further compacted by an additional factor of one hundred.

Impact and Applications

Studies on chromatin packaging continue to reveal the details of the precise chromosomal architecture that results from the progressive coiling of the single DNA molecule into increasingly compact structures. Evidence suggests that the regulation of this coiling and packaging within the chromosome has a significant effect on the properties of the genes themselves. In fact, errors in DNA packaging can lead to inappropriate gene expression and developmental abnormalities. In humans, the blood disease thalassemia, several neuromuscular diseases, and even male sex determination can all be explained by the altered assembly of chromosomal structures.

Chromatin domains, composed of coiled solenoid fibers, may contain several genes, or the boundary of a domain can lie within a gene. These domains have the capacity to influence gene expression, and this property is mediated by specific DNA sequences known as locus control regions (LCRs). An LCR is like a powerful enhancer that activates transcription, thereby turning on gene expression. The existence of such sequences was first recognized from a study of patients with β-thalassemia and a related condition known as hereditary persistence of fetal hemoglobin. In these disorders, there is an error in the expression of a cluster of genes, known as the β-globin genes, that prevents the appearance of adult type hemoglobin. The β-globin genes are linearly arrayed over a 50-kilobase-pair chromatin domain, and the LCR is found upstream from this cluster. Affected patients were found to have normal β-globin genes, but there was a deletion of the upstream LCR that led to failure to activate the genes appropriately. Further investigation led to the conclusion that the variation in expression of these genes observed in different patients was caused by differences in the assembly of the genes into higher-order chromatin structures. In some cases, gene expression was repressed, while in others it was facilitated. Under normal circumstances, a nonhistone protein complex was found to bind to the LCR, causing the chromatin domain to unravel and making the DNA more accessible to transcription factors, thus enhancing gene expression.

DNA sequencing studies have demonstrated a common feature in several genes whose altered expression leads to severe human genetic disease. For example, the gene that causes myotonic dystrophy has a large number of repeating nucleotide triplets in the DNA region immediately adjacent to the protein-encoding segment. Physical studies have shown that this results in the formation of unusually stable nucleosomes, since these repeated sequences create the strongest naturally occurring sites for association with the core histones. It has been suggested that these highly stable nucleosomes are unusually resistant to the unwinding and denaturation of the DNA that must occur in order for gene expression to begin. Ribonucleic acid (RNA) polymerase is the enzyme that makes an RNA transcript of the gene, and its movement through the protein-coding portion of the gene is inhibited if the DNA is unable to dissociate from the nucleosomes. Thus, although the necessary protein product would be normal and functional if it could be made, it is a problem with chromatin unpackaging that leads to reduced gene expression that ultimately leads to clinical symptoms of the disease. Both mild and severe forms of myotonic dystrophy are known, and an increase in the clinical severity correlates exactly with an increased number of nucleotide triplet repeats in the gene. Similar triplet repeats have been found in the genes responsible for Kennedy's disease, Huntington's chorea, spinocerebellar ataxia type I, fragile X syndrome, and dentatorubral-pallidoluysian atrophy.

Fascinating and unexpected recent research results have suggested that a central event in the determination of gender in mammals depends on local folding of DNA within the chromosome. Molecular biologists Peter Goodfellow and Robin Lovell-Badge successfully cloned a human gene from the Y chromosome that determines maleness. This SRY gene (named from the sex-determining region of the Y chromosome) encodes a protein that selectively recognizes a specific DNA sequence and helps assemble a chromatin complex that activates other male-specific genes. More specifically, binding of the SRY protein causes the DNA to bend at a specific angle and causes conformation that facilitates the assembly of a protein complex to initiate the cascade of gene activation leading to male development. If the bend is too tight or too wide, gene expression will not occur, and the embryo will develop as a female.

The unifying lesson to be learned from these examples of DNA packaging and disease is that DNA sequencing studies and the construction of human genetic maps will not by themselves provide all the answers to questions concerning human variation and genetic disease. An understanding of human genetics at the molecular level depends not only on the primary DNA sequence but also on the three-dimensional organization of that DNA within the chromosome. Compelling genetic and biochemical evidence has left no doubt that the packaging process is an essential component of regulated gene expression.

—Jeffrey Knight

See Also: Central Dogma of Molecular Biology; Chromosome Structure; Developmental Genetics; Fragile X Syndrome; Huntington's Chorea.

Further Reading: Roger Kornberg and Anthony Klug, "The Nucleosome," *Scientific American* 244 (1981), provides a somewhat dated but highly readable summary of the primary association of DNA with histone proteins. Peter Russell, *Genetics* (1996), is a college-level textbook with an excellent discussion of chromatin structure and organization. The same topic is adroitly covered from a cellular and biochemical perspective by James Darnell, Harvey Lodish, and David Baltimore in *Molecular Cell Biology* (1990). Alan Wolffe provides a comprehensive review for advanced students and professionals in *Chromatin: Structure and Function* (1995), and he has also written an excellent summary for the general reader of the relationship between gene expression and DNA packaging in "Genetic Effects of DNA Packaging," *Scientific American, Science and Medicine* (November/December, 1995).

Chromosome Structure

Field of study: Classical transmission genetics

Significance: *The separation of the alleles in the production of the reproductive cells is a central feature of the model of inheritance. The realization that the genes are located on chromosomes and that chromosomes occur as pairs that separate during meiosis provides the physical explanation for the basic model of inheritance. When chromosome* *structure is modified, changes in information transmission produce abnormal developmental conditions, most of which contribute to early miscarriages and spontaneous abortions.*

Key terms

DEOXYRIBONUCLEIC ACID (DNA): the chemical molecule that contains the hereditary information

HISTONES: a class of proteins associated with DNA

MEIOSIS: a cell reproduction process in which the chromosome number is reduced from two sets to one set

PROTEINS: molecules composed of amino acids

The Discovery of Chromosome Involvement in Inheritance

The development of the microscope made it possible to study what became recognized as the central unit of living organisms, the cell. One of the most obvious structures within the cell is the nucleus, which is visible because it is surrounded by a membrane that sets it off from the otherwise colorless background. As study continued, dyes were used to stain cell structures to make them more visible. It became possible to see colored structures called "chromosomes" ("color bodies") within the nucleus that became visible when they condensed as the cell prepared for reproduction.

The association of the condensed, visible state of chromosomes with cell reproduction caused investigators to speculate that the chromosomes played a role in the transmission of information. Chromosome counts made before and after cell reproduction showed that the chromosome number remained constant from generation to generation. When it was observed that the nuclei of two cells (the egg and the sperm) fused during sexual reproduction, the association between information transport and chromosome composition was further strengthened. German biologist August Weismann, noting that the chromosome number remained constant from generation to generation despite the fusing of cells, predicted that there must be a cell division that reduced the chromosome number in the egg and sperm cells. The reductional division, meiosis, was described in 1900.

Following the rediscovery of Austrian botanist Gregor Mendel's rules of inheritance in 1900, the work of German zoologist Theodor Boveri and American geneticist Walter Sutton led to the proposal, in 1903, that the character-determining factors (genes) proposed by Mendel were located on the chromosomes and that the factor segregation that was a central part of the model occurred because the like chromosomes of each pair separated during the reduction division in meiosis. This hypothesis, the "chromosome theory of heredity," was confirmed in 1916 by the observations of the unusual behavior of chromosomes and the determining factors located on them by American geneticist Calvin Bridges.

Chromosome Structure and Relation to Inheritance

With the discovery of the nucleic acids came speculation about the roles of deoxyribonucleic acid (DNA) and the associated proteins. During the early 1900's, it was generally accepted that DNA formed a structural support system to hold critical information-carrying proteins on the chromosomes. The identification of the structure of DNA in 1953 by American biologist James Watson and English physicist Francis Crick and the recognition that DNA, not the proteins, contained the genetic information led to study of chromosome structure and the relationships of the DNA and protein components.

It is now recognized that each chromosome contains one DNA molecule. Each plant and animal species has a specific number of chromosomes. Humans have twenty-three kinds of chromosomes, present as twenty-three pairs. Each chromosome can be recognized by its overall length and the position of constrictions that are only visible when the cell is reproducing. At all other stages of the cell's life, the chromosome material is diffuse and is seen only as a general color within the nucleus. When the cell prepares for reproduction, the fibrous DNA molecule tightly coils and condenses into the visible structures. Since there must be information for the two cells that result from the process of reproduction, the chromosomes are present in a duplicated condition when they are visible.

A major feature of the visible, copied chromosomes is a constriction called the "centromere." This constriction may be located anywhere along the chromosome, so its position is useful for identifying chromosomes. In karyotyping, the standard system used to identify human chromosomes, the numbering begins with the longest chromosome with the constriction nearest the center (chromosome 1). Chromosomes with nearly the same length but with the centromere constriction removed from the center position have higher numbers (chromosomes 2 and 3). Shorter chromosomes with a centromere near the middle are next, and the numbering proceeds based on the distance the centromere is removed from the central position. Short chromosomes with a centromere near one end have the highest numbers.

Most of the chromosomes have a centromere that is not centrally located, which results in arms of unequal length. The short arm is referred to as "petite" and is designated the p arm. The long arm is designated the q arm. This nomenclature is useful for referring to features of the chromosome. For example, when a portion of the long arm of chromosome 15 has been lost, the arm is shorter than normal. The loss, a deletion, is designated $15q-$ (chromosome 15, long arm, deletion). The Prader-Willi syndrome, in which the infant has poor sucking ability and poor growth, and later becomes a compulsive eater, results from this deletion. Cri du chat ("cry of the cat") syndrome results from $5p-$. The cry of these individuals is like that of a cat, and they are severely mentally retarded and have numerous physical defects.

Some chromosomes have additional constrictions referred to as "secondary constrictions." The primary centromere constrictions are located where the spindle fibers attach to the chromosomes to move them to the appropriate poles during cell division. The secondary constrictions are sites of specific gene activity. Both of these regions contain DNA base sequence information that is specific to their functions.

Histones

The DNA of the chromosomes is wound around special proteins called "histones." This results in an orderly structure that condenses the DNA mass so that the bulky DNA does not require as much storage space. The wrapped DNA units then fold into second and third levels of organization. The exact processes involved in these higher levels of folding are not fully understood, but the overall condensation reduces the bulk of the DNA nearly one thousandfold. If the DNA is removed from a condensed chromosome, the proteins remain and have nearly the same shape as the chromosome, indicating that it is the proteins that form the chromosome shape. The presence of

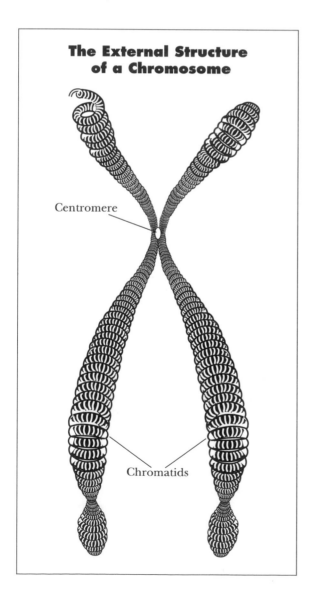

The External Structure of a Chromosome

Centromere

Chromatids

The Internal Structure of a Chromosome

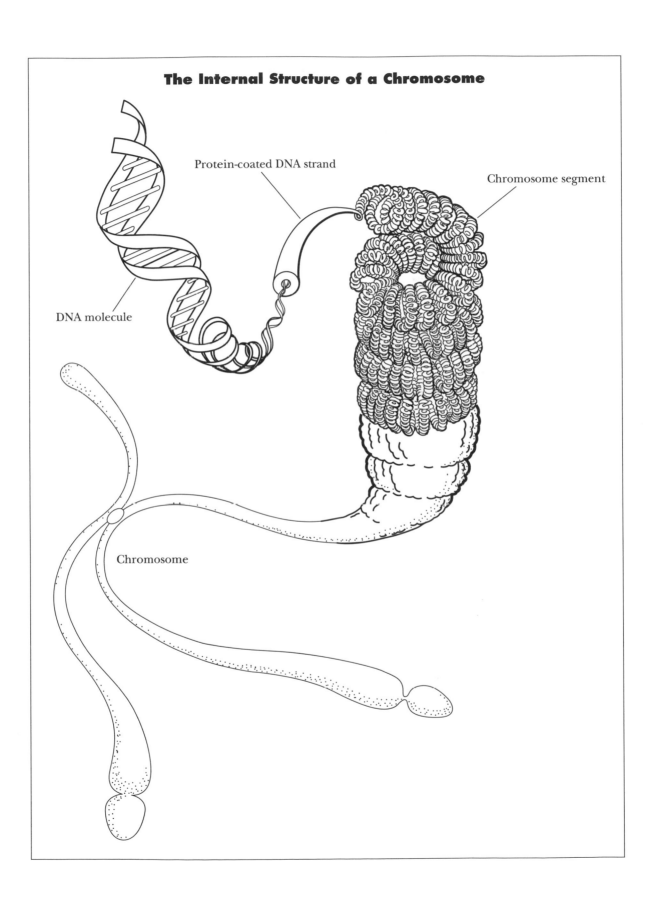

Protein-coated DNA strand

Chromosome segment

DNA molecule

Chromosome

these proteins and the fact that the DNA is wrapped around them raises many questions about how the DNA is copied for reproduction and how the DNA information is read for gene activity. These are areas of active research.

The histone proteins form a structure called a "nucleosome" ("nuclear body"). There are four kinds of histones, and two of each kind join together to form a cylinder-shaped nucleosome structure. The fibrous DNA molecule wraps around each nucleosome approximately two and one-half times with a sequence of unwound DNA between each nucleosome along the entire length of the DNA molecule. The structure, called "chromatin," looks like a string of beads when isolated sections are viewed with an electron microscope. When the chromatin is digested with enzymes that break the DNA backbone in the unwound regions, repeated lengths of chromatin are recovered, showing that the nucleosome wrapping is very regular. These nucleosome regions join together to form the second- and third-level folding as the chromosome condenses when the cell prepares for reproduction.

In addition to the histone proteins, nonhistone proteins attach to the chromatin. With an electron microscope, chromatin loops can be seen extending from a protein matrix. There is evidence that these loops represent replication units along the chromosome, but how the DNA molecule is freed from the histone proteins to be replicated is a major unsolved puzzle.

The condensation of the chromatin is not uniform over the entire chromosome. In the regions immediately adjacent to the centromere, the chromatin is tightly condensed and remains that way throughout the visible cycle. All of the available evidence indicates that this chromatin is not genetically active—it does not appear to contribute to the information content of the cell. It also replicates later than the remaining DNA. Because it has these unexpected characteristics, it is called "heterochromatin" ("the other chromatin"). The remaining chromatin is referred to as "euchromatin" ("true chromatin") because it is active in information transfer, replicates as a unit, and behaves the way one expects genetic information to behave.

Giemsa Stain and Chromosome Painting

When chromosomes are treated with a dye called Giemsa stain, regular banding patterns appear. The bands vary in width, but their positions on the individual chromosomes is consistent. This makes the bands useful in identifying specific chromosome regions. When a chromosome has a structural modification, such as an inversion that results when two breaks occur and the region is inverted during repair, the change in the banding pattern makes it possible to recognize the source of the variation. When a loss of a chromosomal region produces a deficiency disorder, changes in the banding patterns of a chromosome can identify the missing region. Karyotype analysis is a useful tool in genetic counseling because disorders caused by chromosome structure modifications can be identified. If a parent is the source of the modified chromosome, predictions can be made about the possibilities of a second child being similarly affected. Associations between disorders and missing chromosome regions are also useful in identifying what specific functions are associated with specific regions; the loss of function expression gives evidence about the functions and their normal roles. Other stains produce different banding patterns and, when used in combination with the Giemsa banding patterns, allow diagnosis of structure modifications that can be quite complex.

It is also possible to use fluorescent dyes in a process called "chromosome painting" to identify the DNA of individual chromosomes. This allows the recognition of small regions that have been exchanged between chromosomes that are too small to be recognized by measuring the lengths of chromosome arms or produce a noticeable change in the banding pattern. Color differences within chromosomes or at their tips clearly show which chromosomes have exchanged materials, how much material each has exchanged, and where on the chromosomes the exchanges have taken place. Many cancer cells, for example, have multiple chromosome modifications, with material from two or three chromosomes associated in one highly modified chromosome structure.

Chromosome Disorders

Specific DNA sequences called "telomeres" protect the ends of the chromosomes from damage and prevent DNA molecules from sticking together. Research that began in the early 1990's led to the discovery that the telomere regions of the chromosomes are shortened at each DNA replication. When the telomeres have been reduced to some critical point, the cell dies. Other observations indicate that the telomere is returned to its normal length in tumor cells, suggesting that this might contribute to the long life of tumor cells, possibly making them immortal. The relationship of cell age to telomere length and the mechanisms that lead to telomere shortening are not known, but this is an area of active research because it has implications for aging and possible implications for understanding cancer and suggesting new ways to stop the growth of tumors.

The DNA of each chromosome carries a unique part of the information code in the sequence of the bases. The specific sequences are in linear order along the chromosome and form linked sequences of genes called "linkage groups." When the like chromosomes pair and separate during meiosis, one copy of each chromosome is transmitted to the offspring. During meiosis, there may be an exchange of material between the paired chromosomes, but this does not change the information content because the information is basically the same for both chromosomes in any region. There may be differences in the coding sequences, but functionally it is the same informational content. Extreme changes in chromosome structure that result in the moving of information to another chromosome may have consequences on how specific information is expressed; a change in position might result in different regulation or in how the information is transmitted during meiosis.

Position effects result when genes are moved to different regions of the same chromosome or to another chromosome. A normal allele may show a mutant phenotype expression in a new position in the chromosome set. The best-known case occurs when a gene is placed adjacent to a heterochromatic region. The relocated DNA is condensed like the heterochromatic-region DNA, and its normal activity cannot be expressed. Ninety percent of patients with the disorder chronic myelogenous leukemia have an exchange of material (translocation) between chromosome 9 and chromosome 22. Chromosome 22 is shorter than normal and is called the Philadelphia chromosome, after the city in which it was discovered. The placing of a specific gene from chromosome 9 within the broken region adjacent to a gene on chromosome 22 causes the uncontrolled expression of both of the genes and uncontrolled cell reproduction (leukemia). The mechanism is not understood, but the change of regulation has dramatic results.

The separation of like chromosomes during meiosis occurs because the two chromosome arms are attached to a specific centromere. When the centromere is moved to one of the poles, the arms are pulled along, ensuring movement of all of the material of the paired chromosomes to the opposite poles and inclusion in the newly formed cells. Translocations occur when chromosomes are broken and material is placed in the wrong position by the repair system, causing a chromosome region to become attached to a different centromere. This leads to an inability to properly separate the regions of the arm, which can result in duplication of some of the chromosomal regions (when two copies of the same arm move to one cell) or deficiencies (when none of the material from a chromosome arm moves into a cell). This is a common outcome with translocation heterozygotes (individuals with both normal chromosomes and translocated chromosomes in the same cells). Pairing of like chromosome regions occurs, but rather than two chromosomes paired along their entire lengths, the arms of the two translocated chromosomes are paired with the arms of their normal pairing partners. The separation of the chromosomes produces duplications of material from one chromosome arm or a deficiency of that material 50 percent of the time. If these cells are involved in fertilization, the offspring will show duplication or deficiency disorders.

—*D. B. Benner*

See Also: Chromatin Packaging; DNA Replication; DNA Structure and Function; Linkage Maps.

Further Reading: Carol Greider and Elizabeth Blackburn, "Telomeres, Telomerase, and Cancer," *Scientific American* 274 (February, 1996), contains a review of the nature of telomeres and their importance in cell life. *Genetics* (1996), by Peter Russell, provides an introduction to chromosome structure and function. *Biology* (1996), by Neil Campbell, is a college-level biology textbook that provides introductory explanations of chromosomes.

Chromosome Theory of Heredity

Field of study: Classical transmission genetics

Significance: *The chromosome theory of heredity originated with American geneticist Walter Sutton, who first suggested that genes were located on chromosomes. This theory guided much of genetic research in the early twentieth century, including development of the earliest genetic maps based on linkage. In 1931, several experiments confirmed the chromosome theory by demonstrating that certain rearrangements of the heritable traits (or genes) were always accompanied by corresponding rearrangements of the microscopically observable chromosomes.*

Key terms

GENE: a unit of heredity controlling a single trait consisting of a length of deoxyribonucleic acid (DNA) coding for a single protein

CHROMOSOME: one of the paired structures containing genetic information inside the cell's nucleus

LINKAGE: the inheritance of two or more genes together as a unit if they are located close together on the same chromosome

CROSSING-OVER: the breakage of chromosomes followed by the interchange of the resulting fragments; also, the recombination of genes that results from the chromosomal rearrangement

INDEPENDENT ASSORTMENT: the inheritance of genes independently of one another when they are located on separate chromosomes

GENETIC MAPPING: the locating of gene positions along chromosomes

LINKAGE MAPPING: a form of genetic mapping that uses recombination frequencies to estimate the relative distances between linked genes

PHYSICAL MAPPING: a form of genetic mapping that associates a gene with a microscopically observable chromosome location

Mendel's Law of Independent Assortment

In a series of experiments first reported in 1865, Austrian botanist Gregor Mendel established the first principles of genetics. Mendel showed that the units of heredity were inherited as particles that maintained their identity across the generations; these units of heredity are now known as "genes." These genes exist as pairs in all the body's cells except for the egg and sperm cells. When Mendel studied two traits at a time (dihybrid inheritance), he discovered that different genes were inherited independently of one another, a principle that came to be called the law of independent assortment. For example, if an individual inherits genes *A* and *B* from one parent and genes *a* and *b* from the other parent, in subsequent generations the combinations *AB*, *Ab*, *aB*, and *ab* would all occur with equal frequency. Gene *A* would go together with *B* just as often as with *b*, and gene *B* would go with *A* just as often as with *a*. Mendel's results were ignored for many years after he published his findings, but his principles were rediscovered in 1900 by Erich Tschermak von Seysenegg in Vienna, Austria, Carl Erich Correns in Tübingen, Germany, and Hugo de Vries in Amsterdam, Holland. Organized research in genetics soon began in various countries in Europe and also in the United States.

Sutton's Hypothesis

Mendel's findings had left certain important questions unanswered: Why do the genes exist in pairs? Why do different genes assort independently? Where are the genes located? Answers to these questions were first suggested in 1903 by a young American scientist, Walter Sutton, who had read about the rediscovery of

Mendel's work. By this time, it was already well known that all animal and plant cells contain a central portion called the nucleus and a surrounding portion called the cytoplasm. Division of the cytoplasm is a very simple affair: The cytoplasm simply squeezes in two. The nucleus, however, undergoes mitosis, a complex rearrangement of the rod-shaped bodies called chromosomes, which exist in pairs. Sex cells (eggs or sperm) are "haploid," with one chromosome from each pair. All other body cells, called somatic cells, have a "diploid" chromosome number in which all chromosomes are paired. During mitosis, each chromosome becomes duplicated, then the two strands (or chromatids) split apart and separate. One result of mitosis is that the chromosome number of each cell is always preserved. Sutton also noticed that eggs in most species are many times larger than sperm because of a great difference in the amount of cytoplasm. The nuclei of egg and sperm are approximately equal in size, and these nuclei fuse during fertilization, a process in which two haploid sets of chromosomes combine to make a complete diploid set. From these facts, Sutton concluded that the genes are probably in the nucleus, not the cytoplasm, because the nucleus divides carefully and exactly while the cytoplasm divides inexactly. Also, if genes were in the cytoplasm, one would expect the mother's contribution to be much greater than the father's, contrary to the repeated observation that the parental contributions to heredity are usually equal.

Of all the parts of diploid cells, only the chromosomes were known to exist in pairs. If genes were located on the chromosomes, it would explain why they existed in pairs (except singly in eggs and sperm cells). In fact, the known behavior of chromosomes exactly paralleled the postulated behavior of Mendel's genes. Sutton's hypothesis that genes were located on chromosomes came to be called the chromosome theory of heredity. According to Sutton's hypothesis, Mendel's genes assorted independently because they were located on different chromosomes. However, there were only a limited number of chromosomes (eight in fruit flies, fourteen in garden peas, and forty-six in humans), while there were hundreds or thousands of genes. Sutton therefore predicted that Mendel's law of independent assortment would only apply to genes located on different chromosomes. Genes located on the same chromosome would be inherited together as a unit, a phenomenon now known as "linkage."

In 1903, Sutton outlined his chromosomal theory of heredity in a paper entitled "The Chromosomes in Heredity." Many aspects of this theory were independently proposed by Theodor Boveri, a German researcher who had worked with sea urchin embryos at the Naples Marine Station in Italy.

Linkage and Crossing-Over

Sutton had predicted the existence of linked genes before other investigators had adequately described the phenomenon. The subsequent discovery of linked genes lent strong support to Sutton's hypothesis. English geneticists William Bateson and Reginald C. Punnett described crosses involving linked genes in both poultry and garden peas, while American geneticist Thomas Hunt Morgan made similar discoveries in the fruit fly (*Drosophila melanogaster*). Instead of assorting independently, linked genes most often remain in the same combinations in which they were transmitted from prior generations: If two genes on the same chromosome both come from one parent, they tend to stay together through several generations and be inherited as a unit. On occasion, these combinations of linked genes do break apart, and these rearrangements were attributed to "crossing-over" of the chromosomes, a phenomenon in which chromosomes were thought to break apart and then recombine. Some microscopists thought they had observed X-shaped arrangements of the chromosomes that looked like the result of crossing-over, but many other scientists were skeptical about this claim because there was no proof of breakage and recombination of the chromosomes in these X-shaped arrangements.

Genetic Mapping

Sutton had been a student of Thomas Hunt Morgan at Columbia University in New York City. When Morgan began his experiments with fruit flies around 1909, he quickly became convinced that Sutton's chromosome theory would lead to a fruitful line of research. Morgan and his students soon discovered many new mutations in fruit flies, representing many new genes. Some of these mutations were linked to one another, and the linked genes fell into four linkage groups corresponding to the four chromosome pairs of fruit flies. In fruit flies as well as other species, the number of linkage groups always corresponds to the number of chromosome pairs.

One of Morgan's students, Alfred H. Sturtevant, reasoned that the frequency of recombination of linked genes should be small for genes located close together and higher for genes located far apart. In fact, the frequency of crossing-over between linked genes could serve as a rough measure of the distance between them along the chromosome. Sturtevant assumed that the frequency of recombination would be roughly proportional to the distance along the chromosome; recombination between closely linked genes would be a rare

event, while recombination between genes further apart would be more common. Sturtevant first used this technique in 1913 to determine the relative positions of six genes on one of the chromosomes of *Drosophila*. For example, the genes for white eyes and vermilion eyes recombined about 30 percent of the time, and the genes for vermilion eyes and miniature wings recombined about 3 percent of the time. Recombination between white eyes and miniature wings took place 34 percent of the time, close to the sum of the two previously mentioned frequencies (30 percent plus 3 percent). Therefore, the order of arrangement of the genes was:

white ←30 units→ vermilion ←3 units→ miniature

Since the distances were approximately additive (the smaller distances added up to the larger distances), Sturtevant concluded that the genes were arranged along each chromosome in a straight line like beads on a string. In

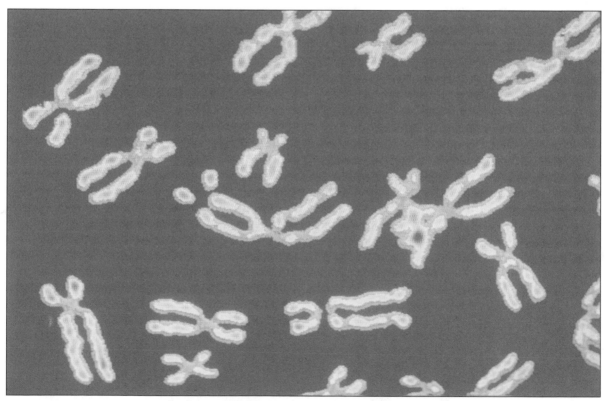

Normal human chromosomes. (Dan McCoy/Rainbow)

all, Sturtevant was able to determine such a linear arrangement among six genes in his initial study (an outgrowth of his doctoral thesis) and many more genes subsequently. Calvin Bridges, another one of Morgan's students, worked closely with Sturtevant. Over the next several years, Sturtevant and Bridges conducted numerous genetic crosses involving linked genes. They used recombination frequencies to determine the arrangement of genes along chromosomes and the approximate distances between these genes, thus producing increasingly detailed genetic maps of several *Drosophila* species.

The use of Sturtevant's technique of making linkage maps was widely copied. As each new gene was discovered, geneticists were able to find another gene to which it was linked, and the new gene was then fitted into a genetic map based on its linkage distance to other genes. In this way, geneticists began to make linkage maps of genes along the chromosomes of many different species. There are now over a thousand genes in *Drosophila* whose locations have been mapped using linkage mapping. Extensive linkage maps have also been developed for mice (*Mus musculus*), humans (*Homo sapiens*), corn or maize (*Zea mays*), and bread mold (*Neurospora crassa*). In bacteria such as *Escherichia coli*, other methods of genetic mapping were developed based on the order in which genes were transferred during bacterial conjugation. These mapping techniques reveal that the genes in bacteria are arranged in a circle or, more precisely, in a closed loop resembling a necklace. This loop can break at any of several locations, after which the genes are transferred from one individual to another in the order of their location along the chromosome. The order can be determined by interrupting the process and testing to see which genes had been transferred before the interruption.

Confirmation of the Chromosome Theory

The first confirmation of the chromosome theory was published in 1916 by Bridges, who studied the results of a type of abnormal cell division. When egg or sperm cells are produced by meiosis, only one chromosome of each chromosome pair is normally included in each of the resultant cells. In a very small proportion of cases, one pair of chromosomes fails to separate (or "disjoin"), so that one of the resultant cells has an extra chromosome while the other cell is missing that chromosome. This abnormal type of meiosis is called "nondisjunction." In fruit flies, as in humans and many other species, females normally have two X chromosomes (XX) and males have two unequal chromosomes (XY). Bridges discovered some female fruit flies that had the unusual chromosome formula XXY; he suspected that these unusual females had originated from nondisjunction, in which two X chromosomes had failed to separate during meiosis. Bridges studied one cross using a white-eyed XXY female mated to a normal, red-eyed male. (The gene for white eyes was known to be sex-linked; it was carried on the X chromosome.) Bridges was able to predict both the genetic and chromosomal anomalies that would occur as a result of this cross. Among the unusual predictions that were verified experimentally was the existence of a chromosome configuration (XYY) that had never been observed before. Using the assumption that the gene for white eyes was carried on the X chromosome in this and other crosses, Bridges was able to make unusual predictions of both genetic and chromosomal results. These studies greatly strengthened the case for the chromosomal theory.

In 1931, Harriet Creighton and Barbara McClintock were able to confirm the chromosomal theory of inheritance much more directly. Creighton and McClintock used corn plants whose chromosomes had structural abnormalities on either end, enabling them to recognize the chromosomes under the microscope. One chromosome, for example, had a knob at one end and an attached portion of another chromosome at the other end, as shown in the accompanying figure on the following page. Creighton and McClintock then crossed plants differing in two genes located along this chromosome. One gene controlled the color of the seed coat while the other produced either a starchy or waxy kernel. The parental gene combinations (C with wx on the abnormal chromosome and c with Wx on the other chromosome) were always

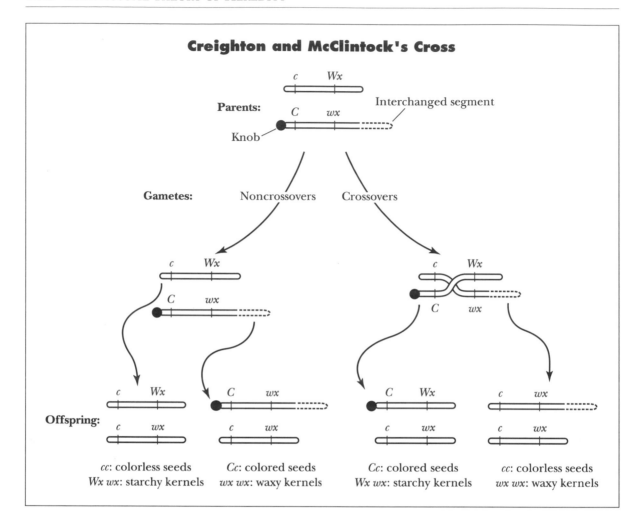

Creighton and McClintock's Cross

Parents:

Knob

Interchanged segment

Gametes: Noncrossovers   Crossovers

Offspring:

cc: colorless seeds
Wx wx: starchy kernels

Cc: colored seeds
wx wx: waxy kernels

Cc: colored seeds
Wx wx: starchy kernels

cc: colorless seeds
wx wx: waxy kernels

preserved in noncrossovers. However, a crossover between the two genes produced two new gene combinations: *C* with *Wx* and *c* with *wx*.

In this cross, Creighton and McClintock observed that the chromosomal appearance in the offspring could always be predicted from the phenotypic appearance: Seeds with colorless seed coats and starchy kernels had normal chromosomes, seeds with colored seed coats and waxy kernels had chromosomes with the knob at one end and the extra interchanged chromosome segment at the other end, seeds with colorless seed coats and waxy kernels had the interchanged segment but no knob, and seeds with colored coats and starchy kernels had the knob but not the interchanged segment. In other words, whenever the two genes showed rearrangement of the parental combi-

nations, a corresponding switch of the chromosomes could be observed under the microscope. The interchange of chromosome segments was always accompanied by the recombination of genes, or, in the words of the original paper, "cytological crossing-over . . . is accompanied by the expected types of genetic crossing-over. . . . Chromosomes . . . have been shown to exchange parts at the same time they exchange genes assigned to these regions." In short, genetic recombination (the rearranging of genes) was always accompanied by crossing-over (the rearranging of chromosomes). This historic finding established firm evidence for the chromosomal theory of heredity. Later that same year, Curt Stern published a paper describing a very similar experiment using fruit flies.

Physical Mapping and Further Confirmation

Other evidence that helped confirm the chromosome theory came from the study of rare chromosome abnormalities. In 1933, Thomas S. Painter called attention to the large salivary gland chromosomes of *Drosophila*. Examination of these large chromosomes made structural abnormalities in the chromosomes easier to identify. When small segments of a chromosome were missing, a gene was often found to be missing also. These abnormalities, called "chromosomal deletions," allowed the first physical maps of genes to be drawn. In all cases, the physical maps were found to be consistent with the earlier genetic maps (or linkage maps) based on the frequency of crossing-over.

When Bridges turned his attention to the "bar eyes" trait in fruit flies, he discovered that the gene for this trait was actually another kind of chromosome abnormality called a "duplication." Again, a chromosome abnormality that could be seen under the microscope could be related to a genetic map based on linkage. Larger chromosome abnormalities included "inversions," in which a segment of a chromosome was turned end-to-end, and "translocations," in which a piece of one chromosome became attached to another. There were also abnormalities in which entire chromosomes were missing or extra chromosome were present. Each of these chromosomal abnormalities was accompanied by corresponding changes in the genetic maps based on the frequency of recombination between linked genes. In cases in which the location of a chromosomal abnormality could be identified microscopically, this permitted an anchoring of the genetic map to a physical location along the chromosome. The correspondence between genetic maps and chromosomal abnormalities provided important additional evidence in support of the chromosomal theory. Other forms of physical mapping were developed decades later in mammals and bacteria. The increasingly precise mapping of gene locations led the way to the development of modern molecular genetics, including techniques for isolating and sequencing individual genes.

The discovery of restriction endonuclease enzymes during the 1970's allowed geneticists to cut DNA molecules into small fragments. In 1980, a team headed by David Botstein measured the sizes of these "restriction fragments" and found many cases in which the length of the fragment varied from person to person because of changes in the DNA sequence. This type of variation is generally called a "polymorphism." In this case, it was a polymorphism in the length of the restriction fragments (known as a restriction fragment length polymorphism, or RFLP). The use of the RFLP technique has allowed rapid discovery of the location of many human genes. The Human Genome Project (an effort by scientists worldwide to determine the location and sequence of every human gene) would never have been proposed had it not been for the existence of this mapping technique.

—Eli C. Minkoff

See Also: Chromosome Structure; Classical Transmission Genetics; Dihybrid Inheritance; *Drosophila melanogaster*; Linkage Maps; Meiosis; Mendel, Gregor, and Mendelism.

Further Reading: Many of the classic papers that contributed to the chromosomal theory are reprinted in James A. Peters, *Classic Papers in Genetics* (1959). Included in this book are the now-famous papers in which Mendel established the principles of genetics, Sutton first proposed the chromosomal theory of heredity, Sturtevant produced the first genetic map based on linkage, and Creighton and McClintock confirmed that the recombination of linked genes always took place by a process that also rearranged the chromosomes. Other important papers by Bridges, Painter, Morgan, Bateson, and Punnett are also included. Botstein's initial paper on the RFLP technique appears in the *American Journal of Human Genetics* 32 (1980). Good books that explain the principles covered in this article include C. P. Swanson, *Cytology and Cytogenetics* (1957); D. T. Suzuki et al., *An Introduction to Genetic Analysis* (1981); M. Strickberger, *Genetics* (1968); H. Rees et al., *Chromosome Genetics* (1977); W. Hexter et al., *The Science of Genetics* (1976), and B. Lewin, *Genes V* (1994). The history of the gene concept is chronicled by E. A. Carlson, *The Gene: A Critical History* (1966).

Classical Transmission Genetics

Field of study: Classical transmission genetics

Significance: *Transmission genetics refers to the passing of the information needed for the proper function of an organism from parents to their offspring as a result of reproduction. In sexual reproduction, parents produce specialized cells (eggs and sperm) that fuse to produce a new individual. Each of these cells contains one copy of each of the required units of information (the "genes"), which provide the blueprint necessary for the offspring to develop into individual, functioning organisms.*

Key terms

CHROMOSOME: a structure in the nucleus of a cell that contains the genetic material

GENE: a portion of deoxyribonucleic acid (DNA) that carries character-determining factors

REPRODUCTION: the production of more of the same kind; in biological systems, the production of a new generation of individuals using the information transmitted by each parent

OFFSPRING: the new individuals resulting from reproduction

CROSS: the mating of parents to produce offspring during sexual reproduction

The Discovery of Transmission Genetics

The desire to improve plant and animal production is as old as agriculture. For centuries, humans have been using selective breeding programs that have resulted in the production of thousands of varieties of plants and breeds of animals. The Greek philosopher-scientist Hippocrates suggested that small bits of the body of the parent were passed to the offspring during reproduction. These small bits of arms, heads, stomachs, and livers were thought to develop into a new individual. Following the development of the microscope, it became possible to see the cells, the small building blocks of living organisms. Study of the cell during the 1800's showed that sexual reproduction was the result of the fusion of specialized cells from two parents (eggs and sperm). It was also observed that these cells contained chromosomes ("color bodies" visible when the cells reproduced) and that the number and kind of these chromosomes was the same in both the parents and the offspring. This suggested that the chromosomes carried the genetic information and that each parent transmitted the same number and kinds of chromosomes. For example, humans have twenty-three kinds of chromosomes. The offspring receives one of each kind from each parent and so has twenty-three chromosome pairs. Since the parents and the offspring have the same number and kinds of chromosomes, and since each parent transmits one complete set of the chromosomes, it was thought that there must be a process of cell division that reduces the parent number from two sets of chromosomes to one set in the production of the egg or sperm cells. The parents would each have twenty-three pairs (forty-six) chromosomes, but their reproductive cells would each contain only one of each chromosome (twenty-three).

In the 1860's, the Austrian botanist Gregor Mendel repeated studies of inheritance in the garden pea and, using the results, developed a model of genetic transmission. The significance of Mendel's work was not recognized during his lifetime, but it was rediscovered in 1900. In that same year, the predicted reductional cell division during reproduction was fully described, and the science of genetics was born.

Classical Transmission Genetics

Genetics is the study of variation. It is recognized that a particular feature of an animal or plant is inherited because there is variation in the expression of that feature, and variation in expression follows a recognizable inheritance pattern. For example, it is known that blood types are inherited, both because there is variation (blood types A, B, and O) and because examination of family histories reveals patterns that show transmission of blood-type information from parents to children.

Variation in character expression may have one of two sources: environmental conditions or inherited factors. If a plant is grown on poor soil, it might be short. The same plant grown on good soil might be tall. A plant that is short because of an inherited factor cannot grow tall even if it is placed on richer soil. From this example, it can be seen that there may be two different ways to determine whether a specific character expression is environmentally or genetically determined: testing for environmental influences and testing for inherited factors. Many conditions are not so easily resolved as this example; there may be many complex environmental factors involved in producing a condition, and it would be impossible to test them all. Knowledge of inheritance patterns can, however, help in determining whether inherited factors play a role in a condition. Cancer-associated genes have been located using family studies that show patterns consistent with a genetic contribution to the disease. There are certainly environmental factors that influence cancer production, but those factors are not as easily recognized.

The patterns of transmission genetics were developed because the experimenters focused their attention on single, easily recognized characteristics. Mendel carefully selected seven simple characteristics of the pea plant, such as height of the plant, color of the flower, and color of the seeds. The second reason for success was the use of carefully controlled crosses. The original parents were selected from families that did not show variations in the characteristic of interest. For example, plants from a pure tall family were crossed with plants from a pure short family. Control of the information passed by the parents allowed the experimenter to follow the variation of expression from parents to offspring through a number of generations.

The classic genetic transmission pattern is the passing of information for each characteristic from each parent to each offspring. The offspring receives two copies of each gene. (The term "gene" is used to refer to a character-determining factor; Mendel's original terminology was "factor.") Each parent also had two copies of each gene, so in the production of the specialized reproductive cells, the number must be reduced. Consider the following example. A tall pea plant has two copies of the information for height, and both copies are for

tall height (tall/tall). This plant is crossed with a plant with two genes for short height (short/short). The information content of each plant is reduced to one copy: The tall plant transmits one tall gene, and the short plant transmits one short gene. The offspring receive both genes and have the information content tall/short.

The situation becomes more complex and more interesting when one or both of the parents in a cross have two different versions of the gene for the characteristic. If, for example, one parent has the height genes tall/short and the other has the genes short/short, the cells produced by the tall/short parent will be of two kinds: $\frac{1}{2}$ carry the tall gene and $\frac{1}{2}$ carry the short gene. The other parent has only one kind of gene for height (short), so all of its reproductive cells will contain that gene. The offspring will be of two kinds: $\frac{1}{2}$ will have both genes (tall/short), and $\frac{1}{2}$ will have only one kind of gene (short/short). Had it been known that the one parent had one copy for each version of the gene, it could have been predicted that the offspring would have been of two kinds and that each would have an equal chance of appearing. Had it not been known that one of the parents had the two versions of the gene, the appearance of two kinds of offspring would have revealed the presence of both genes. The patterns are repeatable and are therefore useful in predicting what might happen or revealing what did happen in a particular cross. For example, blood-type patterns or deoxyribonucleic acid (DNA) variation patterns can be used to identify the children that belong to parents in kidnapping cases or in cases in which children are mixed up in a hospital.

In a second example, the pattern is more complex, because both parents carry both versions of the gene: a tall/short to tall/short cross. Each parent will produce $\frac{1}{2}$ tall-gene-carrying cells and $\frac{1}{2}$ short-gene-carrying cells. Any cell from one parent may randomly join with any cell from the other parent, which leads to the following patterns: $\frac{1}{2}$ tall × $\frac{1}{2}$ tall = $\frac{1}{4}$ tall/tall; $\frac{1}{2}$ tall × $\frac{1}{2}$ short = $\frac{1}{4}$ tall/short; $\frac{1}{2}$ short × $\frac{1}{2}$ tall = $\frac{1}{4}$ short/tall; $\frac{1}{2}$ short × $\frac{1}{2}$ short = $\frac{1}{4}$ short/short. Tall/short and short/tall are the same, yielding totals of ¼ tall/tall; ½ tall/short; and ¼ short/short, or a 1:2:1 ratio. This was the ratio that Mendel recognized and used to develop his model of transmission genetics. Mendel used pure parents (selected to breed true for the one characteristic), so he knew when he had a generation in which all of the individuals had one of each gene. As in the previous example, if it had been known that each of the parents had one of each gene, the ratio could have been predicted; conversely, by using the observed ratio, the information content of the parents could be deduced. Using a blood-type example, if one parent has blood-type genes AO and the other parent has the genes BO, the possible combinations observed in their offspring would be AB, AO, BO, and OO, each with the same probability of occurrence (½A gene-bearing and ½O gene-bearing cells in one parent × ½B gene-bearing and ½O gene-bearing cells in the other parent).

The Patterns of Inheritance

Transmission genetics is useful because it allows researchers to make predictions about specific crosses and explains the occurrence of characteristic expressions in the offspring. In genetic counseling, probabilities of the appearance of a genetic disease can be made when there is an affected child in the family or a family history of the condition. This is possible because, for most inherited characteristics, the pattern is established by the reduction of chromosome numbers that occurs when the reproductive cells are produced and by the random union of reproductive cells from the two parents. The recognition that the genes are located on the chromosomes and the description of the reductional division in which the like chromosomes separate, carrying the two copies of each gene into different cells during the reduction division of meiosis, provide the basis of the regularity of the transmission pattern. It is this regularity that allows the application of mathematical treatments to genetics. Two genes are present for each character in each individual, but only one is passed to each offspring by each parent; therefore, the 50 percent (or ½) probability becomes the basis for

making predictions about the outcome of a cross for any single character.

The classical pattern of transmission genetics occurs because specialized reproductive cells, eggs and sperm, are produced by a special cell reproduction process (meiosis) in which the chromosome number is reduced from two complete sets to one set in each of the cells that result from the process. This reduction results because each member of a pair of chromosomes recognizes its partner, and the chromosomes come together. This joining (pairing) appears to specify that each chromosome in the pair will become attached to a "motor" unit from an opposite side of the cell that will move the chromosomes to opposite sides of the cell during cell division. The result is two new cells, each with only one of the chromosomes of the original pair. This process is repeated for each pair of chromosomes in the set.

Independent Genes

Humans have practiced selective breeding of plants and animals for centuries, but it has only been during the nineteenth and twentieth centuries that the patterns of transmission of inherited characters have been understood. This change occurred because the experimenters focused on a single characteristic and could understand the pattern for that characteristic. Previous attempts had been unsuccessful because the observers attempted to explain a large number of character patterns at the same time. Mendel expanded his model of transmission to show how observations become more complex as the number of characteristics examined is expanded. Consider a plant with three chromosomes and one simple character gene located on each chromosome. In the first parent, chromosome 1 contains the gene for tall expression, chromosome 2 contains the gene for expression of yellow seed color, and chromosome 3 contains the gene for purple flower color. In the other parent, chromosome 1 contains a gene for short height, chromosome 2 contains a gene for green seed color, and chromosome 3 contains a gene for white flower color. Each parent will transmit these genes to their offspring, who will have the genes tall/short, yellow/green, and purple/

white. In the production of reproductive cells, the reductional division of meiosis will pass on one of the character expression genes for each of the three characters. (It is important to remember that the products of the reductional cell division have one of each chromosome. If this did not occur, information would be lost, and the offspring would not develop normally.) The characteristics are located on different chromosomes, and during the division process, these chromosome pairs act independently. This means that the genes that came from any one parent (for example, the tall height, yellow seed, and purple flower expression genes from the one parent) do not have to go together during the division process. Since chromosome pairs act independently, different segregation patterns occur in different cells. The results from one meiosis may be a cell with the tall, green, and purple genes and one with the short, yellow, and white genes. In the same plant, another meiosis might produce a cell with the short, yellow, and purple genes, and the second cell would have the tall, green, and white genes.

Since these genes are independent, height does not influence seed color or flower color, nor does flower color influence seed color or height. The determining gene for each characteristic is located on a different chromosome, so the basic transmission model can be applied to each gene independently, and then the independent patterns can be combined. The tall/short height genes will segregate so that $1/2$ of the cells will contain the tall gene and $1/2$ will contain the short gene. Likewise, the yellow/green seed color genes will separate so that $1/2$ of the cells will contain the yellow gene and $1/2$ will contain a green gene. Finally, $1/2$ of the cells will contain a purple flower gene and $1/2$ will contain a white gene. These independent probabilities can be combined because the probability of any combination is the product of the independent probabilities. For example, the combination tall, purple, white will occur with a probability of $1/2 \times 1/2 \times 1/2 = 1/8$. This means that one should expect eight different combinations of these characters. The possible number of combinations for n chromosome pairs is 2^n. For humans, this means that

any individual may produce 2^{23} different chromosome combinations. This is the same idea as tossing three coins simultaneously. Each coin may land with a head or a tail up, but how each coin lands is independent of how the other coins land. Knowledge of transmission patterns based on chromosome separation during meiosis allows researchers to explain the basic pattern for a single genetic character, but it also allows researchers to explain the great variation that is observed among individuals within a population in which genes for thousands of different characters are being transmitted.

Continuous Variation and Linked Genes

The principles of transmission genetics were established by studying characters with discrete expressions—plants were tall or dwarf, seeds were yellow or green. In 1903, Danish geneticist Wilhelm Johannsen observed that characteristics that showed continuous variation, such as weight of plant seeds, fell into recognizable groups that formed a normal distribution. These patterns could be explained by applying the principles of transmission genetics. Assume a plant has two genes that influence its height and that these genes are on two different chromosomes (for example, 1 and 3). Each gene has two versions. A tall gene stimulates growth (increases the height), but a short gene makes no contribution to growth. A plant with the composition tall-1/tall-1, tall-3/tall-3 would have a maximum height because four genes would be adding to the plant's height. A short-1/short-1, short-3/short-3 plant would have minimum height because there would be no contribution to its height by these genes. Plants could have two contributing genes (tall-1/short-1, tall-3/short-3) or three contributing genes (tall-1/short-1, tall-3/tall-3). The number of offspring with each pattern would be determined by the composition of the parents and would be the result of gene segregation and transmission patterns. Many genes contributing to a single character expression apply to many interesting human characteristics, such as height, intelligence, amount of skin pigmentation, hair color, and eye color.

Mendel's model of the transmission of genes was supported by the observations of chromo-

Alfred Sturtevant's research into the genetics of the fruit fly helped to clarify the role of chromosomes in heredity. (California Institute of Technology)

some pair separation during the reduction division, but early in the twentieth century, it was recognized that some genes did not separate independently. Work in American geneticist Thomas Hunt Morgan's laboratory, especially by an undergraduate student, Alfred Sturtevant, showed that each chromosome contained determining genes for more than one characteristic and established that genes located close together on the same chromosome stayed together during the separation of the paired chromosomes during meiosis. If a pea plant had a chromosome with the tall height gene and, immediately adjacent to it, a gene for high sugar production, and if the other version of this chromosome had a gene for short height and a gene that limited the sugar production, the most likely products from meiosis would be two kinds of cells: one with the genes for tall height and high sugar production and one with the genes for short height and limited sugar production. These genes are said to be "linked," or closely associated on the same chromosome, because they go together as the chromosomes in the pair separate. It is generally accepted that humans contain approximately 100,000 genes, but there are only twenty-three kinds of chromosomes. This means that each chromosome contains many different genes. Each chromosome is considered a linkage group, and one of the goals of genetic study is to locate the gene responsible for each known characteristic to its proper chromosome.

A common problem in medical genetics is locating the gene for a specific genetic disease. Family studies may show that the disease is transmitted in a pattern consistent with the gene being on one of the chromosomes, but there is no way of knowing its location. Variations in DNA structure are also inherited in the classic pattern, and these DNA pattern modifications can be determined using modern molecular procedures. DNA variation patterns are analyzed for linkage to the disease condition. If a specific DNA pattern always occurs in individuals with the disease condition, it indicates that the DNA variation is on the same chromosome and close to the gene of interest because it is transmitted along with

the disease-producing gene. This information locates the chromosome position of the gene, allowing further work to be done to study its structure. Once the postition and structure of the disease-producing gene are understood, researchers can begin to investigate possible treatments using the techniques of gene therapy.

—*D. B. Benner*

See Also: Biotechnology; Chromosome Structure; Congenital Defects; Dihybrid Inheritance; Linkage Maps; Meiosis; Mendel, Gregor, and Mendelism; Monohybrid Inheritance; Multiple Alleles.

Further Reading: *The Cartoon Guide to Genetics* (1991), by Larry Gonick and Mark Wheelis, is an easy-to-read presentation of the basic concepts of genetics. *Human Heredity: Principles and Issues* (1994), by Micheal Cummings, discusses transmission genetic concepts as they apply to human characteristics. *Science as a Way of Knowing* (1993), by John A. Moore, traces the development of scientific thinking with an emphasis on understanding hereditary mechanisms. *Theory and Problems of Genetics* (1991), by William D. Stansfield, provides explanations of basic genetics concepts and an introduction to problem solving, while the *Case Workbook in Human Genetics* (1994), by Ricki Lewis, presents problems based on specific diseases.

Cloning

Field of study: Genetic engineering and biotechnology
Significance: *Cloning, the exact genetic replication of organisms from donor DNA, has broad implications in agriculture and animal husbandry. At the same time, advances in the cloning of complex vertebrates—including, perhaps, humans—have raised nagging bioethical problems.*

Key terms

BIOETHICISTS: scholars who focus on the ethical implications of biological techniques
CYTOPLASM: a viscous substance surrounding the nuclei of cells
DEOXYRIBONUCLEIC ACID (DNA): the material found in all cells that contains the genetic blueprints of organisms

DNA POLYMERASE: an enzyme that creates new DNA by using genetic information found in old DNA

GENE CLONES: exact replications of particular areas of DNA made by the use of microorganisms

GENOME: a complete set of chromosomes; the human genome contains forty-six chromosomes

Cloning and the Replication of DNA

Cloning is almost as old as the human species. Anyone who breaks a leaf or a stem from a plant, puts it in water in which it develops rudimentary roots, and then plants it so that a new plant will develop engages in cloning. Such techniques have been used by farmers for thousands of years and continue to be important in contemporary agriculture. Animal cloning occurs in multiple births in which resulting offspring are genetically identical. In human beings, identical twins are genetic replications of each other (but fraternal twins are not).

In the last half of the twentieth century, sophisticated techniques extended the possibilities of cloning and resulted in the genetic replication of complex vertebrates in laboratories. The potential exists for the cloning of humans, a possibility that causes deep concern among both bioethicists and the general public. The possibilty of cloning vertebrates was considered little more than science fiction until 1953, when American biologist James Watson and English physicist Francis Crick discovered that the deoxyribonucleic acid (DNA) found in every cell of each living organism contains the complete genetic blueprint of that organism. The two researchers posited that DNA's molecular structure is a double-spiral helix chain. The DNA polymerase, an enzyme found in DNA, contains information that is used to create DNA that is genetically identical to the DNA from which it is derived. With the discovery that DNA replication can be achieved under laboratory conditions, the cloning of complex species became a distinct possibility.

Cells are separated from other cells by cell walls. Each cell contains a nucleus surrounded by a thick, jellylike substance called cytoplasm. Individual cells contain between 30,000 and 100,000 genes. Typically, human adults have approximately sixty trillion cells in their bodies. Each cell, with the exception of some specialized reproductive cells, contains all the genes that every other cell contains. DNA molecules are the largest molecules in the cell, although they are still so small that they can be seen only through the most powerful electronic microscopes. Remarkably, if all the cells in a typical human body were stretched out end to end, they would form a line half a billion miles long, or five times the distance from the earth to the sun. Human cells contain forty-six chromosomes, twenty-three from each parent. These chromosomes are called genomes. When the cells divide through mitosis, some of them become differentiated cells that determine such characteristics as hair color, gender, and basic body structure. Even these differentiated cells contain a full blueprint of the organism of which they are part.

Some of the earliest replication experiments took place with the cloning of various bacteria. To prepare the total cell for such cloning, four basic steps are followed. The culture containing the bacterial DNA is first grown and harvested. The cells are then opened to release their contents. Next, the material that has been extracted is treated to draw everything from it except the DNA. Finally, the DNA solution is concentrated so that it can be placed in a hospitable medium where it will multiply.

The Cloning of Dolly

It has long been possible to clone single-cell organisms in laboratories. The process for doing this involves the isolation of a cell in an environment that will sustain it. In time, the cell replicates itself, and each resulting cell will be genetically identical to the original cell. Using such techniques, scientists went one step further in the mid-1970's, when they successfully cloned a tadpole.

In 1997, the world was taken aback when a group of scientists headed by embryologist Ian Wilmut at the Roslin Institute in Scotland announced the successful cloning of a sheep, subsequently named Dolly. Scientists had previously cloned cows and sheep, but they had used embryo cells to accomplish this cloning.

Dolly the sheep, the world's first adult mammal clone, was the subject of intense media interest after the February, 1997, announcement of her birth. (Reuters/Jeff J. Mitchell/Archive Photos)

Dolly was the first vertebrate to be cloned from the cell of an adult vertebrate. Her cloning was accomplished by removing a substantial number of cells from the udder of a six-year-old ewe and placing them in a laboratory dish filled with nutrients, where they were left to grow for five days. At the end of this period, the nutrients were reduced to 5 percent of what the cells needed to continue growing. This withdrawal of sustenance caused the cells to enter a state resembling suspended animation and return to their original, undifferentiated state.

When these cells were placed in the ova of several host sheep, each ovum received a genetic message from them that caused it to begin creating a lamb embryo. Of an initial 277 adult cells introduced into the ova, thirteen resulted in pregnancy, but only one, which resulted in Dolly's delivery, was carried to full term. Dolly is an exact genetic replica of the sheep from whose udder the original cells were extracted. Environmental factors will serve to make Dolly

or any other clone an individual, but genetically, such clones will never have the sort of individuality that an organism produced by more usual reproductive means would possess.

Some doubt has been cast upon the claims of Wilmut and his colleagues about the cloning of Dolly because no one was able to replicate their experiment and produce another clone similar to Dolly. Wilmut himself has conceded that it is remotely possible that Dolly was not the clone of an adult ewe but rather of the fetus the ewe might have been carrying. The ewe from which Dolly was derived could have had some fetal cells circulating in her bloodstream. One of these cells might inadvertently have contaminated the laboratory culture that produced Dolly. This kind of cloning has already been accomplished, so the cloning of Dolly may not represent the breakthrough that was originally supposed.

Whether or not the Dolly experiment is validated, however, other scientists have come close to replicating it. James Robl of the University of Massachusetts and Steven Stice, chief scientist for Advanced Cell Technology in Worcester, Massachusetts, cloned calf embryos from adult cells they removed from the leg of a cow they obtained from a Massachusetts slaughterhouse. Of the hundreds of cells they cloned from this cow, dozens grew into embryos, although none lived longer than sixty days. Regardless of the outcome of the Dolly controversy, scientists will surge ahead in their quest to replicate life through the sort of cloning that is thought to have produced Dolly.

Human Cloning

The announcement of scientific breakthroughs such as the cloning of Dolly have caused deep concern among bioethicists, who see such experimentation as a major step toward human cloning. Such concerns are not without merit. In 1993, human clones were created by biologists at George Washington University. These scientists took cells from seventeen defective human embryos obtained from a fertility clinic that was discarding them. Drawing cells from embryos that ranged from two to eight cells in size, the scientists extracted individual cells and grew them in hospitable cultures. This procedure resulted in the production of some embryos with as many as thirty-two cells. The possibility existed that any one of these cells could be implanted in a human host, who could carry it to full term as any pregnant woman would carry an embryo to delivery. The experiment was terminated before it went that far, but it illustrated without a doubt that the possibility of cloning humans had been catapulted from the realm of science fiction into the arena of stark reality. The scientists who carried out this experiment, Robert J. Stillman and Jerry L. Hall, were officially censured by the faculty and administration of George Washington University. They were cautioned to suspend such experimentation, although such condemnation seemed unlikely to slow progress in the field or discourage future scientists from conducting such experiments.

It has been argued that human cloning offers hope to childless couples who want children who are genetically related to them. In its current state of development, however, the cloning of humans in such situations might prove impractical and highly unlikely. According to Wilmut and his colleagues, it took 277 attempts to create Dolly. The sheer inconvenience of subjecting a human to so many attempts would discourage most people from participating in cloning procedures. Morever, if inconvenience were not a determining factor, cost might be. In 1998, women who sold their eggs to fertility clinics were paid two thousand dollars for each egg; at such rates, it would be prohibitively expensive for a clinic to acquire hundreds of eggs for the impregnation of a single patient.

Of course, such statistics constantly change. Robl and Stice claimed that the success rate of their procedure for cloning cows from adult cells was between 5 and 10 percent. They achieved these results by adding cells that were dividing normally to cow eggs that had their nuclei removed. They then delayed reactivating the DNA longer than Wilmut had in his experiments with the ewe that produced Dolly. Robl and Stice claimed a success rate comparable to the 13 percent rate of success among clinics that use frozen human embryos to impregnate women.

Bioethical Considerations

In December, 1997, physicist Richard Seed announced that he intended to clone human beings. Seed vowed that if he could not carry out his experiments in the United States, where Secretary of Health and Human Services Donna Shalala described him as a "mad scientist," he would carry them out in a country that would not place restrictions upon him. The public expressed alarm at Seed's announcement, and cloning quickly became an issue in both religious and political discussions. The matter was brought before the U.S. Senate in January and February, 1998, with the introduction of a bill sponsored by Republican senator Christopher S. Bond from Missouri. This bill would have made it a felony to use somatic-cell nuclear transfer technology, such as that employed in cloning Dolly, with human cells. Lobbyists from the scientific and medical communities worked strenuously to prevent any legislation that would ban cloning.

When the bill was brought to the Senate floor, opponents used parliamentary tactics to keep it from being debated. Supporters of the bill, seeking to force debate on it, could not marshall the three-fifths majority vote they needed to override the objections to the bill's consideration. Actually, fifty-four senators, a number of them crossing party lines, opposed consideration, while forty-two favored it. The Senate's oldest member, archconservative Strom Thurmond of South Carolina, surprised his colleagues by coming out in opposition to the bill, explaining that his daughter suffered from diabetes and might sometime in the foreseeable future be helped by procedures that could grow out of human cloning. Senators Dianne Feinstein from California and Edward Kennedy from Massachusetts (both Democrats) proposed legislation that would ban the placement of cloned human embryos in the uterus but would permit the continuation of other cloning research for medical purposes.

Nobel-Prize-winning virologist David Baltimore, president of the California Institute of Technology, warned, "Legislation is a very poor way of dealing with issues of scientific and medical progress." He cautioned, "What looks to be revolutionary one day is common practice the next," and urged that rather than enact legislation that would be difficult to change as conditions change, it would be better to encourage federal oversight agencies to enact and enforce regulations that could more easily be changed to meet altered conditions.

Another concern among scientists and legislators is that a human produced through cloning, although a virtual twin to its donor, would suffer a diminution in its human dignity. Psychologists are particularly fearful that such a twin would lack the sense of autonomy that a child produced by conventional means would have. Francis Pizzulli, an attorney who concerns himself with matters related to human cloning, fears that a type of genetic bondage would result from such cloning. Another potential problem would arise from donors who later die of diseases that can be inherited, such as diabetes or Huntington's chorea; humans cloned from these donors would then live in the shadow of their impending fate. Such clones might conceivably be turned down for life insurance, health insurance, and employment if their origins became known.

Richard Seed at a 1998 news conference discussing his plans to clone human beings. (Reuters/Sue Ogrocki/Archive Photos)

The Cloning of Extinct Species

The film *Jurassic Park* (1993) heightened public interest in the possibility of cloning life forms that have been extinct for long periods of time. This possibility interested scientists for years before the popular film attracted public interest. Allan Wilson and his students at the University of California at Berkeley were at the forefront of testing ancient DNA that was available in organisms trapped in amber and preserved in that state for eons. Relatively fresh DNA was also found in organisms that had been buried in ice for long periods so that the organic matter was not drastically different from what it had been at the time the organism died. Using DNA polymerase chain reactions, researchers have been able to create DNA sequences from very old cells, including those from a 17-million-year-old magnolia leaf and a weevil estimated to be 120 million years old. DNA replication of this sort results in the creation of genetically identical organisms.

Such DNA sequences, however, are very short and have offered little hope that extinct organisms can be reproduced through cloning. Creating the DNA sequences is an initial step toward reproduction through cloning. This first step, however, lies quite a distance from the final step, the reproduction or exact replication of a living organism that is a true clone of the original organic matter. A major problem in attempting to clone extinct organisms is that a cell's genomes are activated in the egg cells of the same species. If a species is extinct, such activation is dubious. Furthermore, once a species ceases to exist, a fertilized egg cell, if one could be created, would require a host that could carry it to term. Because cloned offspring are typically larger than offspring produced by the usual reproductive methods, delivery is usually difficult and dangerous. The failure rate is high among clones that result from living cells; it is therefore certain that the failure rate would be even higher if cloning is done with cells that are nonliving but still viable. Such cells clearly contain their

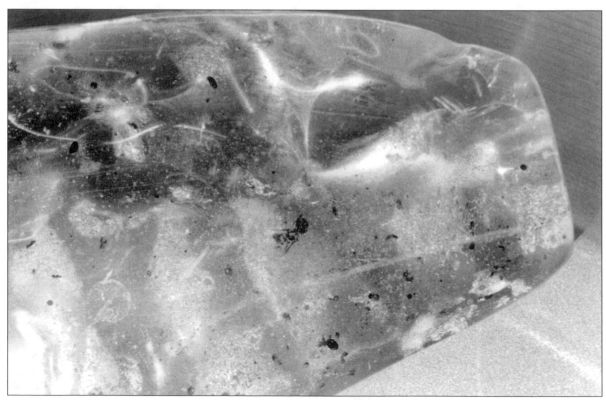

Scientists at the University of California at Berkeley have extracted DNA from ancient insects trapped in amber. (Ben Klaffke)

own DNA, but reconstituting it and replicating it so that cloning can take place is problematic.

The technology does not currently exist to consider cloning such large vertebrates as dinosaurs or mastodons. It is doubtful that any existing animal could carry the fertilized eggs of such creatures to term, particularly in light of the fact that cloned organisms are larger than their counterparts that are not cloned. There is considerable value in studying the genetic composition and DNA of species that no longer exist. The information gained from such study, however, is clearly not dependent upon transforming the DNA into living organisms.

Facilities for Cloning

According to the Centers for Disease Control, the United States had 281 in vitro fertilization clinics in operation at the end of 1997. Half of these clinics had the equipment and informed personnel that would make the cloning of humans possible. Such clinics treat, through intracytoplasmic sperm injection, couples who have been unable to conceive. This procedure involves the use of a microscopic pipette that injects a single sperm cell into the egg of the woman who wishes to conceive. This procedure can easily be adapted to permit the cloning of humans. All that is required is an injection of nucleic DNA from the cell of an adult into an egg whose nucleus has been removed. Although the government has specifically ordered federal funds withheld from any operation that is engaged in cloning, in vitro fertility clinics are largely unaffected by the mandate because federal funds had already been withheld from them.

Recognized U.S. scientists who wish to move in the direction of human cloning can find support outside the United States if they fail to find such support within the country. A Swedish religious group, the Raelians, have already offered to fund Richard Seed's experiments and provide him with laboratory facilities in Sweden if he wishes to move his operation there.

The Use of Bovine Incubation in Cloning

In 1998, reproductive biologists Neal First of the University of Wisconsin and Tanja Dominko of the Oregon Regional Research Center reported an exciting breakthrough at a meeting of the International Embryo Transfer Society that has broad implications for cloning. These two scientist introduced nucleic DNA from a variety of animals that included rats, monkeys, pigs, and sheep into the eggs of cows. The nuclei having been removed from these eggs, they were activated by the nucleic DNA introduced into them so that they produced clones of the donor whose DNA was used.

At the time of their report, the embryos that First and Dominko introduced into the bovine eggs had all miscarried. Their experiments suggest, however, the possibility, theoretically at least, that the eggs of cows might one day be used in the incubation of any adult mammal cells, including those of humans. If such incubation is perfected, human women would not have to endure the risks and high expense of providing their own eggs. The eggs that First used in this experiment were obtained from a slaughterhouse, where they were considered a waste product. With the refinement of this technique, couples who have been unable to reproduce by conventional methods would be able to have offspring that possess the genes of at least one parent. This development would, in the eyes of some, make human cloning a more acceptable procedure than it is currently considered. One remaining caveat in the eyes of many scientists and legislators is that cloned animals tend to suffer from more physical abnormalities than animals produced by usual reproductive means.

—R. Baird Shuman

See Also: Biotechnology; Biotechnology, Risks of; Cloning: Ethical Issues; Cloning Vectors; Genetic Engineering; Genetic Engineering: Social and Ethical Issues; Mitosis; Sheep Cloning.

Further Reading: Arthur L. Caplan's *Moral Matters: Ethical Issues in Medicine and the Life Sciences* (1995) deals with issues related to genetic engineering, as does *The Ethics of Human Gene Therapy* (1997), by LeRoy Walters and Julie Gage Palmer. Ethical and legal considerations related to cloning are the focus of Lori B. Anderson's "Human Cloning: Assessing the

Ethical and Legal Quandaries," *Chronicle of Higher Education* (February 13, 1998) and Vincent Kiernan's "Senate Rejects Bill to Ban Human Cloning," *Chronicle of Higher Education* (February 20, 1998). "Was Dolly a Mistake?" *Time* (March 2, 1998), casts doubt upon how the sheep Dolly was cloned. Other valuable sources are Ira H. Carmen's *Cloning and the Constitution* (1986), Karl Drlica's *Understanding DNA and Gene Cloning* (1992), and R. H. Pritchard and I. B. Holland's *Basic Cloning Techniques* (1985).

Cloning: Ethical Issues

Field of study: Genetic engineering and biotechnology

Significance: *Although cloning of plants has been performed for hundreds of years and cloning from embryonic mammalian cells became commonplace in the early 1990's, the cloning of the sheep Dolly from adult cells in 1996 raised concerns that cloning might be used in a dangerous or unethical manner.*

Key terms

CLONE: an identical genetic twin of any organism; clones can occur naturally or by the intervention of scientists

CLONING: the process of producing a genetic twin in the laboratory by advanced scientific means

BIOETHICS: the study of human actions and goals in a framework of moral standards relating to use and abuse of biological systems

Bioethics and Cloning

Bioethics was founded as a discipline by ethicist Van Rensselaer Potter in the early 1970's as the formal study and application of ethics to biology and biotechnology. The discipline was initially created as an ethical values system to help guide scientists and industrialists in dealing with ecology and the environment. The world has become even more complex since Potter's original vision of a planet challenged by ecological catastrophe. Humans have developed the ability to take genes from one organism and transfer them to another, creating something entirely new to nature, something that can be benign or horrifically dangerous. Moreover, humans have the ability to make indeterminate numbers of genetic copies of these organisms by cloning. Is this technology immoral to use? Is it moral? Can it be both simultaneously? Bioethics examines such questions and suggests ways to define the long-term goals of humanity and, as Potter suggests, "promote the evolution of a better world for future generations."

Cloning is the induction of a genetic twin of an organism. The process of cloning has actually been performed with plants for centuries. Cuttings can be removed from many species and induced to make roots. These cuttings are then grown into full-size genetic copies of the parent plant. This process may seem primitive, but it is cloning. The emergence of crops that cannot be propagated in the standard fashion, such as seedless navel oranges, has led to whole groves of cloned siblings. Few would suggest that such cloning is inherently wrong or unethical.

In 1996, a team of scientists in Scotland headed by Ian Wilmut cloned a mammal—a sheep named Dolly—from adult cells for the first time. While bioethicists had seen no wrong in cloning orange trees and embryonic mammals, they were troubled by the cloning of a sheep. It is important to realize that the cloning of Dolly was not the key bioethical issue. The issue that worried the ethicists was the implication of the clone's existence—that it was only a small step away from cloning a human. If bioethics is concerned with protecting the evolution of future generations of humans, is cloning a potential threat? Could the natural progress of humans toward an unknown evolutionary future be sidetracked or derailed by the intervention and effects of cloning? What would be the social ramifications of human cloning? Does it have the potential to change humanity as it is now known forever? Is cloning simply wrong? These were the questions that immediately beset bioethicists.

Many scientists, including Wilmut, were quick to point out that they would never support human cloning but did not believe that cloning itself was unethical. Most ethicists agreed that cloning animals could help human society in many ways. Genetically engineered animals could be used to create vast quantities of protein-based therapeutic drugs. Commercial animals that are top producers, such as cows with high milk yields, could also be cloned. Human replacement organs could be grown in precisely controlled environments.

Cloning, if misapplied, has frightening potentials. The possibility arises of cloned, genetically programmed soldiers, terrorists, or even workers. An elite society of "perfect" cloned individuals could form and dominate the uncloned "lower" classes. One can easily imagine the cloning of a dying child to comfort grieving parents, the horror of children cloned only to be organ donors for their originals, or the outrage of being cloned without permission.

Human Cloning

Something about human cloning chafes at the human conscience. Bioethicist Karen Rothenberg, in her statements before a United States Congressional subcommittee on public health and safety, suggested why society is made uneasy by the potential implications of human cloning. She broke her argument down into three *I*s. The first *I* is "interdependence." Cloning makes humans uneasy because it only requires one parent. People are humbled because it takes two humans to produce a baby. If part of the definition of humanity is the interdependence upon each other to reproduce, then a cloned human begs the question of just what is human. Rothenberg's second *I* is "indeterminateness." Cloning removes all randomness from human reproduction. With cloning, people predetermine whether they want to reproduce any physical or mental type available. They can control all possible genetic variables in cloning with a predicted outcome. However, does the same genetic variability that decides one's hereditary fate at conception also define some part of humanity? The last *I* is "individuality." It is disconcerting for people to imagine ten or one hundred copies of themselves walking around. Twins and triplets are common now, but what would such a vast change mean to individuality and the concept of the human

soul? In closing, Rothenberg asked whether "the potential benefits of any scientific innovation [are] outweighed by its potential injury to our very concept of what it means to be human."

Andrew Scott of the Urban Institute takes a different view. He believes that bioethics does not apply to cloning but only to what happens after cloning. Cloning does not present a moral dilemma to Scott, assuming that the process does not purposely create "abnormalities." Scott states that "the clone [would] simply be another, autonomous human being . . . carrying the same genes as the donor, and [living] life in a normal, functional way." He suggests that as long as clones are not programmed to be "human-drones" or used in an unethical way, cloning should not be a bioethical worry.

Perhaps the question of whether human cloning is wrong is not well stated. Certainly a great deal of benefit could come from human cloning. Better questions may be: Can humans be trusted not to abuse the technology of cloning? Can those in positions of power be trusted not to use cloning to their advantage and the endangerment of humanity? Carl B. Feldbaum, the president of the Biotechnology Industry Organization, believes that people should be wary of anyone who asks them to allow human cloning and states, "In the future, society may determine that there are sound reasons to clone certain animals to improve the food supply, produce biopharmaceuticals, provide organs for transplantation and aid in research. I can think of no ethical reason to apply this technique to human beings, if in fact it can be applied."

The ethical issues are even more complicated than they first appear. Is the actual process of cloning, as performed by Wilmut, ethical if applied to humans? Wilmut's cloning process produced many failures before Dolly was conceived; only she survived of her 277 cloned sisters. Bioethicists question whether manipulating human embryos to produce clones with only a 0.4 percent success rate is moral; to someone who believes that human life begins at conception, the cloning procedure as performed by Wilmut would almost certainly be unacceptable.

Cloning offers a new and perhaps frightening view of life and the biological universe and brings with it a renewed respect for life. If almost any cell in the body can be used as the basis to clone an entirely new organism, this makes each cell the equivalent of a fertilized egg. While respect for life is renewed from this insight, life is simultaneously cheapened. If each cell contains all the genetic information needed to create a new individual, then what is a single cell worth among millions of copies? The answer may be "very little." When one million or one hundred million potential copies exist, then one copy is worth almost nothing. Therefore, the two contrary feelings of reverence and irreverence linger side by side. The question one must ask is, Which will win out in the end?

—*James J. Campanella*

See Also: Biotechnology, Risks of; Cloning; Eugenics: Nazi Germany; Genetic Engineering: Social and Ethical Issues; Sheep Cloning.

Further Reading: A debate on cloning was published by the U.S. Government Printing Office, *Scientific Discoveries in Cloning: Challenges to Public Policy* (1997), Senate Hearing 105-22. Ian Wilmut et al., "Viable Offspring Derived from Fetal and Adult Mammalian Cells," *Nature* (February 27, 1997), provides details on sheep cloning. For a discussion on bioethics see *Bioethics: Bridge to the Future* (1971), by Van Rensselaer Potter. Martin Johnson's article "Cloning Humans?" *Bioessays* (August, 1997), addresses the biological obstacles to human cloning.

Cloning Vectors

Field of study: Genetic engineering and biotechnology

Significance: *Cloning vectors are one of the key tools required for propagating (cloning) foreign DNA sequences in cells. Cloning vectors are vehicles for the replication of DNA sequences that cannot otherwise replicate. Expression vectors are cloning vectors that provide not only the means for replication but also the regulatory signals for protein synthesis.*

Key terms

FOREIGN DNA: deoxyribonucleic acid (DNA) taken from a source other than the host cell that is joined to the DNA of the cloning vector; also known as "insert DNA"

HOST CELL: the cell (most often the bacteria *Escherichia coli*) in which the cloning vector grows

PLASMID: a small, circular DNA molecule that replicates independently of the host cell chromosome

BACTERIOPHAGE: a virus composed of ribonucleic acid (RNA) or DNA surrounded by a protective protein coat that infects and grows in bacterial cells

RECOMBINANT DNA MOLECULE: a molecule of DNA created by joining DNA molecules from different sources, most often vector DNA joined to insert DNA

The Basic Properties of a Cloning Vector

Cloning vectors were developed in the early 1970's from naturally occurring DNA molecules found in some cells of the bacteria *Escherichia coli* (*E. coli*). These replicating molecules (plasmids) were first used by the American scientists Stanley Cohen and Herbert Boyer as vehicles or vectors, to replicate other pieces of DNA (insert DNA) that were joined to them. Thus the first two essential features of cloning vectors are their ability to replicate in an appropriate host cell and their ability to join to foreign DNA sequences to make recombinant molecules. Plasmid replication requires host-cell-specified enzymes, such as DNA polymerases that act at a plasmid sequence called the "origin of replication." Insert DNA is joined (ligated) to plasmid DNA through the use of two kinds of enzymes: restriction enzymes and DNA ligases. The plasmid DNA sequence must have unique sites for restriction enzymes to cut. Cutting the double-stranded circular DNA at more than one site would cut the plasmid into pieces and would separate important functional parts from one another. However, when a restriction enzyme cuts the circular plasmid at one unique site, it converts it to a linear molecule. Linear, insert DNA molecules (also cut with restriction enzymes) can be joined to cut plasmid molecules similarly through the action of the enzyme DNA ligase. This catalyzes the covalent joining of the insert DNA and plasmid DNA ends to create a circular, recombinant plasmid molecule. Most cloning vectors have been designed to have many unique restriction enzyme cutting sites all in one stretch of the vector sequence. This part of the vector is referred to as the multiple cloning site.

In addition to an origin of replication and a multiple cloning site, most vectors have a third element: a selectable marker. In order for the vector to replicate, it must be present inside an appropriate host cell. Introducing the vector into cells is often a very inefficient process. Therefore, it is very useful to be able to select, from a large population of host cells, those rare cells that have taken up a vector. This is the role of the selectable marker. The selectable marker is usually a gene that encodes resistance to an antibiotic to which the host is normally sensitive. For example, if a plasmid vector has a gene that encodes resistance to the antibiotic ampicillin, only those *E. coli* cells that harbor a plasmid will be able to grow on media containing ampicillin. There are a number of procedures for introducing the plasmid vector into the host cell. Transformation is a procedure in which the host cells are chemically treated so that they will allow small DNA molecules to pass through the cell membrane. Electroporation is a procedure that uses an electric field to create pores in the host cell membrane to let small DNA molecules pass through.

Viruses and Cloning Vectors

In addition to plasmid cloning vectors, some viruses have been modified to serve as cloning vectors. Viruses are infectious agents that are made of a genome, either DNA or ribonucleic acid (RNA), that is surrounded by a protective protein coat. Viruses whose host cells are bacteria are referred to as bacteriophage. Bacteriophage vectors are used in similar ways to plasmid vectors. The vector and insert DNAs are cut by restriction enzymes so that they sub-

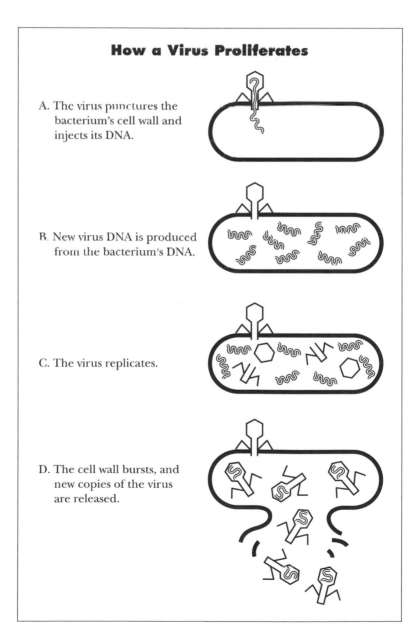

How a Virus Proliferates

A. The virus punctures the bacterium's cell wall and injects its DNA.

B. New virus DNA is produced from the bacterium's DNA.

C. The virus replicates.

D. The cell wall bursts, and new copies of the virus are released.

sequently can be joined by DNA ligase. The newly formed recombinant DNA molecules must enter an appropriate host cell to replicate. In order to introduce the bacteriophage DNA into cells, a whole bacteriophage must be built. This is referred to as "packaging" the DNA. The protein elements of the bacteriophage are mixed with the recombinant bacteriophage DNA and packaging enzymes to create an infectious virus. Appropriate host cells are then infected with the bacteriophage. The natural means for viral replication are used to make many copies of each recombinant molecule. These are then packaged within the infected host cells to make infectious, progeny bacteriophage. In many cases, the final step of viral infection is the lysis of the host cell. This releases the progeny bacteriophage to infect nearby host cells. Bacteriophage vectors have two advantages relative to plasmid vectors: The first advantage is the viral delivery of recombinant DNA to the host cells, which is much more efficient than the transformation or electroporation procedures used to introduce plasmid DNA into host cells; the second advantage is the ability of bacteriophage vectors to clone relatively large fragments of insert DNA compared to plasmid vectors.

Viruses that infect cells other than bacteria have been modified to serve as cloning vectors. This permits cloning experiments using many different kinds of host cells, including human cells. Viral vectors, just like the natural viruses from which they are derived, have specific host and tissue ranges. A particular viral vector will be limited for use in specific species and cell types. The fundamental practice of all virally based cloning vectors involves the covalent joining of the insert DNA to the viral DNA to make a recombinant DNA molecule, introduction of the recombinant DNA into the appropriate host cell, and then propagation of the vector through the natural mechanism of viral replication. There are two fundamentally different ways that viruses propagate in cells. Many viruses, such as the bacteriophage already described, enter the host cell and subvert the cell's biosynthetic machinery to its own reproduction, which ultimately leads to lysis and thereby kills the host cell as the progeny

viruses are released. The second viral life strategy is to enter the host cell and integrate the viral DNA into the host cell chromosome so that the virus replicates along with the host DNA. Such integrating viruses can be stably maintained in the host cell for long periods. The retroviruses, of which the human immunodeficiency virus (HIV) is an example, are a group of integrating viruses that are potentially useful vectors for certain gene therapy applications. Using cloning vectors and host cells other than bacteria allows scientists to produce some proteins that bacteria cannot properly make, permits experiments to determine the function of cloned genes, and is important for the development of gene therapy.

Expression Vectors

Expression vectors are cloning vectors designed to produce proteins from genes foreign to the host cell in a convenient host cell. Expression vectors must do more than simply provide a means for DNA replication; they must also provide the appropriate regulatory signals for the transcription and translation of the foreign gene. Unlike the sequences that code for the amino acids of a protein (coding sequence), the regulatory sequences directing where and when to start and stop transcription are not universal. In general, the regulatory sequences of genes from one species will not be recognized by the protein-synthesizing machinery of cells from another species. Thus, unless the vector provides the appropriate host-cell-specific regulatory sequences, foreign genes will not normally be expressed.

Expression cloning vectors make it possible to produce proteins encoded by genes from one species in the cells of some other species. Furthermore, producing proteins using an expression vector usually results in "overexpression," which is the production of much greater quantities of the protein than are otherwise obtainable. This technology is not only of immense benefit to scientists who study proteins, but it is also used by industry (particularly the pharmaceutical industry) to make valuable proteins. Proteins such as human insulin, growth hormone, and clotting factors that were difficult or impossible to isolate from their natural sources are readily available because the human genes for these proteins have been cloned and expressed in bacteria. In addition to making these therapeutic proteins more abundant and less expensive, an added benefit of cloning and protein expression is that proteins produced by bacteria from their human genes do not provoke an allergic immune response in patients as do proteins such as porcine insulin that are isolated from other species.

Artificial Chromosomes

In 1987, a new type of cloning vector was developed by David Burke, Maynard Olson, and their colleagues. The new vectors, artificial chromosomes, filled the need created by the Human Genome Project (HGP) to clone very large (hundreds of thousands to millions of base pairs in length) insert DNAs. The HGP is the effort to map and ultimately sequence all the chromosomes of humans and a number of other species. Because of the massive scale of the genome projects, a vector capable of propagating much larger DNA fragments than plasmids or bacteriophage is required. The first artificial chromosome vector was developed in the yeast *Saccharomyces cerevisiae*. All the critical DNA sequence elements of a yeast chromosome were identified and isolated, and these were put together to create a yeast artificial chromosome (YAC). The elements of a YAC vector are an origin of replication, a centromere, telomeres, and a selectable marker suitable for yeast cells. A yeast origin of replication (similar to the origin of replication of bacterial plasmids) is a short DNA sequence that the host's replicative enzymes such as DNA polymerase recognize as a site to initiate DNA replication. In addition to replicating, the two new copies of a chromosome must be faithfully partitioned into daughter cells during mitosis. The centromere sequence mediates the partitioning of the chromosomes during cell division because it serves as the site of attachment for the spindle fibers in mitosis. Telomeres are the DNA sequences at the ends of chromosomes. They are required to prevent degradation of the chromosome and for the replication of DNA at the ends of chromosomes.

YAC vectors are used very similarly to the plasmid vectors described earlier. Very large insert DNAs are joined to the YAC vector, and the recombinant molecules are introduced into host yeast cells in which the artificial chromosome is replicated just as the host's natural chromosomes are. YAC cloning technology allows very large chromosomes to be subdivided into a manageable number of pieces that can be organized (mapped) and studied. YAC vectors also provide the opportunity to study DNA sequences that interact over very long distances. Since the development of YAC vectors, artificial chromosome vectors for a number of different host cells have been created.

Impact and Applications

Cloning vectors are one of the key tools enabling recombinant DNA technology. Cloning vectors make it possible to isolate particular DNA sequences from an organism and make many identical copies of this one sequence in order to study the structure and function of that sequence apart from all other DNA sequences. Until the development of the polymerase chain reaction (PCR), cloning vectors and their host cells were the only means to collect many copies of one particular DNA sequence. For long DNA sequences (those over approximately ten thousand base pairs), cloning vectors are still the only means to do this.

Gene therapy is a new approach to treating and perhaps curing genetic disease. Many common diseases are the result of defective genes. Gene therapy aims to replace or supplement the defective gene with a normal, therapeutic gene. One of the difficulties faced in gene therapy is the delivery of the therapeutic gene to the appropriate cells. Viruses have evolved to enter cells, sometimes only a very specific subset of cells, and deliver their DNA or RNA genome into the cell for expression. Thus viruses make attractive vectors for gene therapy. An ideal vector for gene therapy would replace viral genes associated with pathogenesis with therapeutic genes; the viral vector would then target the therapeutic genes to just the right cells. One of the concerns related to the use of viral vectors for gene therapy is the random nature of the viral insertion into the target cell's chromosomes. Insertion of the vector DNA into or near certain genes associated with increased risk of cancer could theoretically alter their normal expression and induce tumor formation.

Plasmid DNA vectors encoding immunogenic proteins from pathogenic organisms are being tested for use as vaccines. DNA immunization offers several potential advantages over traditional vaccine strategies in terms of safety, stability, and effectiveness. Genes from disease-causing organisms are cloned into plasmid expression vectors that provide the regulatory signals for efficient protein production in humans. The plasmid DNA is inoculated intramuscularly or intradermally, and the muscle or skin cells take up some of the plasmid DNA and express the immunogenic proteins. The immune system then generates a protective immune response. There are two traditional vaccination strategies: One uses live, attenuated pathogenic organisms, and the other uses killed organisms as the inoculum. The disadvantage of the former is that, in rare cases, the live vaccine can cause disease. The disadvantage of the latter strategy is that the killed organism does not enter the patient's cells and make proteins like the normal pathogen. Therefore, one part of the immune response, the cell-mediated response, is usually not activated, and the protection is not as good. In DNA immunization, the plasmids enter the patient's cells, and the immunogenic proteins produced there result in a complete immune response. However, unlike a live vaccine, there is no chance of causing disease because the plasmid vector does not carry all of the disease-causing organism's genes.

—*Craig S. Laufer*

See Also: Cloning; Gene Therapy; Genetic Engineering; Genomic Libraries; Restriction Enzymes.

Further Reading: W. French Anderson, "Gene Therapy," *Scientific American* 124 (September, 1995), provides a good review of the promise and problems of gene therapy. Theodore Friedmann, "Overcoming the Obstacles," *Scientific American* 96 (June, 1997), elaborates on the relative merits of different delivery systems for gene therapy. Philip Cohen, "Creators of the Forty-Seventh Chromosome," *New Scientist* 34 (November 11, 1995), describes the efforts to develop human artificial chromosomes. Daniel E. Hassett and J. Lindsay Whitton, "DNA Immunization," *Trends in Microbiology* 307 (August, 1996), reviews the process of DNA immunization and compares it to traditional vaccination strategies. James D. Watson et al., *Recombinant DNA* (1983), is an excellent reference for details on how the different cloning vectors work and to what purposes each is particularly suited.

Complete Dominance

Field of study: Classical transmission genetics

Significance: *Dominance is defined in terms of which version of a particular gene will be observed in heterozygous individuals. Knowing whether the pattern of expression of a characteristic is dominant or recessive helps in making predictions concerning the appearance, or reappearance, of a particular genetic condition or disorder in a family's history.*

Key terms

GENE: the functional unit that produces a specific inherited characteristic

ALLELES: different forms of a specific gene

DEOXYRIBONUCLEIC ACID (DNA): the molecule that contains the hereditary information

PROTEIN: a molecule composed of amino acid subunits whose placement is determined by the information in a specific gene

ENZYME: a protein that speeds a chemical reaction

HETEROZYGOTE: an individual with two different alleles for a gene

HOMOZYGOTE: an individual with two like alleles for a gene

PHENOTYPE: the observed character expression in an individual

The Discovery and Definition of Dominance

Early theories of inheritance were based on the idea that fluids carrying materials for the production of a new individual were transmitted to offspring from the parents. It was assumed that substances in these fluids from the two parents mixed and that the children would therefore show a blend of the parents' characteristics. For instance, individuals with dark hair mated to individuals with very light hair were expected to produce offspring with medium-colored hair. The carefully controlled breeding studies carried out in the 1700's and 1800's did not produce the expected blended phenotypes, but no other explanation was suggested until Austrian botanist Gregor Mendel proposed his model of inheritance. Mendel repeated studies using the garden pea and obtained the same results seen by other investigators, but he counted the numbers of each type produced from each mating and developed his theory based on those observations.

One of the first observations Mendel dealt with was the appearance of only one of the parent expressions in the first generation of offspring (the F_1 generation). For example, a cross of tall plants and dwarf plants resulted in offspring that were all tall. Mendel proposed that the character expression (in this case height) was controlled by a determining factor, later called the "gene." He then proposed that there were different forms of this controlling factor corresponding to the different expressions of the characteristic and termed these "alleles." In the case of plant height, one allele produced tall individuals and the other produced dwarf individuals. He further proposed that in the cross of a tall (*D*) plant and a dwarf (*d*) plant, each parent contributed one factor for height, so the offspring were be *Dd*. (Uppercase letters denote dominant alleles, while lowercase letters denote recessive alleles.) These plants contained a factor for both the tall expression and the dwarf expression, but the plants were all tall, so "tall" was designated the dominant phenotype expression for the height characteristic.

Mendel recognized from his studies that the determining factors occurred in pairs—each sexually reproducing individual contains two alleles for each inherited characteristic. When he made his crosses, he carefully selected pure breeding parents that would have two copies of the same allele. In Mendel's terminology, the parents would be homozygous: A pure tall parent would be designated *DD*, while a dwarf parent would be designated *dd*. His model also proposed that each parent would contribute one factor for each characteristic to each offspring, so the offspring of such a mating should be *Dd* (heterozygous). The tall appearance of the heterozygote defines the character expression (the phenotype) as dominant. Dominance of expression for any characteristic cannot be guessed but must be determined by observation. When there is variation observed in the phenotype, heterozygous individuals must be examined to determine which expression is observed. For phenotypes that are not visible, such as blood types or enzyme activity variations, a test of some kind must be used to determine which phenotype expression is present in any individual.

Mendel's model and the appearance of the dominant phenotype also explains the classic 3:1 ratio observed in the second (F_2) generation. The crossing of two heterozygous individuals (*Dd* × *Dd*) produces a progeny that is $\frac{1}{4}$ *DD*, $\frac{1}{2}$ *Dd*, and $\frac{1}{4}$ *dd*. Because there is a dominant phenotype expression, the $\frac{1}{4}$ *DD* and the $\frac{1}{2}$ *Dd* progeny all have the same phenotype, so $\frac{3}{4}$ of the individuals are tall. It was this numerical relation that Mendel used to establish his model of inheritance.

The Functional Basis of Dominance

The development of knowledge about the molecular activity of genes through the 1950's and 1960's provided information on the nature of the synthesis of proteins using the genetic code passed on in the deoxyribonucleic acid (DNA) molecules. This knowledge allows researchers to explain variations in phenotype expression and to understand the dominance expression of a phenotype at the functional level. An enzyme's function is determined by its structure, and that structure is coded for in the genetic information. The simplest situation is one in which the gene product is an enzyme that acts on a specific chemical reaction that results in a specific chemical product, the phenotype. If that enzyme is not present or if its structure is modified so that it cannot properly perform its function, then the chemical action will not be carried out. The result will be an absence of the normal product and a phenotype expression that varies from the normal expression. For example, melanin is a brown pigment produced by most animals. It is the product of a number of chemical reactions, but one enzyme early in the process is known to be defective in albino animals. Lacking normal enzyme activity, these animals cannot produce melanin, so there is no color in the skin, eyes, or hair. When an animal has the genetic composition cc (c designates colorless, or albino), it has two alleles that are the same, and neither can produce a copy of the normal enzyme. There is none of the chemical action necessary to produce the melanin pigment, so the animal has the albino phenotype. Animals with the genetic composition CC (C designates colored, or normal) have two copies of the allele that produces normal enzymes that result in the production of normal pigment and a color phenotype. When homozygous normal (CC) and albino (cc) animals are crossed, heterozygous (Cc) animals are produced. The c allele codes for production of an inactive enzyme, while the C allele codes for production of the normal, active enzyme. The presence of the normal enzyme promotes pigment production, and the animal has the pigmented phenotype expression. The presence of pigment in the heterozygote leads to the designation that the pigmented phenotype is dominant to albinism or, conversely, that albinism is a recessive phenotype expression because it is seen only in the homozygous (cc) state.

The same absence or presence of an active copy of an enzyme explains why blood types A and B are both dominant to blood type O. When an A allele or a B allele is present, an active enzyme promotes the production of a substance that is identified in a blood test; the blood type A expression or the blood type B expression is seen. When neither of these alleles is present, the individual is homozygous OO. There is no detectable product present, and the blood test is negative; therefore, the individual has blood type O. When the A allele and the B allele are both present in a heterozygous individual, each produces an active enzyme, so both the A and the B product are detected in blood tests; such an individual has blood type AB. The two phenotypes blend and are both expressed in the heterozygote (codominance).

When there are a number of alleles present for the expression of a characteristic, a dominance relation among the phenotype expressions can be established. In some animal coats, very light colors result from enzymes produced by a specific allele that is capable of producing melanin but at a much less efficient rate than the normal version of the enzyme. In the rabbit, chinchilla (c^{ch}) is such an allele. In the Cc^{ch} heterozygote, the normal allele (C) produces a normal, rapidly acting enzyme, and the animal has normal levels of melanin. The normal pigment phenotype expression is observed because the animals are dark in color, so this expression is dominant to the chinchilla phenotype expression. In the heterozygote $c^{ch}c$, the slow-acting enzyme produced by the c^{ch} allele is present and produces pigment, in a reduced amount, so the chinchilla phenotype expression is observed and is dominant to the albino phenotype expression. The result is a dominance hierarchy in which the normal pigment phenotype expression is dominant to both the chinchilla and the albino phenotypes, and the chinchilla expression is dominant to the albino expression.

It is important to note that the dominant phenotype expression is the result of the nature of the action of the specific version of the protein produced by each allele. In other words, dominance is a manifestation of the phenotype expression by the action of the gene product. It is not a property of the allele. In the previous examples, both the albino allele and the chinchilla allele produce a product—a version of the encoded enzyme—but the normal allele produces a version of the enzyme that produces more pigment. The relative ability of the enzymes to carry out the function determines the observed phenotype expression and therefore the dominance association. The C allele does not inhibit the activity of either of the other two alleles or their enzyme products, and the allele does not, therefore, show dominance; rather, its enzyme expression does.

Dominance of Abnormalities

Dominance of a normal phenotype's expression is fairly easy to explain at the level of the functioning protein because the action of the normal product is seen, but dominance of abnormalities is more difficult; however, it is not impossible once the molecular activity of a given situation is understood. Polydactyly, the presence of extra fingers on one hand or extra toes on one foot, is a dominant phenotype. The mechanism that leads to this expression and numerous other developmental abnormalities is not yet understood. One model can be suggested based on an understanding of a regulation defect in bacteria. In this situation, one gene codes for one protein, but it takes two of the protein molecules joined together to form the functional unit required to produce the phenotype expression. In order to carry out the normal function, both of the protein subunits must be normal. A heterozygote with one allele coding for normal protein structure and functional capacity and one allele with a modified code that results in an defective protein will produce both normal and defective protein molecules. These will join together at random to form the required two-protein functional units. The possible combinations will be defective-defective, which results in a nonfunctional unit; defective-normal, which also results in a nonfunctional unit; and normal-normal, which is a normal, functional unit. The majority of the two-protein units will be nonfunctional, and their presence interferes with the action of the few normal units. The normal function will be, at best, greatly reduced, and the overall phenotype expression will be abnormal. One form of hereditary blindness is dominant because the presence of abnormal proteins interferes with the transport of both protein types across a membrane to their proper location in the cells that react to light. The abnormal phenotype appears in the heterozygote, so the abnormal phenotype is the dominant expression. A number of human disease conditions, including some forms of cancer, show the dominant mode of inheritance. For those cancers, this means that a single change in one cell can lead to a tumor-producing phenotype.

Dominant phenotype expression is a relational term—one expression is dominant relative to another expression of the same phenotype—and is not a fixed characteristic. Many genes have influences on more than one phenotype, and the influence of the gene product may have different dominance expressions in different situations. In dogs, for example, most hairless phenotypes appear in heterozygous animals, so the hairless phenotype has the dominant expression. The same allele (*H* for hairless) produces a lethal phenotype in the homozygous (*HH*) state. This means that the lethal effect, resulting from the same *H* allele that causes the dominant hairless condition, is recessive: It appears only in the *HH* homozygote. It therefore cannot be said that the expression of the *H* allele is dominant because, although it has a dominant expression in producing the hairless condition, it is recessive in its effects on the animal's survival phenotype. Conversely, the *h* allele cannot promote the normal growth of hair in the heterozygote, but it does keep the heterozygous animal alive.

Impact and Applications

One of the aims of human genetic research is to find cures for inherited conditions. When a condition shows the recessive phenotype expression, treatment may be effective. The individual lacks a normal gene product, so supplying that product can have a beneficial effect. This is the reason for the success of treatments such as supplying insulin to individuals who are genetically diabetic. There are many technical issues to be considered in such treatments, but they show promise as strategies for dealing with recessive genetic conditions.

Dominant disorders, on the other hand, will be much more difficult to treat. As seen in the previous examples, a heterozygous individual with a disorder already has a normal allele that produces normal gene product. The nature of the interactions between the products results in the defective phenotype expression. Supplying more normal product may not improve the situation. A great deal more knowledge about the nature of the underlying mechanisms will be needed to make treatment effective.

—*D. B. Benner*

See Also: Albinism; Biochemical Mutations; DNA Structure and Function; Monohybrid Inheritance; Protein Structure.

Further Reading: *Biology* (1996), by Neil Campbell, is a college-level biology text that provides an introduction to many topics relating to genetics. Any genetics textbook, such as *Genetics* (1996), by Peter Russell, will include discussions concerning dominance.

Congenital Defects

Field of study: Human genetics
Significance: *Congenital defects are malformations caused by abnormalities in embryonic or fetal development that may interfere with normal life functions or cause a less severe health problem. The defect may be morphological or biochemical in nature. Understanding the causes of birth defects has led to improved means of detection and treatment.*

Key terms

TERATOGEN: any agent that is capable of causing an increase in the incidence of birth defects

TERATOLOGY: the science or study of birth defects

SENSITIVE PERIOD: a critical time during development when organs are most susceptible to teratogens

Normal Development

In order to understand the causes of birth defects, it is necessary to have some understanding of the stages of normal development. If the time and sequence of development of each organ is not normal, an abnormality may result. It has been useful to divide human pregnancy into three major periods: the preembryonic stage, the embryonic stage, and the fetal stage.

The preembryonic stage is the first two weeks after fertilization. During this stage, the fertilized egg undergoes cell division, passes down the fallopian tube, and implants in the uterine wall, making a physical connection with the mother. It is of interest to note that perhaps as many as one-half of the fertilized eggs fail to implant, and another one-half that do implant do not survive the second week. The second stage, the embryonic stage, runs from the beginning of the third week through the end of the eighth week. There is tremendous growth and specialization of cells during this period, as all of the body's organs are formed. The embryonic stage is the time during which most birth defects are initiated.

The fetal stage runs from the beginning of the ninth week to birth. Most organs continue their rapid growth and development during this final period of gestation leading up to birth. By the end of the eighth week, the embryo, although it has features of a human being, is only about 1 inch long. Its growth is amazing during this period, reaching 12 inches by the end of the fifth month and somewhere around 20 inches by birth. It is evident from the description of normal development that the changes the embryo and fetus undergo are very rapid and complicated. It is not unexpected that mistakes can happen, leading to congenital disorders.

Causes of Birth Defects

Ever since humans began recording aspects of their lives, whether in rock art, sculpture, painting, or writing, examples of birth defects have been described by all cultures and ethnic groups. Although the incidence of specific malformations may vary from group to group, the overall incidence of birth defects is probably similar in all people on earth. It is estimated that 3 out of every 100 newborns have some sort of major or minor disorder. An additional 2 to 3 percent have malformations that fully develop sometime after birth. When it is also realized that perhaps another 5 percent of all fertilized eggs have severe enough malformations to lead to an early, spontaneous abortion, the overall impact of birth defects is considerable.

Humans have long sought an explanation for why some couples have babies afflicted with serious birth defects. Such children were long regarded as "omens" or warnings of a bad event to come. The word "teratology" (Greek for "monster causing") was coined by scientists to reflect the connection of "monster" births with warnings. Frequently, ancient people sacrificed

Down syndrome is among the most common severe congenital defects. (Linda K. Moore/Rainbow)

such babies. It was thought that such pregnancies resulted from women mating with animals or evil spirits. Maternal impression has long been invoked as an explanation for birth defects, and from early Greek times until even relatively recent times, stories and superstitions abounded.

Of the birth defects in which a specific cause has been identified, it has been found that some are caused by genetic abnormalities, including gene mutations and chromosomal changes, while others are caused by exposure of the pregnant woman and her embryo or fetus to some sort of environmental toxin such as radiation, viruses, drugs, or chemicals.

Examples of Birth Defects

Many birth defects are caused by changes in the number or structure of chromosomes. The best-known chromosomal disorder is Down syndrome, which results from individuals having an extra chromosome 21, giving them forty-seven chromosomes rather than the normal forty-six. A person with Down syndrome characteristically has a flattened face, square-shaped ears, epicanthal folds of the eye, a short neck, poor muscle tone, slow development, and subnormal intelligence. Cystic fibrosis is an example of a defect caused by a single gene. Affected people inherit a recessive gene from each parent. The disorder is physiological in nature and results in a lack of digestive juices and the production of thick and sticky mucus that tends to clog the lungs, pancreas, and liver. Respiratory infections are common, and death typically occurs by the age of thirty. Cleft lip, or cleft palate, is multifactorial in inheritance (some cases are caused by chromosomal abnormalities or by single-gene mutations). Multifactorial traits are caused by many pairs of genes, each having a small effect, and are usually influenced by factors in the environment. The result is that such traits do not follow precise, predictable patterns in a family.

Genetic factors account for the great majority (perhaps 85 to 90 percent) of the birth defects in which there is a known cause. The remaining cases of known cause are attributed to maternal illness; congenital infections; exposure to chemicals, drugs, and medicines; and

physical factors such as X rays, carbon dioxide, and low temperature. The "government warning" on liquor bottles informs pregnant women that if they drink alcohol during a sensitive period of prenatal development, they run the risk of having children with the fetal alcohol syndrome. There is a wide variation in the effects of alcohol on a developing fetus. Alcohol exposure can lead to an increased frequency of spontaneous abortion, and it depresses growth rates, both before and after birth. Facial features of a child exposed to alcohol may include eye folds, a short nose, small mid-face, a thin upper lip, a flat face, and a small head. These characteristics are likely to be associated with mental retardation. Frequently, however, otherwise normal children have learning disorders and only a mild growth deficiency. Variation in the symptoms of prenatal alcohol exposure has made it difficult to estimate the true incidence of the fetal alcohol syndrome. Estimates for the United States range from 1 to 3 per 1,000 newborns.

In 50 to 60 percent of babies born with a major birth disorder, no specific cause can be identified. Because of this rather large gap in knowledge, nonscientific explanations about the causes of birth defects flourish. What is known is that most congenital defects, whether caused by a genetic factor or an environmental factor, are initiated during the embryonic period. It is also known that some disorders, such as learning disorders, frequently result from damage to the fetus during the last three months of pregnancy. Knowledge about what can be done by parents to avoid toxic exposure and activity that could cause birth defects is critical.

—*Donald J. Nash*

See Also: Cystic Fibrosis; Down Syndrome; Hereditary Diseases.

Further Reading: "Tis the Season," *Psychology Today* 28 (1995), describes how certain birth defects may be related to different seasons of the year and also discusses some of the biological and environmental conditions that cause birth defects. Robina Riccitiello and Jerry Adler, "Your Baby Has a Problem," *Newsweek* Spring/Summer (1997), discusses how advances in medicine have reduced the number of birth defects and how surgeries have been designed to correct some birth defects before babies are born. Anne E. Rossen, "Understanding Congenital Disorders," *Current Health* 18 (May, 1992), is a useful article describing some congenital disorders, including some that are not apparent early in life such as Huntington's chorea. Some of the environmental factors causing congenital defects are also covered.

Criminality

Field of study: Human genetics

Significance: *The pursuit of genetic links to criminality is a controversial field of study that has produced several intriguing examples of the apparent contribution of genetic defects to criminal behavior. However, most of these defects involve major alterations in metabolic pathways that, in turn, affect numerous characteristics. Experts disagree on the validity and significance of these data. These research efforts have also come under strong criticism by opponents who fear that such discoveries will be used to charge that certain ethnic or racial groups are genetically predisposed to deviant behaviors such as criminality.*

Key terms

METABOLIC PATHWAY: a biochemical process that converts certain chemicals in the body to other, often more useful, chemicals with the help of proteins called enzymes

NEUROTRANSMITTER: a chemical that carries messages between nerve cells

Biochemical Abnormalities

Early attempts to identify the root of the tendency for criminal behavior fell under the auspices of biological determinism, which sought to explain and justify human society as a reflection of inborn human traits. For example, Italian physician Cesare Lombroso reported in *L'uomo delinquente* (1876; criminal man) that certain "inferior" groups, by virtue of their apish appearance, were evolutionary throwbacks with criminal tendencies. Since that time, more sophisticated scientific methods have been employed to seek the "root

causes" of criminality. Among the most prominent findings are those that indicate that certain biochemical imbalances, particularly in neurotransmitters, may lead to a range of "abnormal" behaviors. For example, levels of the neurotransmitter serotonin have been found to be low in many people who have attempted suicide and in people with poor impulse control, such as children who torture animals and impulsive arsonists. However, the environment itself may lower or raise serotonin levels, calling into question the importance of genetic influence. The psychological effects of serotonin are also far-reaching, with antidepressant drugs such as Prozac functioning by increasing the amount of time serotonin remains in the system after its release (latency).

Abnormalities in dopamine levels (the primary neurotransmitter in the brain's "pleasure center") have also been implicated in aggressive, antisocial behavior. In 1995, researchers reported that increased latency of dopamine might be associated with a tendency toward violence among alcoholics. A genetic abnormality on the X chromosome that causes a defect in the enzyme monoamine oxidase A was reported by researchers in 1993. This enzyme is responsible for degrading certain neurotransmitters, including dopamine and epinephrine. This defect was linked to a heritable history of low intelligence quotient (IQ) and violent acts in one Dutch family. Males who possess an extra Y chromosome (XYY syndrome males) also demonstrate a variety of behavioral difficulties and are overrepresented in prisons and mental institutions. However, no link to criminal behavior has been established.

In all cases, such genetic abnormalities affect numerous characteristics (often including mental capabilities) and manifest themselves as any number of unassociated antisocial behaviors ranging from exhibitionism to arson. Since criminality simply refers to a violation of the law and since there are numerous types of crimes and motivations for them (such as anger, revenge, and financial gain), it is difficult to make claims of definitive, nonenvironmental links between biochemical disorders and criminal behavior. Poorly defined, multifaceted social descriptors (for example, violence, aggression,

Studies that have looked for links between criminal behavior and genetic factors have proved controversial and inconclusive.
(Ben Klaffke)

and intelligence) are usually used to represent such behaviors and, as such, cannot be considered true "characters." As child psychiatrist Michael Rutter has said, to claim that there is a gene for crime is "like saying there's a gene for Roman Catholicism."

Impact and Applications

Genetic links to criminality entered the public spotlight in the early 1990's as part of the U.S. government's "Violence Initiative" championed by Secretary of Health and Human Services Louis Sullivan. The uproar began in 1992 when Frederick Goodwin, director of the Alcohol, Drug Abuse, and Mental Health Administration, made comments comparing urban youth to aggressive jungle primates. The public feared that research on genetic links to criminality would be used to justify the disproportionate numbers of African Americans and Hispanics in the penal system. Psychiatrist Peter Breggin also warned that unproved genetic links would be used as an excuse to screen minority children and give them sedating drugs to intervene in their impending aggression and criminality. After all, forced sterilization laws had been enacted in thirty U.S. states in the 1920's to prevent reproduction by the "feebleminded" and "moral degenerate." In 1993, public protest led to the temporary cancellation of "Genetic Factors in Crime," a federally funded conference organized by philosopher David Wasserman. A similar "Symposium on Genetics of Criminal and Antisocial Behaviour" was held in London in 1995. However, the public remains highly suspicious of the motivation for such research.

In an era in which genes have been implicated in everything from manic depression to the propensity to change jobs, the belief that genes are responsible for criminal behavior is very enticing. However, such a belief may have severe ramifications. To the extent that society accepts the view that crime is the result of pathological and biologically deviant behavior, it is possible to ignore the necessity to change social conditions such as poverty and oppression that are also linked to criminal behavior. Moreover, this view may promote the claim by criminals themselves that their "genes" made them do it. While biochemical diagnosis and treatment with drugs may be simpler and therefore more appealing than social intervention, it is reminiscent of the days when frontal lobotomy (surgery of the brain) was the preferred method of biological intervention for aggressive mental patients. In the future, pharmacological solutions to social problems may be viewed as similarly inhumane.

—*Lee Anne Martínez*

See Also: Aggression; Behavior; Inborn Errors of Metabolism; Sterilization Laws; XYY Syndrome.

Further Reading: Robert Sapolsky gives an entertaining account of the complex interaction between genes and the environment in "A Gene for Nothing," *Discover* 18 (October, 1997). Francis Fukuyama compares the philosophies of cultural versus biological effects on behavior in "Is It All in the Genes?" *Commentary* 104 (September, 1997). Juan Williams, "Violence, Genes, and Prejudice," *Discover* 15 (November, 1994), gives an excellent account of the controversy and debate that accompanied the U.S. government's funding of research on genetic links to violence and crime. In "The Biology of Violence," *The New Yorker* 71 (March 15, 1995), Robert Wright discusses evolutionary psychology's view that violent responses to oppressive environments may be adaptive rather than genetically inflexible. "The Rise of Neurogenetic Determinism," by Steven Rose, *Nature* 373 (February, 1995), comments on how technological advances have revived genetic explanations for behavior.

Cystic Fibrosis

Field of study: Human genetics
Significance: *Cystic fibrosis, although a rare disease, is the most common lethal inherited disease among Caucasians in the United States and the United Kingdom. Advances in genetic screening and treatment may someday result in the prevention or elimination of this disease.*

Key terms

EPITHELIAL CELLS: cells responsible for transporting salt and water

GENETIC DISEASE: a disease that is inherited rather than caused by an injury or infection

RECESSIVE GENE: a weaker gene whose characteristics are not expressed unless it is paired with another recessive gene

Causes and Symptoms

Cystic fibrosis is caused by an abnormal recessive gene that must be inherited from both parents. If both parents carry the gene, their child has a 25 percent chance of inheriting the abnormal gene from both parents and thus having the disease. The child has a 50 percent chance of having one normal and one abnormal gene, thus becoming a carrier of the disease. Carriers do not develop cystic fibrosis but can transmit the disease to their children if the other parent is also a carrier.

This disease is chronic and has no known cure. Generally, symptoms are apparent shortly after birth and become progressively more serious. Abnormally thick mucus blocks the ducts in sweat glands and glands in the lungs and pancreas. Diagnosis usually involves a simple test that measures excessive sodium and chloride (salt) in the person's sweat. Ten percent of newborns with cystic fibrosis cannot excrete undigested material in their intestines and so develop a blocked intestine. Symptoms of this condition include a swollen abdomen, vomiting, constipation, and dehydration. The child may become severely undernourished because the digestive system is not functioning properly. In turn, the malnutrition causes poor weight gain, impaired blood clotting, slow bone growth, and poor overall growth. Digestive problems usually increase as the child ages.

Although cystic fibrosis can damage other organs, the most serious complications from the disease involve the lungs. Respiratory problems develop as the affected person ages. The lungs may appear normal in early life but will later malfunction. Normal lungs fight off infection by secreting mucin, the primary component of mucus. The mucin helps trap germs and foreign particles in the lungs. The mucus is swept toward the throat by hairlike projections known as "cilia" and is then expelled from the respiratory tract. Lungs also produce a natural antibiotic called "defensin," which destroys germs.

For people with cystic fibrosis, the defense mechanisms are crippled. The lungs produce mucin that is too thick and sticky to be flushed away by the cilia. The mucus stays in the lungs and provides an ideal breeding ground for bacteria. The person suffers from repeated respiratory infections such as bronchitis and pneumonia. In addition, cystic fibrosis pre-

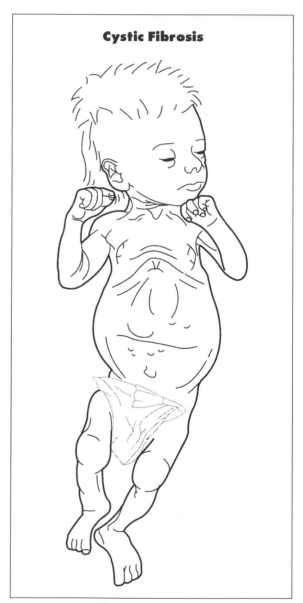

Cystic Fibrosis

Approximately 10% of newborns with cystic fibrosis have a puttylike plug of undigested material in their intestines called the meconium ileus, which results in emaciation with a distended abdomen.

vents the body from properly absorbing salt into the epithelial cells. The lungs cannot absorb the salt, causing a buildup of salt outside the cells. This buildup disables the natural antibiotic defensin, which in turn leads to bacterial infections that cause the increased production of mucus. The damage from infections and the mucus in the airways eventually make breathing impossible.

Treatment and Outlook

Approximately thirty thousand Americans have cystic fibrosis, and one in twenty-five Caucasians are carriers of the disease. In the 1950's, an infant with cystic fibrosis seldom survived past the first year or two. With aggressive medical treatment and therapy to relieve the severe symptoms, the average life span of a person with cystic fibrosis increased to twenty-nine years by the early 1990's; a person with the disorder may live as long as forty years.

Treatment depends on the organs involved. People with cystic fibrosis follow special diets and take dietary supplements so that they have proper nutrition and receive the necessary digestive enzymes, salts, and vitamins that cannot pass through the blocked ducts. The disease makes the digestion and absorption of fats and proteins difficult because certain enzymes are depleted as a result of blocked ducts in the pancreas.

The affected person also undergoes daily backslapping designed to break up the mucus in the lungs. Antibiotics can reduce infections of the lungs. In the 1990's, therapy involving the inhalation of a special enzyme began. This enzyme helps break down the thick mucus so that it is easier to cough out. In late 1997, a potent antibiotic was recommended to the Food and Drug Administration. This antibiotic, called Tobi (tobramycin for inhalation) was the first inhaled antibiotic and appeared to increase the lung function of some cystic fibrosis patients.

Research continues in an effort to determine the nature of the genetic defect that causes cystic fibrosis so that the normal function of the defective gene can be replaced by specially designed therapies. Genetic testing for cystic fibrosis can locate the defect responsible for the disease in about 75 percent of the afflicted people. Since cystic fibrosis can cause more than one hundred other genetic mutations, however, a simple test to detect the variations may be very difficult to develop. In addition, the symptoms of the disease can vary from severe to extremely mild. Research focuses on the development of inexpensive and accurate diagnosis as well as sound genetic counseling in order to reduce the occurrence of the disease. In early 1997, the National Institutes of Health (NIH) recommended that all couples planning to have children should be offered the option of testing for the cystic fibrosis gene mutations.

—*Virginia L. Salmon*

See Also: Genetic Counseling; Genetic Screening; Genetic Testing; Hereditary Diseases; Human Genetics.

Further Reading: *Everything You Need to Know About Diseases* (1996), edited by Michael Shaw, provides a general overview of cystic fibrosis and how it is inherited. A controversial discussion of treatment for cystic fibrosis is presented by John Travis in "Cystic Fibrosis Controversy," *Science News* 151 (May 10, 1997). An informative overview of the lung symptoms is presented in S. Sternberg, "Cystic Fibrosis Puzzle Coming Together," *Science News* 151 (February 8, 1997). Lori Oliwenstien, "How Salt Can Kill," *Discover* 18 (January, 1997), discusses the link between salt and the lungs.

Developmental Genetics

Field of study: Developmental genetics

Significance: *The discovery of the genes responsible for the apparently miraculous conversion of a single homogeneous egg cell into a fully formed organism provides an explanation of how this mysterious process functions. Geneticists and developmental biologists forged connections between readily observed morphological changes in a developing egg and the workings of many unseen genes and proteins. The results of their efforts suggested that common developmental mechanisms exist for diverse organisms and that the experimental manipulation of particular genes could lead to treatments or cures for cancers and developmental abnormalities in humans.*

Key terms

INDUCTION: an easily observed event in which a cell or group of cells signals an adjacent cell to pursue a different developmental pathway and so become differentiated from its neighboring cells

MORPHOGEN: a chemical compound or protein made by cells in an egg that creates a concentration gradient affecting the developmental fate of surrounding cells by altering their gene expression or their ability to respond to other morphogens

DIFFERENTIATION: the process by which a cell changes its phenotype, or outward appearance, and becomes different from its parent cell, usually by altering its gene expression

EPIGENESIS: the formation of differentiated cell types and specialized organs from a single, homogeneous fertilized egg cell without any preexisting structural elements

DEOXYRIBONUCLEIC ACID (DNA): the genetic material of chromosomes, composed of two intertwined strands of sugar molecules, each molecule with one of four different bases attached to it (adenine, guanine, thymine, or cytosine); the sequence of the DNA strand's bases determines its informational content

GENE: a discrete section of DNA with a control region (an "on-off" switch) followed by a much longer sequence of bases coding for a protein molecule; genes are arranged in tandem to form the chromosomes

GENE EXPRESSION: the combined biochemical processes, called "transcription" and "translation," that convert the linearly encoded information in the bases of DNA into the three-dimensional structures of proteins

Early Hypotheses of Development in Diverse Organisms

Scientific investigations of complex phenomena such as animal and plant development often begin with simple observations that lead to questions about causes and effects. These questions prompt the creation of hypotheses to explain the observations, and experiments are devised to test their validity. The history of developmental genetics is full of such observations and hypotheses, some more correct or better supported by experimental data than others. From the earliest times, people noted two inconsistencies in the ways that plants and animals developed: A particular organism produced offspring very much like itself in structure and function, and the fully formed adult consisted of numerous cell types and other highly specialized organs and structures, yet it came from one simple egg cell. Researchers wondered how such simplicity, observed in the egg cell, could give rise to such complexity in the adult and always reproduce the same structures.

In the seventeenth century, the "preformationism" hypothesis was advanced to answer these questions by asserting that a miniature organism existed in the sperm or eggs. After fertilization, this miniature creature simply grew into the fully formed adult. Some microscopists of the time claimed to see a "homunculus," or little man, inside each sperm cell. That the preformationism hypothesis was ill-conceived became apparent when others noted that developmental abnormalities could not be explained satisfactorily, and it became clear that another, more explanatory hypothesis was needed to account for these inconsistencies.

In 1767, Kaspar Friedrich Wolff published

his "epigenesis" hypothesis, in which he stated that the complex structures of chickens developed from initially homogeneous, structureless areas of the embryo. Many questions remained before this new hypothesis could be validated, and it became clear that the chick embryo was not the best experimental system for answering them. Other investigators focused their efforts on the sea squirt, a simpler organism with fewer differentiated tissues.

Work with the sea squirt, a tiny sessile marine animal often seen stuck to submerged rocks, led to the notion that development followed a mosaic pattern. The key property of mosaic development was that any cell of the early embryo, once removed from its surroundings, grew only into the structure for which it was destined or determined. Thus the early embryo consisted of a mosaic of cell types, each determined to become a particular body part. The determinants for each embryonic cell were found in the cell's cytoplasm, the membrane-bound fluid surrounding the nucleus. Other scientists, most notably Hans Driesch in 1892 and Theodor Boveri (working with sea urchin embryos) in 1907, noted that a two-cell-stage embryo could be teased apart into separate cells, each of which grew into a fully formed sea urchin. These results appeared to disagree with the mosaic developmental mechanism, and a new mechanism was postulated: regulative development.

The key property of regulative development was that any cell separated from its embryo could regulate its own development into a complete organism. In contrast to mosaic development, the determinants for regulative development were found in the nuclei of embryonic cells, and Boveri hypothesized that gradients of these determinants, or morphogens, controlled the expression of certain genes. Both Driesch and Boveri worked from an earlier bold hypothesis proposed by German biologist August Weismann in 1883, the "germ-plasm" theory, in which chromosomes assumed an all-important role in controlling development. How the chromosomes contributed to development was not known, and Weismann mistakenly implied that genes were lost from differentiated cells as more and more specific structures formed.

In spite of the apparent disagreements among the several hypotheses advanced to explain the observed phenomena of developing embryos, a grand synthesis was soon formed. Working with roundworm, mollusc, sea urchin, and frog embryos, investigators realized that both mosaic and regulative mechanisms operate during development, with some organisms favoring one mechanism over the other. The most important conclusion coming from these early experiments suggested that certain genes on the chromosomes interacted with both the cytoplasmic and nuclear morphogenetic determinants to control the proliferation and differentiation of embryonic cells. What exactly were these morphogens, where did they originate, and how did they form gradients in the embryo? How did they interact with genes?

The Morphology of Development

Before the "how and why" mechanistic questions of morphogens could be answered, more answers to the "what happens when" questions were needed. Using new, powerful microscopes in conjunction with cell-specific stains, many biologists were able to precisely map the movements of cells during embryogenesis and to create "fate maps" of such cell migrations. Fate maps were constructed for sea squirt, roundworm, mollusc, sea urchin, and frog embryos, and it was clearly shown that specific, undifferentiated cells in the early embryo gave rise to complex body structures in the adult.

In addition, biologists observed an entire stepwise progression of intervening cell types and structures that could be grouped into various stages and that were more or less consistent from one organismal type (species) to another. Soon after fertilization, during the very start of embryogenesis, specific zones with defining, yet structureless, characteristics were observed. These zones consisted of gradients of different biochemical compounds, some of which were morphogens, and they seemed to function by an induction process. Some of these morphogen gradients existed in the egg before fertilization; thus it became evident that the egg was not an entirely amorphous, homogeneous cell but one with some amount of preformation. This preformation took the form of specific morphogen gradients forming polar axes extending from the egg cell's top (dorsal) to its bottom (ventral), from its front (anterior) to its end (posterior), and from one side to the other (medio-lateral).

After these early embryonic events and more cell divisions, in which loosely structured patterns of morphogen gradients were established to form the embryo's polar axes, the cells aggregated into a structure called a "blastula," a hollow sphere of cells. The next stage involved the migration of cells from the surface of the blastula to its interior, a process called "gastrulation." This stage is important in all animals because it forms three tissue types: the ectoderm (for skin and nerves), the mesoderm (for muscle and heart), and the endoderm (for other internal organs). Continued morphogenesis generates a "neurula," an embryo with a developing nervous system and backbone. During axis formation and cell migrations, the embryonic cells are continually dividing to form more cells that are undergoing differentiation into specialized tissue types such as skin or muscle. Eventually, processes referred to as "organogenesis" transform a highly differentiated embryo into one with distinct body structures that will grow into a fully formed adult.

Experimental Systems for Studying Developmental Genes

The biologists studying development faced an important issue: to choose an appropriate experimental organism for elucidating the relationship of the observed embryonic differentiation and morphological changes to the underlying genetic activity. When faced with such a choice, biologists normally favor organisms with the simplest developmental program, ones with the fewest differentiated cell types that will still allow them to answer fundamental questions about the underlying processes. Sea squirts and roundworms satisfied this criterion, but they exhibited a predominantly mosaic form of development and were not the best systems for studying morphogen-dependent induction. Frog embryogenesis, with both mosaic and regulative processes, was well described and contributed greatly to answering the "what and when" questions of sequential events, but no effective genetic system existed for examining the role of genes in differentiation necessary for answering the "why and how" questions of mechanism.

The issue was finally resolved by focusing once again on the morphogens. These mediators of cellular differentiation were found only in trace amounts in developing embryos and thus were difficult if not impossible to isolate in pure form for experimental investigation. An alternative to direct isolation of morphogens was to isolate the genes that make the morphogens. The organism deemed most suitable for such an approach was the fruit fly, *Drosophila melanogaster*, even though its development was more complex than that of the roundworm. Fruit flies could be easily grown in large numbers in the laboratory, and many mutants could be generated quickly; most important, an effective genetic system already existed in *Drosophila*, making it easier to create and analyze mutants. The person who best used the fruit fly system and greatly contributed to the understanding of developmental genetics was Christiane Nüsslein-Volhard, who shared a 1995 Nobel Prize in Physiology or Medicine with Edward B. Lewis and Eric Wieschaus.

The Genes of Development

The first important developmental genes discovered in *Drosophila* were the ones acting the latest in morphogenesis, but their discovery led to the isolation of the gene for the morphogen controlling the anterior-posterior axis of the embryo, the bicoid gene. Researchers also identified the homeotic gene, which determined body shape during organogenesis. The study of mutants with legs in place of antennae allowed the isolation of the responsible genes. Once isolated, these homeotic genes were used, through powerful genetic engineering and cloning techniques, to find other genes in different mutants created by Nüsslein-Volhard.

The bicoid gene was the first of thirty genes discovered that determined the initial polarities of the early, undifferentiated embryo. The bicoid gene's discovery validated the gradient hypothesis originally proposed by Boveri because its gene product functioned as a "typical" morphogen. It was a protein that existed in the highest concentration at the egg's anterior pole and diffused to lower concentrations toward the posterior pole, thus forming a gradient. Through the use of more fruit fly mutants, geneticists showed that the bicoid protein stimulated the gene expression of another early gene, called "hunchback gene," which in turn affected the expression of other genes called "kruppel gene" and "knirps gene." Molecular geneticists showed that the bicoid protein controlled the hunchback gene by binding directly to the gene's control region. After more than one hundred years of unanswered questions, biologists understood the molecular properties of a morphogen and demonstrated, through experiments with Nüsslein-Volhard's mutants, how it functioned. Answers were being found to the "how and why" mechanistic questions.

Soon after these initial discoveries, a virtual explosion of experimental findings linking specific gene activities (genotypes) to specific embryonic pathways and structures (phenotypes) poured into scientific publications. It soon became clear that some fifty genes were involved in differentiating an amorphous fruit fly egg into a highly structured larva, with compartments of differentiated cells poised to undergo further development through yet more genes into a fully formed fly. These genes were grouped into three major categories, with each category of genes responsible for the formation of specific, differentiated cell groups, mainly along the anterior-posterior axis of the developing embryo. The first group of genes, the "maternal-effect" genes, included the bicoid gene. These genes, located in special "nurse" cells of the mother, made proteins that contributed to the initial preformed character of the embryo by establishing loosely structured morphogen gradients along the egg's axes before fertilization occurred. The second category of genes, the "segmentation" group, consisted of three subgroups called "gap," "pair-rule," and "polarity." The hunchback, kruppel, and knirps genes were all gap genes. The last category contained the "homeotic" genes, which ultimately determined the body part identity of the previously differentiated cell groups.

Pattern Formation

It was thus firmly established that genes made morphogens and operated in sequence to control development. Exactly how the genes interacted with each other to orchestrate the egg's transformation into a complex organism was less clear. Through the use of highly specific stains to track the morphogens in mutant embryos, a fascinating picture of these interactions emerged. Even before fertilization, shallow, poorly defined gradients were established by genes of the mother, such as the bicoid gene. These morphogen gradients established the anterior-to-posterior and dorsal-to-ventral axes. After fertilization, these maternally generated morphogens bound to the control regions of the gap genes, which caused the formation of four differently colored zones extending from the anterior to the posterior of the *Drosophila* egg cell; in other words, the colors changed in a sequence moving from left to right across an oblong egg resting on its "belly." The gap gene products then bound to the pair-rule gene's control regions and caused their activation; seven colored bands could be seen across the egg cell. When the next set of genes in the segmentation category, the polarity genes, was activated, fourteen bands appeared across the embryo. The first three bands (going from anterior to posterior) eventually developed into the head structures, but only after the last category of genes was activated, the homeotic genes. The next three bands, also under the influence of the homeotic genes, gave rise to the thorax, including the legs and wings of the fly, and the last eight bands produced the abdomen. In all, fourteen distinctly colored bands or segments were produced by the cooperative yet complex interactions of the maternal-effect, segmentation, and homeotic genes. Mutations (heritable changes in the base sequence induced by X rays) in any of these many developmental genes caused distinct and easily observed changes in the developing segment pattern as it proceeded from the initial axis polarities, through the four-band and seven-band stages, to the final fourteen segments. Genes such as hunchback, giant, gooseberry, and hedgehog were all named with reference to the specific phenotypic changes caused in the banding patterns by mutation.

Later, homeotic gene action transformed each segment into a specific body part of the adult fly. The homeotic genes were called the "master" genes because they controlled large numbers of other genes required to make a whole wing or leg. Two clusters of homeotic genes were discovered on a *Drosophila* chromosome, with three genes in one and five in the other. The three-gene cluster controlled the development of those segments destined to become the abdomen; the five-gene cluster controlled the segments for thorax structures such as legs and wings.

A general principle applying to developmental processes in all organisms emerged from the elegant work with *Drosophila* mutants: Finer and finer patterns of differentiated cells are progressively formed in the embryo along its major axes by morphogens acting on genes in a cascading manner, in which one gene set controls the next in the sequence until a highly complex pattern of differentiated cells arises through the changing and intersecting waves of gene expression. Each cell within its own patterned zone then responds to the homeotic gene products and contributes to the formation of distinct, identifiable body parts.

Another important corollary principle was substantiated by the genetic analysis of development in *Drosophila* and other organisms. In direct contrast to Weismann's implication about gene loss during differentiation, convincing evidence showed that genes were not systematically lost as egg cells divided and acquired distinguishing features. Even though a muscle cell was highly differentiated from a skin cell or a blood cell, each cell type retained the same numbers of chromosomes and genes as the original, undifferentiated, but fertilized egg cell. What changed in each cell was the pattern of gene expression, so that different proteins were made by specific genes while other genes were turned off. The morphogens, working in complex combinatorial patterns during the course of development, determined which genes would stay "on" and which would be turned "off."

Impact and Applications

The discovery and identification of the developmental genes in *Drosophila* and other lower organisms led to the discovery of similarly functioning genes in higher organisms, including humans. The base-pair sequences of many of the developmental genes, especially shorter subregions coding for sections of the morphogen that bind to the control regions of target genes, are conserved, or remain the same, across diverse organisms. This gene sequence conservation allowed researchers to find similar genes in humans. For example, some forty homeobox genes were eventually found in mice and humans, even though only eight were initially discovered in *Drosophila*. Some of the late-acting human homeobox genes are responsible for such developmental

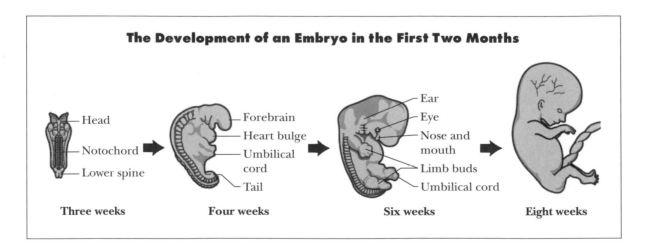

The Development of an Embryo in the First Two Months

Head — Notochord — Lower spine

Three weeks

Forebrain — Heart bulge — Umbilical cord — Tail

Four weeks

Ear — Eye — Nose and mouth — Limb buds — Umbilical cord

Six weeks

Eight weeks

abnormalities as fused fingers and extra digits on the hands and feet. One of the most interesting abnormalities is craniosynostosis, a premature fusion of an infant's skull bones that can cause mental retardation. In 1993, developmental biologist Robert Maxson and his research group at the University of Southern California's Norris Cancer Center were the first to demonstrate that a mutation in a human homeobox gene (the *MSX2* gene) was directly responsible for craniosynostosis and other bone/limb abnormalities requiring corrective surgeries. Maxson made extensive use of "knockout" mice, genetically engineered mice lacking particular genes, to test his human gene isolates. He and his research group made great progress in understanding the role of the *MSX2* gene as inducer of surrounding cells in the developing embryo. When this induction process fails because of defective *MSX2* genes, the fate of cells destined to participate in skull and bone formation and fusion changes, and craniosynostosis occurs.

A clear indication of the powerful cloning methods developed in the late 1980's was the discovery and isolation in 1990 of an important mouse developmental gene called "brachyury" ("short tails"). The gene's existence in mutant mice had been inferred from classical genetic studies sixty years prior to its isolation. In 1997, Craig Basson, Quan Yi Li, and a team of coworkers isolated a similar gene from humans and named it "T-box brachyury" (*TBX5*). Discovered first in mice, the "T-box" is one of those highly conserved subregions of a gene, and it allowed Basson and Li to find the human gene. When mutated or defective in humans, *TBX5* causes a variety of heart and upper limb malformations referred to as Holt-Oram syndrome. *TBX5* codes for an important morphogen affecting the differentiation of embryonic cells into mesoderm, beginning in the gastrulation phase of embryonic development. These differentiated mesodermal cells are destined to form the heart and upper limbs.

One of the important realizations emerging from the explosive research into developmental genetics in the 1990's was the connection between genes that function normally in the developing embryo but abnormally in an adult,

causing cancer. Cancer cells often display properties of embryonic cells, suggesting that cancer cells are reverting to a state of uncontrolled division. Some evidence indicates that mutated developmental genes participate in causing cancer. Taken together, the collected data from many isolated human developmental genes, along with powerful reproductive and cloning technologies, promise to lead to cures and preventions for a variety of human developmental abnormalities and cancers.

—Chet S. Fornari

See Also: DNA Structure and Function; Genetic Engineering; Homeotic Genes; Knockout Genetics and Knockout Mice; Protein Synthesis; RNA Structure and Function; Sheep Cloning; Transgenic Organisms.

Further Reading: A very readable and well-illustrated but somewhat technical account of developmental genetics, especially in *Drosophila*, can be found in James Watson's *Recombinant DNA* (1992). The developmental biologist Scott Gilbert presents a detailed description of all aspects of development, especially the basic biology, in his book *Developmental Biology* (1988). The 1996 fourth edition of Neil Campbell's text *Biology* is an excellent source of general information on many aspects of development. Benjamin Lewin's *Genes* (1997) is a comprehensive, clear discussion of developmental genes, with excellent illustrations. A popular but still scholarly account of development, evolution, and shape formation in animals can be found in Rudolf Raff's *The Shape of Life* (1996). Christiane Nüsslein-Volhard's article "Gradients That Organize Embryo Development," *Scientific American* (August, 1996), provides an incisive, clear history of development and key experiments.

Diabetes

Field of study: Human genetics
Significance: *Diabetes mellitus (sugar diabetes) is one of the most common chronic disorders and is estimated to affect 5 to 10 percent of the adult population in the United States and other countries of the Western world. The medical and socio-*

economic costs of diabetes are considerable, and it is a major risk factor for coronary artery disease, blindness, and kidney failure. An understanding of the genetic and environmental factors that cause different forms of diabetes will aid in the development of successful therapies and prevention of the disorder.

Key terms

AUTOIMMUNE DISEASE: a disease in which an organism's immune system attacks its own cells

COMPLEX DISEASE: a disease that does not show a clear Mendelian pattern of inheritance; the disease shows a tendency to run in families and is often caused by the influence of multiple genetic and nongenetic factors

GENETIC HETEROGENEITY: a situation in which two or more genes can produce the same disease symptoms

INSULIN DEPENDENT DIABETES MELLITUS (IDDM, TYPE I): a type of diabetes that usually appears before the age of forty in which patients must receive exogenous insulin to survive

NONINSULIN DEPENDENT DIABETES MELLITUS (NIDDM, TYPE II): a type of diabetes that usually appears after the age of forty, commonly among people who are obese

Types of Diabetes

Diabetes mellitus is a complex, heterogeneous group of disorders in which elevated blood sugar is a common characteristic. The body of a diabetic cannot metabolize dietary sugars and starches in the normal way. The hormone insulin, produced in the beta cells of the pancreas, normally regulates carbohydrate metabolism; in the diabetic, however, insulin activity is either blocked or insufficient. As a result, the level of sugar in the blood increases and is eliminated in the urine. The sugar in the urine was noted centuries ago by a physician who described it as tasting like honey or sugar (the word "mellitus" means "honey-sweet" in Latin).

The two major types of diabetes are insulin-dependent diabetes mellitus (type I) and non-insulin-dependent diabetes mellitus (type II). Type I diabetes accounts for about 10 percent of the total cases. It is characterized by the destruction of the insulin-producing beta cells of the pancreas and usually occurs before the age of forty. The observation that patients frequently possess antibodies that attack their own pancreatic beta cells indicates that the disease may be an autoimmune disorder. Since there is a lack of insulin, the patient must receive exogenous insulin to survive. Type II diabetes makes up about 90 percent of cases and usually occurs in people who are over forty and frequently in those who are obese. Patients usually produce some of their own insulin, and the symptoms often can be treated with dietary management or oral drugs.

It should be noted that although type I is referred to as "juvenile-onset" diabetes and type II as "adult or maturity-onset" diabetes, either type may occur at any age. Regardless of the type, the long-term complications of high blood sugar include blindness, heart disease, kidney disease, poor wound healing, and poor circulation.

Genetics and Diabetes

Since diabetes is a heterogeneous group of disorders, underlying genetic factors are also likely to be heterogeneous. The familial aggregation of diabetes has been observed for a long time. The risk for type I diabetes in the general population is around 0.3 to 0.5 percent, but the risk to a sibling of a patient is about 6 percent. Although both sexes are affected about equally, the risk for offspring varies with the sex of the patient, ranging from 1 to 3 percent for children of diabetic mothers to 4 to 6 percent for children of diabetic fathers. The role of nongenetic factors in the causation of type I diabetes is underscored by the observation that if one identical twin has diabetes, the likelihood that the other twin will have diabetes is 50 percent or less, despite the fact that identical twins have all of their genes in common.

In contrast to type I diabetes, an identical twin of someone with type II diabetes will also have diabetes over 90 percent of the time. The chance of a sibling of a diabetic having type II diabetes is approximately 10 to 15 percent. Major risk factors include age, a family history of diabetes, and obesity. Regular exercise is known to decrease the risk. A subtype of type II diabetes, maturity-onset diabetes of the young

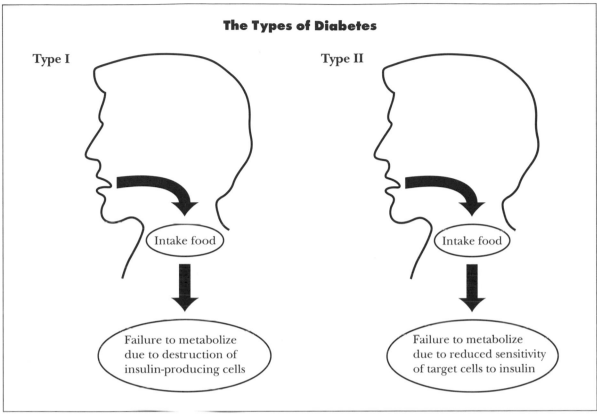

The Types of Diabetes

Type I

Intake food

Failure to metabolize
due to destruction of
insulin-producing cells

Type II

Intake food

Failure to metabolize
due to reduced sensitivity
of target cells to insulin

Diabetes mellitus results from the body's failure to metabolize sugar properly. Type I diabetes is an autoimmune process that destroys insulin-producing cells and is often found in children. Type II diabetes results from the reduced sensitivity of target cells to insulin, requiring increased secretion of insulin to maintain blood sugar levels; this type generally affects adults and is often caused by poor eating habits and lack of exercise over years.

(MODY), appears to be inherited as an autosomal dominant disorder. MODY often appears in patients under the age of twenty-five, and there is a 50 percent chance that siblings and children of an affected individual will also have the disorder.

Studies of obesity in mice and humans have also yielded information on how specific genes may cause obesity and diabetes. A gene identified as the leptin gene regulates the hunger center in the brain, resulting in decreased food intake. The same gene causes the body to increase metabolism. Animals with an abnormal leptin gene eat too much, metabolize food less efficiently, and become obese. A second gene, identified as the beta-3 adrenergic gene, increases fat cells to metabolize fat. Defects in this gene are frequently associated with people who are obese and diabetic.

Impact and Applications

Although there is still no cure for diabetes, increased understanding of the physiology and genetics of the disorder has led to improvements in the therapies involving insulin and in more effective counseling of families. Genetic technology has produced cheaper and more efficient ways of manufacturing insulin. Risk factors associated with the different types of diabetes have been elucidated more clearly, which has increased the effectiveness of genetic counseling.

—Donald J. Nash

See Also: Autoimmune Disorders; Hereditary Diseases; Human Genetics; Twin Studies.

Further Reading: Soumitra Ghosh and Nicholas J. Schork, "Genetic Analysis of NIDDM: The Study of Quantitative Traits," *Diabetes* (January, 1996), provides a useful bibliography and glossary. Chapter 7 in *Living with Our Genes* (1998), by Dean Hamer and Peter Copeland, has useful information concerning obesity and diabetes. *Clinical Genetics Handbook* (1993), by Arthur Robinson and Mary G. Linden, is a useful source for practical genetic and clinical information on many genetic disorders, including diabetes.

Diamond v. Chakrabarty

Field of study: Genetic engineering and biotechnology

Significance: *In 1980, the U.S. Supreme Court upheld the right to patent a live, genetically altered organism. The decision was opposed by many scientists and theologians who believed that such organisms would pose a threat to the future of humanity.*

Key term

PATENT: a grant made by the government that gives the creator or inventor the sole right to make, use, or sell that invention for a specific period of time, usually seventeen years in the United States

Patent on Life Form Upheld

On June 16, 1980, the U.S. Supreme Court voted 5 to 4 that living organisms could be patented under federal law. The case involved Ananda M. Chakrabarty, a scientist who, while working for General Electric in 1972, had created a new form of bacteria, *Pseudomona originosa*, that was able to break down crude oil and that therefore could be used to clean up oil spills. Chakrabarty filed for a patent, but an examiner for the Patent Office rejected the application on the ground that living things are not patentable subject matter under existing patent law. Commissioner of Patents and Trademarks Sidney A. Diamond supported this view. Federal patent law provided that a patent could be issued only to a person who invented or discovered any new and useful "manufacture" or "composition of matter." The U.S. Court of Customs and Patent Appeals reversed that decision in 1979, concluding that the fact that microorganisms are alive has no legal sig-

nificance. It held that a live, human-made bacterium is a patentable item since the microorganism was manufactured by cross-breeding four existing strains of bacteria and had never existed in nature.

Writing for the majority, Chief Justice Warren Burger upheld the patent appeals court judgment, making a distinction between the new bacterium and "laws of nature, physical phenomena and abstract ideas," which are not patentable. In the Court majority's view, Chakrabarty had invented a form of life that did not exist in the natural world, so it could not be considered part of nature. Instead, it was a product of human "ingenuity and research" that deserved patent protection. Items not patentable include new minerals that are discovered in the earth or a new species of plant found in a distant forest. These things occur naturally and are not created by humans. Burger also stressed that physicist Albert Einstein could not have patented his formula $E=mc^2$, since it is a law of nature, nor could Sir Isaac Newton have received a patent for the law of gravity. Discoveries such as these are part of the natural world and cannot be owned by a single individual.

Chakrabarty, on the other hand, had not found an unknown, natural species, nor had he discovered a law of nature. His new bacterium had a distinctive name and was developed in the laboratory for a specific purpose. None of the characteristics of the new organism could be found in nature. His discovery, Burger reemphasized, was patentable because he had created it.

Opposition to the Ruling

The Court majority refused to consider arguments made in friend-of-the-court briefs filed by opponents of genetic engineering. The briefs were presented by groups representing scientists, including several Nobel Prize winners, and religious organizations. One brief suggested that genetic research posed a dangerous and serious threat to the future of humanity and should, therefore, be prohibited. Possible dangers included the spread of pollution and disease by newly created bacteria, none of which would have any natural enemies.

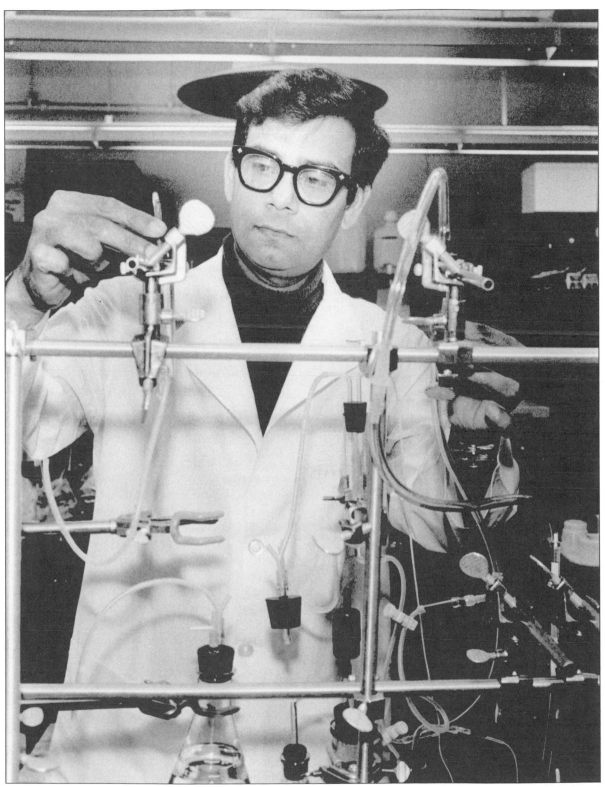

Ananda M. Chakrabarty in 1980, after the Supreme Court affirmed his right to patent a living organism he had invented. (AP/Wide World Photos)

Other threats involved the possible loss of genetic diversity, if, for instance, only the "best" form of laboratory-created plant seeds were grown. Research into human genetics could lead to newly designed gene material that could be used to build a "master race," thereby devaluing other human lives. Justice Burger concluded, however, that humans could be trusted not to create such horrible things. Quoting William Shakespeare's *Hamlet*, the chief justice asserted that it is sometimes better "to bear those ills we have than to fly to others that we know not of." People can try to guess what genetic manipulation could lead to, but it would also be a good idea to expect good things from science rather than "a gruesome parade of horribles." Besides, he then said, it did not matter whether a patent was granted in this case; in either case, scientific research would continue into the nature of genes.

The People's Business Commission, a nonprofit educational foundation, had argued that granting General Electric and Chakrabarty a patent would give corporations the right "to own the processes of life in the centuries to come" through genetic manipulation. Chief Justice Burger wrote that the Court was "without competence to entertain these arguments." They did not have enough information available to determine whether to ignore such fears "as fantasies generated by fear of the unknown" or accept them. Such a determination was not the responsibility of the Court, however. Questions of the morality of genetic research and manipulation were better left to Congress and the political process. How to proceed in these matters could only be resolved "after the kind of investigation, examination, and study that legislative bodies can provide and courts cannot."

Justice William J. Brennan, Jr., presented a brief dissenting opinion. He noted that Congress had twice, in 1930 and 1970, permitted new types of plants to be patented. However, those laws made no mention of bacteria. Thus, Brennan argued, Congress had indicated that only plants could receive patents and that the legislators had thus clearly indicated that other life forms were excluded from the patent process. The Court majority rejected this view, arguing that Congress had not specifically excluded other life forms.

—*Leslie V. Tischauser*

See Also: Biotechnology; Cloning; Genetic Engineering: Social and Ethical Issues.

Further Reading: *Diamond, Commissioner of Patents and Trademarks v. Chakrabarty* (1980), 447 U.S. 303, is the official citation of the Supreme Court decision. *Bioremediation for Marine Oil Spills* (1991), by the U.S. Office of Technology Assessment, provides a nontechnical overview of the use of genetic organisms, such as the bacterium patented by Chakrabarty, for bioremediation.

Dihybrid Inheritance

Field of study: Classical transmission genetics

Significance: *The simultaneous analysis of two different hereditary traits may produce more information than the analysis of each trait separately. In addition, many important hereditary traits are controlled by more than one gene. Traits controlled by two genes serve as an introduction to the more complex topic of traits controlled by many genes.*

Key terms

ALLELES: different forms of the same gene; any gene may exist in several forms having very similar but not identical DNA sequences

DIHYBRID: an organism that is heterozygous for both of two different genes

HETEROZYGOUS: a condition in which the two copies of a gene (one inherited from each of the parents) in an individual are different alleles

HOMOZYGOUS: a condition in which the two copies of a gene in an individual are the same allele; synonymous with "purebred"

Mendel's Discovery of Dihybrid Inheritance

Austrian botanist Johann Gregor Mendel was the first person to describe both monohybrid and dihybrid inheritance. When he crossed purebred round-seed garden peas with purebred wrinkled-seed plants, they produced only monohybrid round seeds. He planted the monohybrid round seeds and allowed them to fertilize themselves; they subsequently produced ³⁄₄ round and ¹⁄₄ wrinkled seeds. He concluded correctly that the monohybrid generation was heterozygous for an allele that produces round seeds and another allele that produces wrinkled seeds. Since the monohybrid seeds were round, the round allele must be dominant to the wrinkled allele. He was able to explain the 3:1 ratio in the second generation by assuming that each parent contributes only one copy of a gene to its progeny. If W represents the round allele and w the wrinkled allele, then the original true-breeding parents are WW and ww. When eggs and pollen are produced, they each contain only one copy of the gene. Therefore the monohybrid seeds are heterozygous Ww. Since these two alleles will separate during meiosis when pollen and eggs are produced, ¹⁄₂ of the eggs and pollen will be W and ¹⁄₂ will be w. Mendel called this "segregation." When the eggs and pollen combine randomly during fertilization, ¹⁄₄ will produce WW seeds, ¹⁄₂ will produce Ww seeds, and ¹⁄₄ will produce ww seeds. Since W is dominant to w, both the WW and Ww seeds will be round, producing ³⁄₄ round and ¹⁄₄ wrinkled seeds. When Mendel crossed a purebred yellow-seed plant with a purebred green-seed plant, he observed an entirely analogous result in which the yellow allele (G) was dominant to the green allele (g).

Once Mendel was certain about the nature of monohybrid inheritance, he began to experiment with two traits at a time. He crossed purebred round, yellow pea plants with purebred wrinkled, green plants. As expected, the dihybrid seeds that were produced were all round and yellow, the dominant form of each trait. He planted the dihybrid seeds and allowed them to fertilize themselves. They produced ⁹⁄₁₆ round, yellow seeds; ³⁄₁₆ round, green seeds; ³⁄₁₆ wrinkled, yellow seeds; and ¹⁄₁₆ wrinkled, green seeds. Mendel was able to explain this dihybrid ratio by assuming that in the dihybrid flowers, the segregation of W and w was independent of the segregation of G and g. Mendel called this "independent assortment." Thus, of the ³⁄₄ of the seeds that are round, ³⁄₄ should be yellow and ¹⁄₄ should be green, so that ³⁄₄ × ³⁄₄ = ⁹⁄₁₆ should be round and yellow, and ³⁄₄ × ¹⁄₄ = ³⁄₁₆ should be round and green. Of the ¹⁄₄ of the seeds that are wrinkled, ³⁄₄ should be yellow and ¹⁄₄ green, so that ¹⁄₄ × ³⁄₄ = ³⁄₁₆ should be wrinkled and yellow, and ¹⁄₄ × ¹⁄₄ = ¹⁄₁₆ should be wrinkled and green. This relationship can be seen in table 1 (semicolons indicate that the two genes are on different chromosomes).

Sex Chromosomes

Humans and many other species have sex chromosomes. In humans, normal females have two X chromosomes and normal males have one X and one Y chromosome. Therefore,

Table 1. Dihybrid Inheritance and Sex Linkage

		Pollen		
		$W;G$	$W;g$	$w;G$	$w;g$
Eggs $W;G$		$W\,W;G\,G$ round, yellow	$W\,W;G\,g$ round, yellow	$W\,w;G\,G$ round, yellow	$W\,w;G\,g$ round, yellow
$W;g$		$W\,W;G\,g$ round, yellow	$W\,W;g\,g$ round, green	$W\,w;G\,g$ round, yellow	$W\,w;g\,g$ round, green
$w;G$		$W\,w;G\,G$ round, yellow	$W\,w;G\,g$ round, yellow	$w\,w;G\,G$ wrinkled, yellow	$w\,w;G\,g$ wrinkled, yellow
$w;g$		$W\,w;G\,g$ round, yellow	$W\,w;g\,g$ round, green	$w\,w;G\,g$ wrinkled, yellow	$w\,w;g\,g$ wrinkled, green

Table 2.

Sperm

Eggs		A;R	a;R	A;Y	a;Y
	A;R	A A;R R normal female	A a;R R normal female	A A;R Y normal male	A a;R Y normal male
	A;r	A A;R r normal female	A a;R r normal female	A A;r Y red-green color-blind male	A a;r Y red-green color-blind male
	a;R	A a;R R normal female	a a;R R albino female	A a;R Y normal male	a a;R Y albino male
	a;r	A a;R r normal female	a a;r r albino female	A a;r Y red-green color-blind male	a a;r Y albino, red-green color-blind male

sex-linked traits, which are controlled by genes on the X or Y chromosome, are inherited in a different pattern than the genes that have already been described. Since there are few genes on the Y chromosome, most sex-linked traits are controlled by genes on the X chromosome.

Every daughter gets an X chromosome from each parent, and every son gets an X from his mother and a Y from his father. Human red-green color blindness is controlled by the recessive allele (*r*) of an X-linked gene. A red-green color-blind woman (*rr*) and a normal man (*RY*) will have normal daughters (all heterozygous *Rr*) and red-green color-blind sons (*rY*). Conversely, a homozygous normal woman (*RR*) and a red-green color-blind man (*rY*) will have only normal children, since their sons will get a normal X from the mother (*RY*) and the daughters will all be heterozygous (*Rr*). A heterozygous woman (*Rr*) and a red-green color-blind man (*rY*) will have red-green color-blind sons (*rY*) and daughters (*rr*), and normal sons (*RY*) and daughters (*Rr*) in equal numbers.

A dihybrid woman who is heterozygous for red-green color blindness and albinism (a recessive trait that is not sex linked) can make four kinds of eggs with equal probability: *R;A*, *R;a*, *r;A*, and *r;a*. A normal, monohybrid man who is heterozygous for albinism can make four

kinds of sperm with equal probability: *R;A*, *R;a*, *Y;A*, and *Y;a*. By looking at table 2 above, it is easy to predict the probability of each possible kind of child from this mating.

The probabilities are $^6/_{16}$ normal female, $^2/_{16}$ albino female, $^3/_{16}$ normal male, $^3/_{16}$ red-green color-blind male, $^1/_{16}$ albino male, and $^1/_{16}$ albino, red-green color-blind male. Note that the probability of normal coloring is $^3/_4$ and the probability of albinism is $^1/_4$ in both sexes. There is no change in the inheritance pattern for the gene that is not sex linked.

Other Examples of Dihybrid Inheritance

A hereditary trait may be controlled by more than one gene. To one degree or another, almost every hereditary trait is controlled by many different genes, but often one or two genes have a major effect compared with all the others, so they are called "single-gene" or "two-gene" traits. Dihybrid inheritance can produce traits in various ratios, depending on what the gene products do. A number of examples will be presented, but they do not exhaust all of the possibilities.

The comb of a chicken is the fleshy protuberance that lies on top of the head. There are four forms of the comb, each controlled by a different combination of the two genes that control this trait. The first gene exists in two forms (*R* and *r*), as does the second (*P* and *p*). In each case, the form represented by the uppercase letter is dominant to the other form. Since there are two copies of each gene (with the exception of genes on sex chromosomes), the first gene can be present in three possible combinations: *RR*, *Rr*, and *rr*. Since *R* is dominant, the first two combinations produce the same trait, so the symbols *R_* and *P_* can be used to represent either of the two combinations. Chickens with *R_;P_* genes have what is called a walnut comb, which looks very much like the meat of a walnut. The gene combinations *R_;pp*, *rr;P_*, and *rr;pp* produce combs that are called rose, pea, and single, respectively. If two chickens that both have the gene combination *Rr;Pp* mate, they will produce progeny that are $^9/_{16}$ walnut, $^3/_{16}$ rose, $^3/_{16}$ pea, and $^1/_{16}$ single (see table 1 for an explanation of these numbers).

White clover synthesizes small amounts of cyanide, which gives clover a bitter taste. There are some varieties that produce very little cyanide (sweet clover). When purebred bitter clover is crossed with some varieties of purebred sweet clover, the progeny are all bitter. However, when the hybrid progeny is allowed to fertilize itself, the next generation is $^9/_{16}$ bitter and $^7/_{16}$ sweet. This is easy to explain if it is assumed that bitter/sweet is a dihybrid trait.

The bitter parent would have the gene combination *AA;BB* and the sweet parent *aa;bb*, where *A* and *B* are dominant to *a* and *b*, respectively. The bitter dihybrid would have the gene combination *Aa;Bb*. When it fertilizes itself, it would produce $^9/_{16}$ *A_;B_*, which would be bitter, and $^3/_{16}$ *A_;bb*, $^3/_{16}$ *aa;B_*, and $^1/_{16}$ *aa;bb*, all of which would be sweet. Clearly, both the *A* allele and the *B* allele are needed in order to synthesize cyanide. If either is missing, the clover will be sweet.

Absence of Dominance

In all of the previous examples, there was one dominant allele and one recessive allele. Not all genes have dominant and recessive alleles. If a purebred snapdragon with red flowers (*RR*) is crossed with a purebred snapdragon with white flowers (*rr*), all the monohybrid progeny plants will have pink flowers (*Rr*). The color depends on the number of *R* alleles present: two *R*'s will produce a red flower, one *R* will produce a pink flower, and no *R*'s will produce a white flower. This is an example of partial dominance or additive inheritance.

Consider a purebred red wheat kernel (*AA;BB*) and a purebred white wheat kernel (*aa;bb*) (see table 3 below). If the two kernels are planted and the resulting plants are crossed with each other, the progeny dihybrid kernels will be light red (*Aa;Bb*). If the dihybrid

Table 3.

		Pollen			
		A;B	*A;b*	*a;B*	*a;b*
Eggs	*A;B*	*A A;B B* red	*A A;B b* medium red	*A a;B B* medium red	*A a;B b* light red
	A;b	*A A;B b* medium red	*A A;b b* light red	*A a;B b* light red	*A a;b b* very light red
	a;B	*A a;B B* medium red	*A a;B b* light red	*a a;B B* light red	*a a;B b* very light red
	a;b	*A a;B b* light red	*A a;b b* very light red	*a a;B b* very light red	*a a;b b* white

Dihybrid ratios may change if both genes are on the same chromosome.

plants grown from the dihybrid kernels are allowed to self-fertilize, they will produce $^{1}/_{16}$ red (*AA;BB*), $^{4}/_{16}$ medium red (*AA;Bb* and *Aa;BB*), $^{6}/_{16}$ light red (*AA;bb, Aa;Bb,* and *aa;BB*), $^{4}/_{16}$ very light red (*Aa;bb* and *aa;Bb*), and $^{1}/_{16}$ white (*aa;bb*). The amount of red pigment depends on the number of alleles (*A* and *B*) that control pigment production. Although it may appear that this is very different than the example in table 1, they are in fact very similar.

All of the inheritance patterns that have been discussed are examples of "independent assortment," in which the segregation of the alleles of one gene is independent of the segregation of the alleles of the other gene. That is exactly what would be expected from meiosis if the two genes are not on the same chromosome. If two genes are on the same chromosome and sufficiently close together, they will not assort independently and the progeny ratios will not be like any of those described. In that case, the genes are referred to as "linked" genes.

—James L. Farmer

See Also: Epistasis; Linkage Maps; Mendel, Gregor, and Mendelism; Monohybrid Inheritance; Quantitative Inheritance.

Further Reading: Thomas Brock et al., *Biology of Microorganisms* (1988), contains a college-level overview of various inheritance patterns. Raul Cano et al., *Essentials of Microbiology* (1988), is a well-illustrated, nontechnical book with useful information on dihybrid inheritance. Slightly more technical and also well-illustrated is Gerard Tortora et al., Microbiology (1989).

Diphtheria

Field of study: Bacterial genetics
Significance: *Diphtheria is an acute bacterial disease known best for damaging the respiratory system. Afflicted individuals die from this as well as from damage to the heart, nerves, and kidney. Genetic research has led to better understanding of diphtheria's cause, action, and treatment.*

Key terms

ANAPHYLAXIS: a severe, sometimes fatal allergic reaction

ANTITOXIN: a vaccine containing antibodies
ANTIBODIES: proteins that neutralize poisons in the body
CUTANEOUS: related to or affecting skin
PROTEIN: a complex, chainlike substance composed of many conjoined amino acids

Diphtheria Symptoms and Cure

The acute bacterial disease diphtheria is caused by rod-shaped *Corynebacterium diphtheriae* (*C. diphtheriae*), discovered in 1883 by Edwin Klebs and Friedrich Löffler. Diphtheria involves the respiratory tract, nerves, and heart in ways that can be lethal. After 1950, the disease became uncommon in industrialized nations because of immunization by vaccination with antitoxin originally isolated from horses by Emil Adolf von Behring in the 1880's. In such nations, diphtheria is mostly caught from travelers coming from developing nations, where it is much more common, who may be asymptomatic carriers or who may have active diphtheria.

The usual body entry point of *C. diphtheriae* is the mouth or the nose, though it also enters via the skin (cutaneous route). After infection and a two- to five-day incubation period, diphtheria's first symptoms are localized inflammation that kills cells in the respiratory tract or in the skin. Respiratory diphtheria initially appears as a sore throat in which a dirty gray membrane (diphtheria pseudomembrane) forms and spreads through the respiratory system. The pseudomembrane (made mostly of dead cells, bacteria, and white blood cells) causes a husky voice and is accompanied by swollen glands. In severe cases, diphtheria kills by heart failure or throat paralysis as little as one day after its initial appearance as a sore throat. Fortunately, such lethality occurs mostly in unimmunized individuals. Cutaneous *C. diphtheriae* infections most often produce only skin lesions, though they can cause death if the bacteria spreads widely through the blood and damages the heart, nerves, and kidneys. Damage depends upon the bacterial entry site, individual immunization status, and the amount of toxin made.

Although most people in industrialized nations are immunized, the consequences of

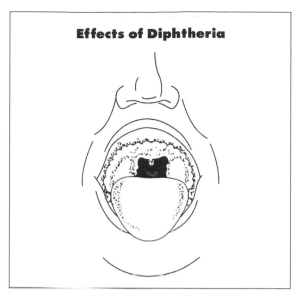

Effects of Diphtheria

Diphtheria causes a thick, grayish green membrane to form over the larynx, tonsils, pharynx, and sinus cavities.

diphtheria can be so severe that therapy by diphtheria antitoxin should begin as soon as symptoms suggest the disease. Cure of diphtheria requires, in addition to the antitoxin, destruction of all *C. diphtheriae* in afflicted individuals. Immunization is the first line of defense, so it is crucial to ensure that the suspected diphtheria sufferer is not sensitive to antitoxin because incautious antitoxin administration may cause lethal anaphylaxis in sensitive people. Individual sensitivity is identified by scratch tests with diluted antitoxin. In sensitive people, desensitization is achieved through the sequential administration of increasing doses of antitoxin in an intensive care unit until effective doses are safely reached.

Diphtheria is so dangerous that all patient contacts are tested for *C. diphtheriae*. Afflicted individuals are given penicillin, erythromycin, and/or antitoxin, depending on the presence, absence, and severity of diphtheria symptoms. Though adequate universal immunization is a sure diphtheria control, booster shots—like those for tetanus prophylaxis—should be given every ten years in addition to childhood shots. Because of the extremely infectious and fatal nature of diphtheria, all people positive for *C. diphtheriae* must be kept in bed, isolated, and treated until symptoms and bacteria are absent

after antibiotic therapy stops. This may require four to six weeks.

Genetics and Diphtheria

Diphtheria symptoms are caused by diphtheria toxin, a protein so lethal that 6 micrograms will kill a 150-pound human. Most often, the toxin first localizes in respiratory mucosa cells or cutaneous sites, where it causes diphtheria pseudomembrane or skin lesions by interacting with the protein translocase. Translocase is essential to synthesis of proteins needed for body cell growth, survival, and reproduction. Diphtheria toxin and translocase interact through a process called adenine ribosylation, similar to that in cholera. Diphtherial adenine ribosylation inactivates translocase, preventing its action and killing affected cells. Dead respiratory cells form diphtheria pseudomembrane, which closes off the throat. In skin, toxin-killed cells cause skin lesions. Destruction of nerve, heart, and kidney cells leads to damage in those tissues.

The diphtheria pseudomembrane may cut off breathing. In such cases, suffocation is prevented by a tracheotomy (a surgical incision in the neck that creates an airway). Major causes of quick diphtheria fatality are damage to nerves and the heart. Study of diphtheria genetics and biochemistry has revealed that the toxin is a protein made by specific genes and that *C. diphtheriae* strains that do not produce the toxin are harmless. In addition, genetic studies have identified interaction of the toxin with respiratory mucosa cell translocase as well as similar action in many other tissues. Use of bacterial genetics has also enabled more scientific production of diphtheria antitoxin. The antitoxin is useful to visitors of regions where the disease is common. Its universal use has led to a worldwide decrease in diphtheria fatalities to fewer than five deaths per million people. The immunization is effective for ten years.

Impact and Applications

Diphtheria has long been a serious, worldwide threat. During the twentieth century, its danger greatly diminished in industrialized nations with the advent of antitoxin and the wide use of antibiotics to kill *C. diphtheriae*. In poorer nations, diphtheria still flourishes and is a severe threat, partly because of less advanced medical practices and the public's fear of immunization.

Prevention of diphtheria relies mostly on immunization via antitoxin. The isolation and identification of diphtheria toxin and the development of antitoxin have depended on genetic methods that now protect most people from the disease. Wherever it afflicts people, diphtheria treatment also requires the use of antibiotics. Hence, advanced diphtheria prevention and treatment will be best effected by using genetic, immunologic, and biochemical methods to produce vaccines effective for more than ten years and to produce more potent antibiotics. Efforts toward these ends will most likely utilize molecular genetics to clearly define why diphtheria is intractable to lifelong vaccination and to produce better antibiotics. Especially valuable will be DNA sequence analysis, gene amplification, and DNA fingerprinting.

—*Sanford S. Singer*

See Also: Cholera; DNA Fingerprinting; Genetic Engineering; Molecular Genetics.

Further Reading: *Conn's Current Therapy* (1995), edited by Robert E. Rakel, provides a succinct overview of diphtheria and its treatment for general readers. *Fishbein's Medical and Health Encyclopedia* (1987) contains solid information on diphtheria and useful illustrations. *Merck Manual, 16th Edition* (1991), edited by Robert Berkow and Andrew J. Fletcher, contains medical details on epidemiology, symptoms, diagnosis, treatment, outbreak management, and antitoxin sensitivity.

DNA Fingerprinting

Field of study: Genetic engineering and biotechnology

Significance: *DNA fingerprinting is a method used to analyze deoxyribonucleic acid (DNA) found in biological materials such as blood, semen, bone, skin, hair, and saliva. The results are used to identify criminals and to settle paternity and maternity disputes based on the uniqueness of the DNA patterns in these genetic materials.*

Key terms

DEOXYRIBONUCLEIC ACID (DNA): the biological structure that stores the genetic information in most living organisms; two linear molecules made of nucleotides entwine to form the DNA double helix

ELECTROPHORESIS: the process of using electrical forces to move DNA strands through a sieving device to separate them by size

POLYMERASE CHAIN REACTION (PCR): an in vitro (outside the body) method for reproducing DNA in the laboratory

RESTRICTION ENZYMES: a class of enzymes obtained from microorganisms that cut the DNA strands at specific places

RESTRICTION FRAGMENT LENGTH POLYMORPHISM (RFLP): the most common method of DNA fingerprinting based on variations in length of DNA fragments from one person to another

Background and Basis for DNA Fingerprinting

The concept of deoxyribonucleic acid (DNA) fingerprinting was pioneered by Alec Jeffreys in 1984 at his laboratory at England's Leicester University. He found repetitive patterns of DNA that varied from one person to another and concluded that every individual has a unique sequence of building blocks or genetic code in specific regions on a strand of their DNA. Jeffreys realized that this variation could be used to determine identity. Because his approach provided such definite results, it was characterized as a "fingerprint," and the technique is now known as DNA fingerprinting or DNA profiling.

The power of DNA fingerprinting lies in its ability to determine unique matches in the lengths of repetitive DNA patterns for a given person. These variations can be found in all the DNA in the human body. The DNA in one cell of an individual is identical to the DNA in all other cells of that individual, whether they be blood cells, semen cells, skin cells, or hair cells. As an example, the DNA in the semen cells that a rapist may leave behind at a crime scene can be compared with the DNA in the blood cells taken from a suspect under court order. If the suspect is the rapist, the DNA profiles in the semen and the blood will be identical, which is referred to as a DNA match. In paternity disputes, the biological father of a child can be determined by comparing the father's DNA fingerprints with the child's. In all but the rarest cases, the results of the DNA fingerprinting provide a level of certainty so high that paternity will be proved or disproved.

Methods Used in DNA Fingerprinting

The most common type of DNA fingerprinting is called restriction fragment length polymorphism (RFLP). In RFLP, DNA is extracted from individuals or from the crime scene, and technicians use chemicals called restriction enzymes to divide the DNA into fragments at specific places. The fragments are then placed in a gel and separated according to size by a laboratory technique called electrophoresis. The analysis of the length of the DNA fragments at one genetic location (polymorphic locus) is reflected on a piece of X-ray film by producing bands on the photographic film that resemble a supermarket bar code. The

difference between individuals is revealed by the unique spacing between the fifteen or twenty bands in the bar code. The power and significance of this procedure grows as more locations on the DNA strands are analyzed and matches are found on more pieces of film. Finding a pair of people with the same DNA pattern in four places (the minimum number of DNA segments that labs commonly analyze) seldom occurs, except in the case of identical twins. The band structures for six DNA segments are as unique as a person's fingerprints.

In many crime cases, the amount of DNA evidence recovered is too small to run a successful DNA test. This problem was first addressed in late 1984 by genetic chemist Kary Mullis. His idea was to find a particular part of the DNA to work with, copy it again and again, and run the DNA fingerprinting tests on the copies. This process is known as the polymerase chain reaction (PCR) since a chemical called polymerase generates the reaction that reproduces the DNA. Advancements in PCR and RFLP procedures during the late 1980's and early 1990's

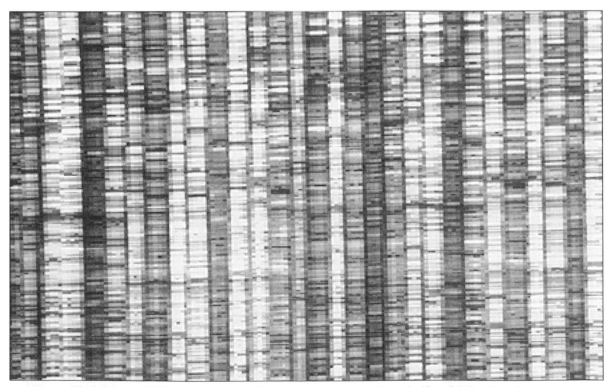

The bands on a DNA fingerprint resemble those on a supermarket bar code. (Dan McCoy/Rainbow)

have made the chances of two DNA profiles matching randomly to be infinitesimally small, even when based on the most minute of samples.

Since the totality of all the DNA involved in a crime case or a paternity case cannot be analyzed, numerical statements of the significance of a DNA match are cited. When a DNA match is introduced in court, it is almost always presented with a number known as a random-match probability, which is an estimate of the chance that a randomly selected person will share the DNA pattern present in both the defendant and the sample from the crime scene. With improvements in both the reliability and the turnaround time of PCR and RFLP results, the random-match probability is typically one in one million, one in one billion, or even less. These numbers do not account for possible mishandling of evidence, laboratory errors, or planting of evidence at a crime scene, all of which could also explain a DNA match. However, numerous studies have shown that laboratory errors are relatively rare and tend to be so easy to spot when they do happen that they need not be considered in most cases.

—Alvin K. Benson

See Also: Forensic Genetics; Gel Electrophoresis; Paternity Tests; Polymerase Chain Reaction; Restriction Enzymes.

Further Reading: *And the Blood Cried Out* (1996), Harlan Levy, gives an account of the power of DNA profiling illustrated with many case histories, including the O. J. Simpson case. *DNA Fingerprinting* (1994), Michael Krawczak, discusses medical perspectives of DNA fingerprinting. *DNA Fingerprinting: State of the Science* (1993), edited by S. D. Pena, describes PCR and RFLP procedures. *DNA Fingerprinting: An Introduction* (1992), Lorne T. Kirby, provides a good overview of the DNA fingerprinting process.

DNA Isolation

Field of study: Genetic engineering and biotechnology
Significance: *The isolation of DNA is the basis for all biotechnology and genetic engineering. Before* *it can be used for such work, DNA must be isolated from other substances such as complex carbohydrates, proteins, and RNA.*

Key terms

CHLOROFORM/ISOAMYL ALCOHOL (CIA): a mixture of two chemicals used in DNA isolation to rid the extract of the contaminating compound phenol

DEOXYRIBONUCLEIC ACID (DNA): the genetic material for most organisms, a double-stranded substance composed of units called "nucleotides"

PHENOL: a simple chemical used in DNA extraction to precipitate proteins and aid in their removal

PROTEIN: a complex substance composed of many amino acids joined by chemical bonds

RIBONUCLEIC ACID (RNA): another form of genetic material, usually a single-stranded substance consisting of units called "ribonucleotides"

DNA Discovery and Extraction

Deoxyribonucleic acid (DNA) was discovered in 1869 by the Swiss physician Friedrich Miescher, who studied white blood cells in pus obtained from a surgical clinic. Miescher found that when bandages that had been removed from the postoperative wounds of injured soldiers were washed in a saline solution, the cells on the bandages swelled into a "gelatinous mass" that consisted largely of DNA. Miescher had isolated a depolymerized form of DNA—that is, DNA not in the normal double-stranded conformation. After a series of experiments, Miescher concluded that the substance he had isolated originated in the nucleus of the blood cells; he first called the substance "nuclein" and later "nucleic acid."

The first problem when extracting DNA is "lysing," or breaking open, the cell. Bacteria, yeast, and plant cells usually have a thick cell wall protecting their inner cell membranes. This cell wall complicates the opening process. Bacteria, such as *Escherichia coli*, are the easiest of these cells to open by a process called "alkaline lysis," in which cells are treated with a solution of sodium hydroxide and detergent that degrades both the cell wall and the cell membrane. Yeast cells are often broken open

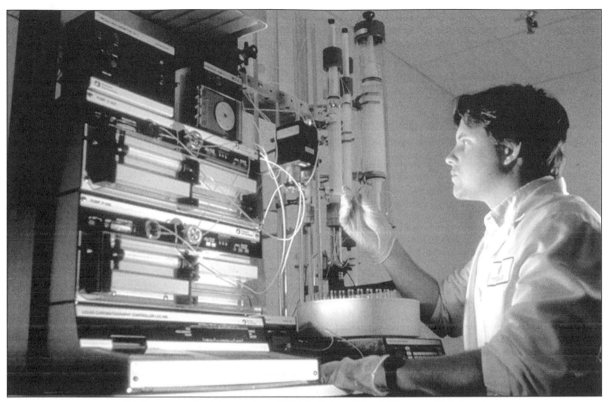

A laboratory technician performs DNA isolation. (Hank Morgan/Rainbow)

with enzymes such as lysozyme that degrade cell walls or by using a "French press," a piston in an enclosed chamber that forces cells open under high pressure. Plant tissue is usually mechanically broken into a fine cell suspension before extraction by grinding frozen tissue in a mortar and pestle. Once the suspension of cells is obtained, the tissue may be treated with a variety of enzymes to break down cell walls or with strong detergents, such as sodium lauryl sarcosine, that disrupt and dissolve both cell walls and cell membranes. Animal cells, such as white blood cells, do not have cell walls and can generally be opened by "osmotic shock," the lysing of cells by moving them from a liquid environment with a high solute concentration to an environment with a very low solute concentration.

Isolation and Purification

Although lysis methods differ between cell types, the process of DNA isolation and purification is consistent. The isolation process may be imagined as a series of steps designed to remove either naturally occurring biological contaminants from the DNA or contaminants added by the scientist during the extraction process. The biological contaminants already present in cells are proteins and ribonucleic acid (RNA); additionally, plant cells have high levels of complex carbohydrates. Contaminants intentionally added by scientists may include salts and various chemicals.

After cells are lysed, a high-speed centrifugation is performed to form large-scale, insoluble cellular debris, such as membranes and organelles, into a pellet. The liquid extract remaining still contains dissolved proteins, RNA, and DNA. If salts are not present in the extract, they are added; salt must be present later for the DNA to precipitate efficiently. Proteins must be removed from the extract since some not only degrade DNA but also inhibit enzymatic reactions with DNA. Protein is precipitated by mixing the extract with a chemical called "phenol." When phenol and the extract are mixed in a test tube, they separate into two parts like oil and water. If these fluids are centrifuged, precipitated proteins will actually collect between the two liquids at a spot called the "interphase." The liquid layer containing the dissolved DNA is then drawn up and away from the precipitated protein.

The protein-free solution still contains DNA, RNA, salts, and traces of phenol dissolved into the extract. To remove the contaminating phenol, the extract is mixed with a chloroform/ isoamyl alcohol solution (CIA). Again like oil and water, the DNA extract and CIA separate into two layers. If the two layers are mixed vigorously and separated by centrifugation, the phenol will move from the DNA extract into the CIA layer. At this point the extract—removed to a new test tube—contains RNA, DNA, and salt.

The extract is next mixed with 100 percent ethanol, inducing the DNA to precipitate out in long strands. The DNA strands may be isolated by either spooling the sticky DNA around a glass rod or by centrifugation. If spooled, the DNA is placed in a new test tube; if centrifuged, the liquid is decanted from the pellet of DNA. The precipitated DNA, with salt and RNA present, is still not pure. It is washed for a final time with 70 percent ethanol, which does not dissolve the DNA but forces salts present to go into solution. The DNA is then reisolated by spooling or centrifugation and dried to remove all traces of ethanol. At this point, only DNA and RNA are left; this mixture can be dissolved in a low-salt buffer containing the enzyme RNase, which degrades any RNA present, leaving pure DNA.

Technological advances have allowed deproteinization by the use of "spin-columns" without the employment of toxic phenol. The raw DNA extract is placed on top of a column containing a chemical matrix that binds proteins but not DNA; the column is then centrifuged in a test tube. The raw extract passes through the chemical matrix and exits protein-free into the collection tube.

—James J. Campanella

See Also: Biotechnology; Cloning Vectors; Gel Electrophoresis; Restriction Enzymes.

Further Reading: The fascinating story of Friedrich Miescher's work can be found in Al- fred Mirsky, "The Discovery of DNA," *Scientific American* (June, 1968). *Recombinant DNA* (1992), by James Watson et al., gives a concise background on cells and DNA. *Molecular Cloning: A Laboratory Manual* (1989), by Joseph Sambrook et al., is the best practical source for methods and background in DNA extraction.

DNA Repair

Field of study: Molecular genetics
Significance: *To protect the integrity of their genetic material, cells of all living things have developed mechanisms to correct damage to deoxyribonucleic acid (DNA) that may occur as a result of internal processes or external agents. Many of these mechanisms are found in organisms ranging from very simple to very complex, indicating that they evolved early in the history of life. Disruption of DNA repair mechanisms in humans has been associated with the development of cancers.*

Key terms

DEOXYRIBONUCLEIC ACID (DNA): the genetic material for most organisms, a double-stranded substance composed of units called nucleotides

NUCLEOTIDE: the basic unit of DNA, consisting of a five-carbon sugar, a base containing nitrogens, and a phosphate group

BASE: the component of a nucleotide that gives it its identity and special properties

ENZYME: a molecule, usually a protein, that assists and accelerates cellular reactions without itself being altered by the reaction

PROTEIN: a macromolecule composed of many amino acid subunits held together by chemical bonds

DNA Structure and DNA Damage

All living things are exposed every day to agents such as radiation or certain chemicals that injure their genetic material. In addition, damage to deoxyribonucleic acid (DNA) occurs as a result of spontaneous breakdown or interaction with products of cellular metabolism. While these forces also damage other cellular components, DNA is the most critical since it serves as the blueprint to direct the functions of the cell; it must be accurately maintained to do this correctly and to serve as the source of information for all daughter cells produced by cellular replication. Thus DNA is the only biological molecule that cells repair. Everything else can be replaced. There are many ways in which DNA can be damaged; the consequences of DNA damage may be a change in the meaning of a gene (a mutation), a break in the DNA molecule, or the abnormal joining of two DNA molecules. To a single-celled creature such as a bacterium, DNA damage may mean death. To a multicellular organism such as a person or an animal, damaged DNA in some of its cells may mean loss of function of organs or tissues or can lead to cancer development.

A brief overview of the structure of DNA will help in understanding how it can be damaged. DNA is assembled from nucleotides, of which there are four types defined by the base they contain. If the DNA double helix is pictured as a twisted ladder, the outside chains (sometimes referred to as the "backbone" of the DNA) are alternating units of sugar and phosphate, and the "rungs" of the ladder are bases. There are four types of bases found in DNA: the double-ring purines, adenine and guanine, and the single-ring pyrimidines, cytosine and thymine. The structure of each base allows it to pair with only one other base: adenine pairs with thymine, and cytosine pairs with guanine. Base pairing holds the two strands of the double helix together and is essential for the synthesis of new DNA molecules (DNA replication) and for the transfer of information from DNA to ribonucleic acid (RNA) in the process of transcription. DNA replication is carried out by an enzyme called DNA polymerase, which reads the information (the sequence of bases) on a single strand of DNA, brings the appropriate nucleotide to pair with the existing strand one nucleotide at a time, and joins it to the end of a growing chain. In addition to copying entire long strands of DNA every time a cell divides, DNA polymerases are also responsible for repairing short, damaged regions of DNA. Transcription occurs through a process similar to DNA replication, except that an RNA polymerase copies only a portion of one of the strands of DNA (a gene), making an RNA copy. The RNA can then direct the production of a particular protein, which is the product of the gene.

One of the most frequent forms of DNA damage is loss of a base. Purines are particularly unstable, and many are lost each day in human cells. If a base is absent, the DNA cannot be copied correctly during DNA replication because transcription and mutation or death may result. Another common type of DNA damage is a pyrimidine dimer, an abnormal linkage between two cytosines, two thymines, or a cytosine and thymine next to one another in a DNA strand. These are caused by the action of ultraviolet light on DNA. A pyrimidine dimer creates a distortion in the double helix that interferes with the processes of DNA replication and transcription. Another form of DNA damage is a break in the backbone of one or both strands of the double helix. Breaks can block DNA replication, create problems during cell division, or cause rearrangements of the chromosomes. DNA replication itself can cause problems by inserting an incorrect base or an additional or too few bases in a new strand. While DNA replication errors are not DNA damage as such, they can also lead to mutations or death and are subject to repair.

DNA Repair

DNA repair ability is found in most organisms. Even some viruses such as bacteriophages, which infect bacteria, and herpes viruses, which infect animals, are capable of repairing some damage to their genetic material. The DNA repair systems of single-celled organisms, including bacteria and yeasts, have been extensively studied for many years. In the 1980's and 1990's, techniques including the use of recombinant DNA methods revealed that DNA repair systems of larger organisms such as humans, animals, and plants are quite similar to those of microorganisms.

Scientists generally classify DNA repair systems into three categories on the basis of complexity, mechanism, and the fate of the damaged DNA. "Damage reversal" systems are the simplest: They usually require only a single enzyme to directly act on the damage and restore it to normal, usually in a single step. "Damage removal" systems are somewhat more complicated: These involve cutting out and replacing a damaged or inappropriate base or section of nucleotides and require several proteins to act together in a series of steps. "Damage tolerance" systems are those that respond to and act on damaged DNA but do not actually repair the original damage. Instead, they are ways for cells to cope with DNA damage in order to continue growth and division.

Damage Reversal Systems

Photoreactivation is one of the simplest and perhaps oldest known repair systems: It consists of a single enzyme that can split pyrimidine dimers in the presence of light. An enzyme called photolyase catalyzes this reaction; it is found in many bacteria, lower eukaryotes, insects, and plants but seems to be absent in mammals (including humans). A similar gene is present in mammals but may code for a protein that functions in another type of repair.

X rays and some chemicals such as peroxides can cause breaks in the backbone of DNA. Simple breaks in one strand are rapidly repaired by the enzyme DNA ligase. Mutant strains of microorganisms with reduced DNA ligase activity tend to have high levels of recombination since DNA ends are very reactive and since there are enzymes that promote interaction of broken DNA molecules. While recombination is important in generating genetic diversity during sexual reproduction, it can also be dangerous if DNA molecules are joined inappropriately. The result can be aberrant chromosomes that do not function properly.

Damage Removal Systems

Damage removal systems are accurate and efficient but require the action of several enzymes and are more energetically "expensive" to the cell. There are three types of damage removal systems that work in the same general way but act on different forms of DNA damage. In "base excision" repair, an enzyme called a DNA glycosylase recognizes a specific damaged or inappropriate base and cuts the base-sugar linkage to remove the base. Next, the backbone is cut by another protein that removes the baseless sugar; then a new nucleotide is in-

serted to replace the damaged one by a DNA polymerase enzyme. Finally, the break in the backbone is sealed by DNA ligase. There are a number of specific glycosylases for particular types of DNA damage caused by radiation and chemicals.

The "nucleotide excision" repair system works on DNA damage that is "bulky" and that creates a block to DNA replication and transcription, such as ultraviolet-induced pyrimidine dimers and some kinds of DNA damage created by chemicals. It probably does not recognize a specific abnormal structure but sees a distortion in the double helix. Several proteins joined in a complex scan the DNA for helix distortions. When one is found, the complex binds to the damage and creates two cuts in the DNA strand containing the damaged bases on either side of the damage. The short segment with the damaged bases (around twenty-seven nucleotides in humans) is removed from the double helix, leaving a short gap that can be filled by DNA polymerase using the intact nucleotides in the other DNA strand as a guide. In the last step, DNA ligase rejoins the strand. Mutants that are defective in nucleotide excision repair have been isolated in many organisms and are sensitive to killing and mutation induction by ultraviolet light and similar-acting chemical mutagens. Humans with the hereditary disease xeroderma pigmentosum are sunlight-sensitive and have very high risks of skin cancers on sun-exposed areas of their bodies. These individuals have defective copies of genes that code for proteins involved in nucleotide excision repair. A comparison of the genes defective in xeroderma pigmentosum patients and those involved in nucleotide excision repair in simpler organisms reveals a great deal of similarity, indicating that this repair system evolved early in the history of life.

"Mismatch" repair occurs after DNA replication as a last "spell check" on its accuracy. By comparing mutation rates in *Escherichia coli* bacteria that either have or lack mismatch repair ability, scientists have estimated that this process adds between one hundred and one thousand times more accuracy to the replication process. It is carried out by a group of proteins that can scan DNA and look for incorrectly paired bases (or unpaired bases) that will have aberrant dimensions in the double helix. The incorrect nucleotide is removed as part of a short stretch, and then the DNA polymerase gets a second try to insert the correct sequence. In 1993, Richard Fishel and his colleagues and Bert Vogelstein and his colleagues isolated the first genes for human mismatch repair proteins and showed that they are very similar to those of the bacterium *Escherichia coli* and the simple eukaryote baker's yeast. Further studies in the 1990's revealed that mismatch repair genes are defective in people with hereditary forms of colon cancer; this deficiency is the reason that cancers develop in these individuals.

Damage Tolerance Systems

Not all DNA damage is or can be removed immediately; some of it may persist for a while. If a DNA replication complex encounters DNA damage such as a pyrimidine dimer, it will normally act as a block to further replication of that DNA molecule. In eukaryotes, however, DNA replication initiates at multiple sites and may be able to resume downstream of a damage site, leaving a "gap" of single-stranded, unreplicated DNA in one of the two daughter molecules resulting from DNA replication. The daughter-strand gap is potentially just as dangerous as the original damage site, if not more so. The reason for this is that if the cell divides with a gap in a DNA molecule, there will be no way to accurately repair that gap or the damage in one of its two daughter cells. Such a cell will either acquire a mutation or die. To avoid this problem, cells have developed a way to repair daughter-strand gaps by recombination with an intact molecule of identical or similar sequence. The "recombinational" repair process, which requires a number of proteins, yields two intact daughter molecules, one of which still contains the original DNA damage. In addition to dealing with daughter-strand gaps, recombinational repair systems can also repair single- and double-strand breaks caused by the action of X rays and similar chemicals on DNA. Many of the proteins required for recombinational repair are also involved in the genetic recombination that occurs in meiosis, the sexual division process of higher cells. In 1997, it was shown that the products of the breast cancer susceptibility genes *BRCA1* and *BRCA2* participate in both recombinational repair and meiotic recombination.

An alternative choice for a DNA polymerase blocked at a DNA damage site is to change its specificity so that it can insert any nucleotide opposite the normally nonreadable damage and continue DNA replication. This type of "damage bypass" is very likely to cause a mutation, but if the cell cannot replicate its DNA, it will not be able to divide. In *Escherichia coli* bacteria, there is a set of genes that are turned on when the bacteria have received a large amount of DNA damage. Some of these gene products alter the DNA polymerase and allow damage bypass. This system has been termed the "SOS response" to indicate that it is a system of last resort. Other organisms, including humans, seem to have similar damage bypass mechanisms that allow a cell to continue growth despite DNA damage at the price of mutations. For this reason, damage bypass systems are sometimes referred to as "error-prone" or mutagenic repair systems.

Impact and Applications

DNA repair systems are an important component of the metabolism of cells. Studies in microorganisms have shown that as little as one unrepaired site of DNA damage per cell can be lethal or lead to permanent changes in the genetic material. The integrity of DNA is normally maintained by an elaborate series of interrelated checks and surveillance systems. The greatly increased risk of cancer suffered by humans with hereditary defects in DNA repair shows how important these systems are in avoiding the genetic changes that can cumulatively lead to the development of cancers. As the relationship between mutations in DNA repair genes and cancer susceptibility becomes clearer, this information will be used in directing the course of cancer therapy and possibly in providing gene therapy to individuals with cancer.

—*Beth A. Montelone*

See Also: Chemical Mutagens; DNA Replication; DNA Structure and Function; Mutation and Mutagenesis.

Further Reading: *Science* magazine declared the DNA repair enzyme "Molecule of the Year" in 1994 and published three short reviews in the December 23, 1994, issue that discuss three repair processes: "Mechanisms of DNA Excision Repair" by Aziz Sancar, "Transcription-Coupled Repair and Human Disease" by Philip C. Hanawalt, and "Mismatch Repair, Genetic Stability, and Cancer" by Paul Modrich. Robert A. Weinberg discusses cancer and the roles of DNA repair genes in "How Cancer Arises," *Scientific American* 197 (September, 1996).

DNA Replication

Field of study: Molecular genetics
Significance: *Cells and organisms pass hereditary information from generation to generation. To assure that offspring contain the same genetic information as their parents, the genetic material must be accurately reproduced. DNA replication is the molecular basis of heredity and is one of the most fundamental processes of all living cells.*

Key terms

DEOXYRIBONUCLEIC ACID (DNA): the genetic material, a two-stranded molecule consisting of a chain of nucleotide subunits

ORIGIN: the site at which replication begins on a DNA molecule

ENZYME: a molecule, usually a protein, that accelerates chemical reactions in a cell

PROTEIN: a complex molecule composed of a folded chain of amino acids

DNA Structure and Function

The importance of chromosomes in heredity has been known since early in the twentieth century. Chromosomes consist of both deoxyribonucleic acid (DNA) and protein, and in the early twentieth century there was considerable controversy concerning which component was the hereditary molecule. Early evidence favored the proteins. In 1944, however, a series of classic experiments by Oswald Avery, Maclyn McCarty, and Colin MacLeod lent strong support to the proponents favoring DNA as the genetic material. They showed that a genetic transforming agent of bacteria was DNA and

not protein. In experiments reported in 1952, Alfred Hershey and Martha Chase provided evidence that DNA was the genetic material of bacteriophages (viruses of bacteria). Combined with additional circumstantial evidence from many sources, DNA became favored as the hereditary molecule, and a heated race began to determine its molecular structure.

In 1953, James Watson and Francis Crick published a model for the atomic structure of DNA. Their model was based on known chemical properties of DNA and X-ray diffraction data obtained from Rosalind Franklin and Maurice Wilkins. The structure itself made it clear that DNA was indeed the molecule of heredity and provided evidence for how it might be copied. The molecule resembles a ladder. The "rails" are composed of repeating units of sugar and phosphate, forming a backbone for the molecule. Each "rung" consists of a pair of bases, one attached to each of the two rails and held together in the middle through weak bonds. Since there are thousands to hundreds of millions of units on a DNA molecule, the weak interactions for each pair of bases add up to a strong force to hold the two strands together. DNA, then, consists of two strands, each consisting of a repeating sugar-phosphate backbone and nitrogenous bases with the two strands held together by base-pair interactions. The two strands are oriented in opposite directions. The ends of a linear DNA molecule can be distinguished by which part of the backbone sugar is exposed and are referred to as the 5′ (five prime) end and the 3′ end, named for a particular carbon atom on the ribose sugar. If one DNA strand is oriented 5′ to 3′, its complementary partner is oriented 3′ to 5′. This organization has important implications for the mechanism of DNA synthesis.

There are four different bases: adenine (A), guanine (G), cytidine (C), and thymine (T). They can be arranged in any order on a DNA strand, allowing the enormous diversity necessary to encode the blueprint of every organism. A key feature of the double-stranded DNA molecule is that bases have strict pairing restrictions: A can only pair with T; G can only pair with C. Thus if a particular base is known on one strand, the corresponding base is automatically known on the other. Each strand can serve as a template, or mold, dictating the precise sequence of bases on the other. This feature is fundamental to the process of DNA replication.

The genome (the complete DNA content of an organism) stores all the genetic infor-mation that determines the features of that organism. The features are expressed when the DNA is transcribed to a messenger mole-cule, ribonucleic acid (RNA), then into pro-tein. The proteins encoded by the organism's genes in its DNA carry out all of the activities of the cell.

The Replication Process

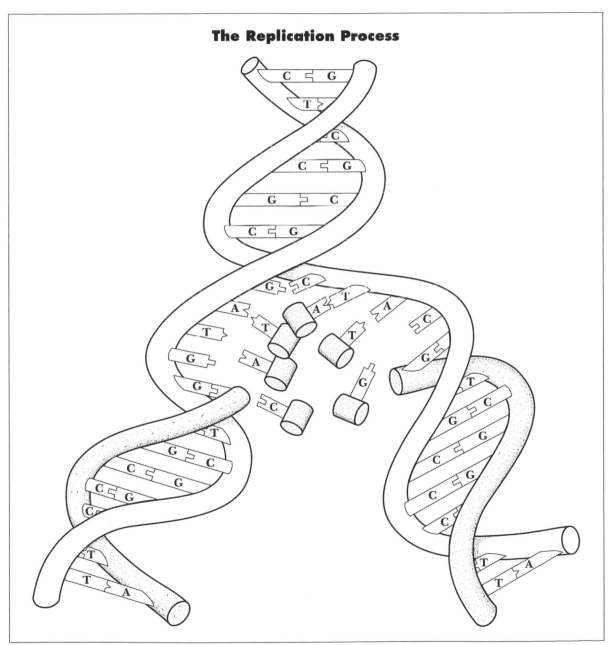

In DNA replication, a double-standard parent helix splits apart and reassembles into two identical daughter helixes. The amino acid base pairs are reproduced exactly.

The Cell Cycle

In eukaryotic organisms (most organisms other than bacteria), cells progress through a series of four stages between cell divisions. The stages begin with a period of growth (G1 phase), followed by replication of the DNA (S phase). A second period of growth (G2 phase) is followed by division of the cell (M phase). The two cells resulting from the cell division each go through their own cell cycle or may enter a dormant stage (G0 phase). The passage from one stage to the next is tightly regulated and directed by internal and external signals to the cell.

The transition from G1 into S phase marks the beginning of DNA replication. In order to enter S phase, the cell must pass through a checkpoint or restriction point in which the cell determines the quality of its DNA: If there is any damage to the DNA, entry into S phase will be delayed. This prevents the potentially lethal process of beginning replication on a DNA molecule that has damage that would prevent completion of replication. If conditions are determined to be acceptable, a "molecular switch" is thrown, triggering the initiation of DNA replication. What is the nature of this molecular switch? There are many differ-

ent proteins that participate in the process of DNA replication, and they can have their activity turned off and on by other proteins. Addition or removal of a chemical group called a phosphate is a common mechanism of chemical switching. This reaction is catalyzed by a class of enzymes called kinases. Certain key proteins are phosphorylated at the boundary of the G1 and S phases of the cell cycle by kinases, switching on DNA replication.

Origins and Initiation

If the human genome were replicated from one end to the other, it would take several years to complete the process. The DNA molecule is simply too large to be copied end to end. Instead, replication is initiated at many different sites called origins of replication, and DNA synthesis proceeds from each site in both directions until regions of copied DNA merge. The region of DNA copied from a particular origin is called a replicon. Using hundreds to tens of thousands of initiation sites and replicons, the genome can be copied in a matter of hours. The structure of replication origins has been difficult to identify in all but a few organisms, most notably yeast. Origins consist of several hundred base pairs of DNA comprising sequences that attract and bind a set of proteins called the origin recognition complex (ORC). The exact mechanism by which the origin is activated is still under investigation, but a favored model is supported by all of the available evidence.

The ORC proteins are believed to be bound to the origin DNA throughout the cell cycle but become activated at the G1/S boundary through the action of kinases. Kinases add phosphate groups to one or more of the six ORC proteins, activating them to initiate DNA replication. Different replicons are initiated at different times throughout S phase. It is unclear how the proposed regulatory system distinguishes between replicons that have been replicated in a particular S phase and those that have not, since each must be used once and only once during each cell division cycle.

A number of different enzymatic activities are required for the initiation process. The two strands of DNA must be unwound or separated,

Life Cycle of a Cell

G₂ phase (growth)

Mitosis

Prophase

Metaphase

Anaphase

Telophase

Interphase

S phase (DNA synthesis)

G₁ phase (growth)

exposing each of the parent strands so they can be used as templates for the synthesis of new, complementary strands. This unwinding is mediated by an enzyme called a helicase. Once unwound, the single strands are stabilized by the binding of proteins called single-strand binding proteins (SSB). The resulting structure resembles a "bubble" or "eye" in the DNA strand. This structure is recognized by the DNA synthesis machinery that is recruited to the site, and DNA synthesis begins. As replication proceeds, the DNA continues to unwind through the action of the helicase. The site at which unwinding and DNA synthesis are occurring is at either end of the expanding eye or bubble, called a replication fork.

DNA Synthesis

The DNA synthesis machinery is not able to synthesize a strand of DNA from scratch; rather, a short stretch of RNA is used to begin the new strands. The synthesis of the RNA is catalyzed by an enzyme called primase. This short piece of RNA, or primer, is extended using DNA nucleotides by the enzymes of DNA synthesis, called DNA polymerases. The RNA primer is later removed and replaced by DNA. Nucleotide monomers align with the exposed template DNA strand one at a time and are joined by the DNA polymerase. The joining of nucleotides into a growing DNA chain requires energy. This energy is supplied by the nucleotide monomers themselves. A high-energy phosphate bond in the nucleotide is split, and the breakage of this high-energy bond provides the energy to drive the polymerase reaction.

The two strands of DNA are not synthesized in the same way. The two strands are oriented opposite one another, but DNA synthesis only occurs in one direction: 5′ to 3′. Therefore, one strand, called the leading strand, is synthesized continuously in the same direction that the replication fork is moving, while the lagging strand is synthesized away from the direction of fork movement. Since the lagging-strand DNA synthesis and fork movement are in opposite directions, this strand of DNA must be made in short pieces that are later joined. Lagging-strand synthesis is therefore said to be discontinuous. These short intermediates are called

Okazaki fragments, named for their discoverer, Reiji Okazaki. Overall, DNA synthesis is said to be semidiscontinuous.

The DNA synthesis machine operating at the replication fork is a complex assembly of proteins. Many different activities are necessary to efficiently carry out the process of DNA replication. Several proteins are necessary to recognize the unwound origin and assemble the rest of the complex. Primase must function to begin both new strands and is then required periodically throughout synthesis of the lagging strand. A doughnut-shaped clamp called PCNA functions as a "processivity factor" to keep the entire complex attached to the DNA until the job is completed. Helicase is continuously required to unwind the template DNA and move the fork along the parent molecule. As the DNA is unwound, strain is created on the DNA ahead of the replication fork. This strain is alleviated through the action of topoisomerase enzymes. Single-strand binding proteins are needed to stabilize the regions of unwound DNA that exist before the DNA is actually copied. Finally, an enzyme called ligase is necessary to join the regions replicated from different origins and to attach all of the Okazaki fragments of the lagging strand. All of these proteins are part of a well-orchestrated, efficient machine ideally suited to its task of copying the genetic material.

DNA polymerases are not perfect. At a relatively low frequency, they can add an incorrect nucleotide to a growing chain, one that does not match the template strand as dictated by the base-pairing rules. However, because the DNA molecules are so extremely large, novel mechanisms for proofreading have evolved to ensure that the genetic material is copied accurately. DNA polymerases can detect the misincorporation of a nucleotide and use an additional enzymatic activity to correct the mistake. Specifically, the polymerase can "back up" and cut out the last nucleotide added, then try again. With this and other mechanisms to correct errors, the observed error rate for DNA synthesis is a remarkable one error in every billion nucleotides added.

Impact and Applications

DNA replication is a fundamental cellular process: Proper cell growth cannot occur without it. It must be carefully regulated and tightly controlled. Despite its basic importance, the details of the mechanisms that regulate DNA replication are poorly understood. Even with all of the checks and balances that have evolved to ensure a properly replicated genome, occasional mistakes do occur. Attempting to replicate a genome damaged by chemical or other means may simply lead to death of a single cell. Far more ominous are genetic errors that lead to loss of regulating mechanisms. Without regulation, cell growth and division can proceed without normal limits, resulting in cancer. Much of the focus for the study of cell growth and regulation is to set a foundation for the understanding of how cancer cells develop. This knowledge may lead to new techniques for selective inhibition or destruction of cancer cells.

Manipulation of DNA replication and cell cycle control are the newest tools for progress in genetic engineering. In early 1997, the first successful cloning of an adult mammal, Dolly the sheep, raised important new issues about the biology and ethics of manipulating mammalian genomes. The technology now exists to clone human beings, although such experiments are not likely to be carried out. More relevant is the potential impact on agriculture. It is now possible to select for animals that have the most desirable traits, such as lower fat content or disease resistance, and create herds of genetically identical animals. Of direct relevance to humans is the potential impact on the understanding of fertility and possible new treatments for infertility.

A new class of genetic diseases was discovered in the 1980's called triplet repeat diseases. Regions of DNA consist of copies of three nucleotides (such as CGG) that are repeated up to fifty times. Through unknown mechanisms related to DNA replication, the number of repeats may increase from generation to generation, at some point reaching a threshold level at which disease symptoms appear. Diseases found to conform to this pattern include fragile X syndrome, Huntington's chorea, and Duchenne muscular dystrophy.

The process of aging is closely related to DNA replication. Unlike bacteria, eukaryotic organisms have linear chromosomes. This poses problems for the cell, both in maintaining intact chromosomes (ends are unstable) and in replicating the DNA. The replication machinery cannot copy the extreme ends of a linear DNA molecule, so organisms have evolved alternate mechanisms. The ends of linear chromosomes consist of telomeres (short, repeated DNA sequences that are bound and stabilized by specific proteins), which are replicated by a separate mechanism using an enzyme called telomerase. Telomerase is inactivated in mature cells, and there may be a slow, progressive loss of the telomeres that ultimately leads to the loss of important genes, resulting in symptoms of aging. Cancer cells appear to have reactivated their telomerase, so potential anticancer therapies are being developed based on this information.

—Michael R. Lentz

See Also: Cancer; DNA Structure and Function; *Escherichia coli*; Molecular Genetics; Mutation and Mutagenesis.

Further Reading: Arthur Kornberg discovered the enzymes that replicate DNA. His autobiography, *For the Love of Enzymes: The Odyssey of a Biochemist* (1989) is a rich history of the process of science and discovery. The same can be said for *The Double Helix* (1968), an account by James Watson of the race to solve the structure of the DNA molecule. *Understanding DNA and Gene Cloning: A Guide for the Curious* (1997) by Karl Drlica is an excellent introduction to the basic properties of DNA and its modern applications.

DNA Structure and Function

Field of study: Molecular genetics
Significance: *Structurally, DNA is a relatively simple molecule; functionally, however, it has wide-ranging effects on the cell. It functions primarily as a stable repository of genetic information in the cell and as a source of genetic information for the production of proteins. Greater knowledge of the*

characteristics of DNA has led to advances in the fields of genetic engineering, gene therapy, and molecular biology in general.

Key terms

ENZYME: a molecule, usually a protein, used by cells to facilitate and speed up a chemical reaction

GENE: a relatively short portion of deoxyribonucleic acid (DNA) that contains genetic information, usually to produce a protein

GENE EXPRESSION: the processes (transcription and translation) by which the genetic information in DNA is converted into protein

NUCLEIC ACIDS: molecules, such DNA and ribonucleic acid (RNA), that are involved in the genetic processes of the cell

PROTEIN: a molecule composed of amino acids linked in a long chain

TRANSCRIPTION: the process by which genetic information in DNA is converted into messenger RNA (mRNA)

TRANSLATION: the process by which the genetic information in a messenger RNA molecule is converted into protein

Chemical and Physical Structure of DNA

Deoxyribonucleic acid (DNA) is the genetic material found in all cells. Chemically, it is classified as a nucleic acid, a relatively simple molecule composed of nucleotides. A nucleotide consists of a sugar (deoxyribose), a phosphate group, and one of the nitrogenous bases: adenine (A), cytosine (C), guanine (G), or thymine (T). In fact, nucleotides differ only in the particular nitrogenous base that they contain. Ribonucleic acid (RNA) is the other type of nucleic acid found in the cell; however, it contains ribose as its sugar instead of deoxyribose and has the nitrogenous base uracil (U) instead of thymine. Nucleotides can be assembled into long chains of nucleic acid via connections between the sugar on one nucleotide and the phosphate group on the next, thereby creating a sugar-phosphate "backbone" in the molecule. The nitrogenous base on each nu-

The Four Nucleotides That Compose DNA

cleotide is positioned such that it is perpendicular to the backbone, as shown in the following diagram:

```
sugar — phosphate — sugar — phosphate — sugar — phosphate — sugar
  |                   |                   |                   |
base                base                base                base
```

Any one of the four DNA nucleotides (A, C, G, or T) can be used at any position in the molecule; it is therefore the specific sequence of nucleotides in a DNA molecule that makes it unique and able to carry genetic information. The genetic information is the sequence itself.

In the cell, DNA exists as a double-stranded molecule; this means that it consists of two chains of nucleotides side by side. The double-stranded form of DNA can most easily be visualized as a ladder, with the sugar-phosphate backbones being the sides of the ladder and the nitrogenous bases being the rungs of the ladder, as shown in the following diagram:

```
sugar — phosphate — sugar — phosphate — sugar — phosphate — sugar
  |                   |                   |                   |
base                base                base                base
  |                   |                   |                   |
base                base                base                base
  |                   |                   |                   |
sugar — phosphate — sugar — phosphate — sugar — phosphate — sugar
```

This ladder is then twisted into a spiral shape. Any spiral-shaped molecule is called a "helix," and since each strand of DNA is wound into a spiral, the complete DNA molecule is often called a "double helix." This molecule is extremely flexible and can be compacted to a great degree, thus allowing the cell to contain large amounts of genetic material.

The Discovery of DNA as the Genetic Material

Nucleic acids were discovered in 1869 by the physician Friedrich Miescher. He isolated these molecules, which he called "nuclein," from the nuclei of white blood cells. This was the first association of nucleic acids with the nucleus of the cell. In the 1920's, experiments performed by other scientists showed that DNA could be located on the chromosomes within the nucleus. This was strong evidence for the role of DNA in heredity, since at that time there was already a link between the activities of chromo-somes during cell division and the inheritance of particular traits, largely because of the work of the geneticist Thomas Hunt Morgan about ten years earlier.

However, it was not immediately apparent, based on this evidence alone, that DNA was the genetic material. In addition to DNA, proteins are present in the nucleus of the cell and are an integral part of chromosomes as well. Proteins are also much more complex molecules than nucleic acids, having a greater number of building blocks; there are twenty amino acids that can be used to build proteins, as opposed to only four nucleotides for DNA. Moreover, proteins tend to be much more complex than DNA in terms of their three-dimensional structure as well. Therefore, it was not at all clear in the minds of many scientists of the time that DNA had to be the genetic material, since proteins could not specifically be ruled out.

In 1928, the microbiologist Frederick Griffith supplied some of the first evidence that eventually led to the identification of DNA as the genetic material. Griffith's research involved the bacterium *Streptococcus pneumoniae*, a common cause of lung infections. He was working primarily with two different strains of this bacterium: a strain that was highly virulent (able to cause disease) and a strain that was nonvirulent (not able to cause disease). Griffith noticed that if he heat-killed the virulent strain and then mixed its cellular debris with the living, nonvirulent strain, the nonvirulent strain would be "transformed" into a virulent strain. He did not know what part of the heat-killed virulent cells was responsible for the transformation, so he simply called it the "transforming factor" to denote its activity in his experiment. Unfortunately, Griffith never took the next step necessary to reveal the molecular identity of this transforming factor.

That critical step was taken by another microbiologist, Oswald Avery, and his colleagues in 1944. Avery essentially repeated Griffith's experiments with two important differences: Avery partially purified the heat-killed virulent strain preparation and selectively treated this preparation with a variety of enzymes to see if the transforming factor could be eliminated, thereby eliminating the transformation itself.

Avery showed that transformation was prevented only when the preparation was treated with deoxyribonuclease, an enzyme that specifically attacks and destroys DNA. Other enzymes that specifically destroy RNA or proteins could not prevent transformation from occurring. This was extremely strong evidence that the genetic material was DNA.

Experiments performed in 1952 by molecular biologists Alfred Hershey and Martha Chase using the bacterial virus *T2* finally demonstrated conclusively that DNA was indeed the genetic material. Hershey and Chase studied how *T2* infects bacterial cells to determine what part of the virus, DNA or protein, was responsible for causing the infection, thinking that whatever molecule directed the infection would have to be the genetic material of the virus. They found that DNA did directly participate in infection of the cells by entering them, while the protein molecules of the viruses stayed outside the cells. Most strikingly, they found that the original DNA of the "parent" viruses showed up in the "offspring" viruses produced by the infection, directly demonstrating inheritance of DNA from one generation to another. This was an important element of the argument for DNA as the genetic material.

The Watson-Crick Double-Helix Model of DNA

With DNA conclusively identified as the genetic material, the next step was to determine the structure of the molecule. This was finally accomplished when the double-helix model of DNA was proposed by molecular biologists James Watson and Francis Crick in 1953. This model has a number of well-defined and experimentally determined characteristics. For example, the diameter of the molecule, from one sugar-phosphate backbone to the other, is 20 angstroms. (As a reference, there are 10 million angstroms in one millimeter, which itself is one-thousandth of a meter.) There are 3.4 angstroms from one nucleotide to the next, and the entire double helix makes one turn for every ten nucleotides, a distance of about 34 angstroms. These measurements were determined by the physicists Maurice Wilkins and Rosalind Franklin around 1951 using a process called "X-ray diffraction," in which crystals of DNA are bombarded with X rays; the resulting patterns captured on film gave Wilkins and Franklin, and later Watson and Crick, important clues about the physical structure of DNA.

Another important aspect of Watson and Crick's double-helix model is the interaction between the nitrogenous bases in the interior of the molecule. Important information about the nature of this interaction was provided by molecular biologist Erwin Chargaff in 1950. Chargaff studied the amounts of each nitrogenous base present in double-stranded DNA from organisms as diverse as bacteria and humans. He found that no matter what the source of the DNA, the amount of adenine it contains is always roughly equal to the amount of thymine; there are also equal amounts of guanine and cytosine in DNA. This information led Watson and Crick to propose an interaction, or "base pairing," between these sets of bases so that A always base pairs with T (and vice versa), and G always base pairs with C. Another name for this phenomenon is "complementary base pairing": A is said to be the "complement" of T, and so on.

The force that holds complementary bases, and therefore the two strands of DNA, together is a weak chemical interaction called a "hydrogen bond," which is created whenever a hydrogen atom in one molecule has an affinity for nitrogen or oxygen atoms in another molecule. The affinity of the atoms for each other draws the molecules together in the hydrogen bond. A-T pairs have two hydrogen bonds between them because of the chemical structure of the bases, whereas G-C pairs are connected by three hydrogen bonds, making them slightly stronger and more stable than A-T pairs. The entire DNA double helix, although it is founded upon the hydrogen bond, one of the weakest bonds in nature, is nonetheless an extraordinarily stable structure because of the combined force of the millions of hydrogen bonds holding most DNA molecules together. However, these hydrogen bonds can be broken under certain conditions in the cell. This usually occurs as part of the process of the replication of the double helix, in which the two strands of DNA must come apart in order to be duplicated. In the cell, the hydrogen bonds are broken with the help of enzymes. Under artificial conditions in the laboratory, hydrogen bonds in the double helix can easily be broken just by heating a solution of DNA to high temperatures (close to the boiling point).

Other Features of the Watson-Crick Model

Watson and Crick were careful to point out that their double-helix model of DNA was the first model to immediately suggest a mechanism by which the molecule could be duplicated. They knew that this duplication, which must occur before the cell can divide, would be a necessary characteristic of the genetic material of the cell and that an adequate model of DNA must help explain how this duplication could occur. Watson and Crick realized that the mechanism of complementary base pairing that was an integral part of their model was a potential answer to this problem. If the double helix is separated into its component single-strand molecules, each strand will be able to direct the replacement of the opposite, or complementary, strand by base pairing properly with only the correct nucleotides. For example, if a single-strand DNA molecule has the sequence TTAGTCA, the opposite complementary strand will always be AATCAGT; it is as if the correct double-stranded structure is "built in" to each single strand. Additionally, as each of the single strands in a double-strand DNA molecule goes through this addition of complementary nucleotides, two new DNA double helices are produced where there was only one before. Further, these new DNA molecules are completely identical to each other, barring any mistakes that might have been made in the replication process.

A strand of DNA also has a certain direction

James Watson (left) and Francis Crick pose with a model of the double-helical structure of DNA. (Archive Photos)

built into it; the DNA double helix is often called "antiparallel" in reference to this aspect of its structure. Antiparallel means that although the two strands of the DNA molecule are essentially side by side, they are oriented in different directions relative to the position of the deoxyribose molecules on the backbone of the molecule. To help keep track of the orien-

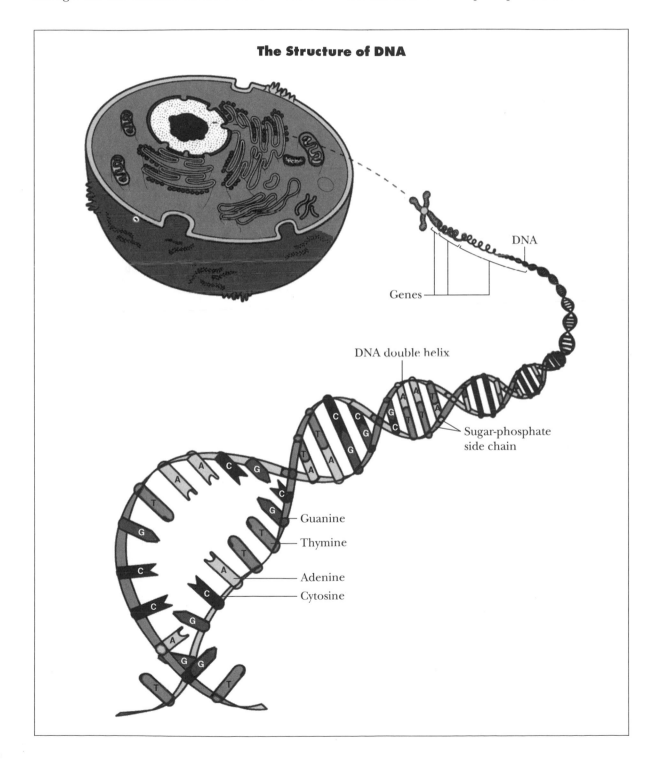

The Structure of DNA

DNA

Genes

DNA double helix

Sugar-phosphate side chain

Guanine

Thymine

Adenine

Cytosine

tation of the DNA molecule, scientists often refer to a "5 prime to 3 prime" direction. This designation comes from numbering the carbon atoms on the deoxyribose molecule (from "1 prime" to "5 prime") and takes note of the fact that the deoxyribose molecules on the DNA strand are all oriented in the same direction in a head-to-tail fashion. If it were possible to stand on a DNA molecule and walk down one of the sugar-phosphate backbones, one would encounter a 5 prime carbon atom on a sugar, then the 3 prime carbon, and so on all the way down the backbone. If one were walking on the other strand, the 3 prime carbon atom would always be encountered before the 5 prime carbon. The concept of an antiparallel double helix has important implications for the ways that DNA is produced and used in the cell. Generally, the cellular enzymes that are involved in processes concerning DNA are restricted to recognizing just one direction. For example, DNA polymerase, the enzyme that is responsible for making DNA in the cell, can only make DNA in a 5 prime to 3 prime direction, never the reverse.

Watson and Crick postulated a right-handed helix as part of their double-helix model; this means that the strands of DNA turn to the right, or in a counterclockwise fashion. This is now regarded as the "biological" (B) form of DNA because it is the form present inside the nucleus of the cell and in solutions of DNA. It is not the only possible form of DNA, however. In 1979, an additional form of DNA was discovered by molecular biologist Alexander Rich that exhibited a zigzag, left-handed double helix; he called this form of DNA "Z-DNA." Stretches of alternating G and C nucleotides most commonly give rise to this conformation of DNA, and scientists think that this alternative form of the double helix is important for certain processes in the cell in which various molecules bind to the double helix and affect its function.

The Function of DNA in the Cell

There are two major roles that DNA plays in the cell. The first is to serve as a storehouse of the cell's genetic information. Normally, cells have only one complete copy of their DNA molecules, and this copy is, accordingly, highly protected. DNA is a chemically stable molecule; it resists damage or destruction under normal conditions, and, if it is damaged, the cell has a variety of mechanisms to ensure the molecule is rapidly repaired. Furthermore, when the DNA in the cell is duplicated in a process called "DNA replication," this duplication occurs in a regulated and precise fashion so that a perfect copy of DNA is produced. Once the genetic material of the cell has been completely duplicated, the cell is ready to divide in two in a process called "mitosis." After cell division, each new cell of the pair will have a perfect copy of the genetic material; thus these cells will be genetically identical to each other. DNA thus provides a mechanism by which genetic information can be transferred easily from one generation of cells (or organisms) to another.

The second role of DNA is to serve as a blueprint for the ultimate production of proteins in the cell. This process occurs in two steps. The first step is the conversion of the genetic information in a small portion of the DNA molecule, called a gene, into messenger RNA (mRNA). This process is called "transcription," and here the primary role of the DNA molecule is to serve as a template for synthesis of the mRNA molecule. The second step, translation, does not involve DNA directly; rather, the mRNA produced during transcription is in turn used as genetic information to produce a molecule of protein. However, it is important to note that genetic information originally present in the DNA molecule indirectly guides the synthesis and final amino acid sequence of the finished protein. Both of these steps, transcription and translation, are often called "gene expression." A single DNA molecule in the form of a chromosome may contain thousands of different genes, each providing the information necessary to produce a particular protein. Each one of these proteins will then fulfill a particular function inside or outside the cell.

Impact and Applications

Knowledge of the physical and chemical structure of DNA and its function in the cell has undoubtedly had far-reaching effects on the science of biology. However, one of the biggest effects has been the creation of a new scientific discipline: molecular biology. With the advent of Watson and Crick's double-helix model of DNA, it became clear to many scientists that, perhaps for the first time, many of the important molecules in the cell could be studied in detail and that the structure and function of these molecules could also be elucidated. Within fifteen years of Watson and Crick's discovery, a number of basic genetic processes in the cell had been either partially or completely detailed, including DNA replication, transcription, and translation. Certainly the seeds of this revolution in biology were being planted in the decades before Watson and Crick's 1953 model, but it was the double helix that allowed scientists to investigate the important issues of genetics on the cellular and molecular levels.

An increased understanding of the role DNA plays in the cell has also provided scientists with tools and techniques for changing some of the genetic characteristics of cells. This is demonstrated by the rapidly expanding field of genetic engineering, in which scientists can precisely manipulate DNA and cells on the molecular level to achieve a desired result. Additionally, more complete knowledge of how the cell uses DNA has opened windows of understanding into abnormal cellular processes such as cancer, which is fundamentally a defect involving the cell's genetic information or the expression of that information.

Through the tools of molecular genetics, many scientists hope to be able to correct almost any genetic defect that a cell or an organism might have, including cancer or inherited genetic defects. The area of molecular biology that is concerned with using DNA as a way to correct cellular defects is called "gene therapy." This is commonly done by inserting a normal copy of a gene into cells that have a defective copy of the same gene in the hope that the normal copy will take over and eliminate the effects of the defective gene. It is hoped that this sort of technology will eventually be used to overcome even complex problems such as Alzheimer's disease and acquired immunodeficiency syndrome (AIDS).

One of the most unusual and potentially rewarding applications of DNA structure was introduced by computer scientist Leonard Adleman in 1994. Adleman devised a way to use short pieces of single-stranded DNA in solution as a rudimentary "computer" to solve a relatively complicated mathematical problem. By devising a code in which each piece of DNA stood for a specific variable in his problem and then allowing these single-stranded DNA pieces to base pair with each other randomly in solution, Adleman obtained an answer to his problem in a short amount of time. Soon thereafter, other computer scientists and molecular biologists began to experiment with other applications of this fledgling technology, which represents an exciting synthesis of two formerly separate disciplines. It may be that this research will prove to be the seed of another biological revolution with DNA at its center.

—Randall K. Harris

See Also: Biotechnology; Central Dogma of Molecular Biology; Chromosome Structure; DNA Replication; Genetic Engineering: Medical Applications; Molecular Genetics; RNA Structure and Function.

Further Reading: *The Transforming Principle* (1985), by Maclyn McCarty, gives an insider's view of the circumstances surrounding Oswald Avery's pivotal experiments. James D. Watson, *The Double Helix* (1968), gives another insider's view of the fascinating race to find the structure of DNA. Watson is also one of the authors of *Recombinant DNA* (1992), a detailed and readable guide to DNA and its use in genetic engineering. For a more general discussion of DNA and its basic structure and function, consult *Unraveling DNA* (1993), by Maxim D. Frank-Kamenetskii. *DNA for Beginners* (1983), by Israel Rosenfield et al., provides an entertaining, yet factual, cartoon account of basic DNA structure and function as well as more advanced topics in molecular biology.

Down Syndrome

Field of study: Human genetics
Significance: *Down syndrome is one of the most studied genetic diseases. The discovery that the*

syndrome is usually caused by the presence of an extra chromosome was a landmark in the understanding of the causes of genetic defects.

Key terms

ANEUPLOID: a cell or individual with one or more missing or extra chromosomes

MEIOSIS: the process of cell division that produces cells that contain half the number of chromosomes as the original cell

NONDISJUNCTION: the failure of homologous chromosomes to separate correctly during cell division

TRISOMY: the condition of having an extra chromosome present

Discovery and Cause

Down syndrome is one of the most common chromosomal defects in human beings. According to some studies, it occurs in one in seven hundred live births; other studies place the number at one in nine hundred. Further, it occurs in about one in every two hundred conceptions. This syndrome (a pattern of characteristic abnormalities) was first described in 1866 by the English physician John Langdon Down. While in charge of an institution housing the profoundly mentally retarded, he noticed that almost one in ten of his patients had a flat face and slanted eyes causing Down to use the term "mongolism" to describe the syndrome; this term, however, is misleading. Males and females of every race can and do have this syndrome. To remove the unintentional racist implications of the term "mongolism," Lionel Penrose and his colleagues changed the name to Down syndrome.

Although Down syndrome was observed and reported in the 1860's, it was almost one hundred years before the cause was discovered. In 1959, the French physician Jérôme Lejeune and his associates realized that the presence of an extra chromosome 21 was the apparent cause of this condition. This fact places Down syndrome in the broader category of aneuploid conditions. All human cells have forty-six chromosomes or strands made up of the chemical called deoxyribonucleic acid (DNA). The sections or subdivisions along these forty-six strands, called genes, are responsible for giving humans all their chemical and physical traits,

The Cause of Down Syndrome

Down syndrome, or trisomy 21, is caused by the presence of an abnormal third chromosome in pair 21.

as well as contributing to their emotional and personality traits. An aneuploid is a cell with forty-five or forty-seven or more chromosomes, with the missing or extra strands of DNA leaving the individual with too few or too many genes. This aneuploid condition then results in significant alterations in one's traits and a great number of potential abnormalities.

In a normal individual, the forty-six strands are actually twenty-three pairs of chromosomes that are referred to as homologous because each pair is the same size and contains the same genes. In most cases of Down syndrome, there are three copies of chromosome 21. An aneuploid with three of a particular chromosome is called trisomic; thus Down syndrome is often called trisomy 21. The extra chromosome is gained because either the egg or sperm that came together at fertilization contained an extra one. This error in gamete (egg or sperm) production is called nondisjunction and occurs during the process of cell division called meiosis. During correctly occurring meiosis, the homologous chromosome pairs are separated from each other, forming gametes with twenty-three chromosomes, one from each pair. If nondisjunction occurs, a pair fails to separate, producing a gamete with twenty-two or twenty-four chromosomes. If the pair that has failed to separate is chromosome 21, then the potential exists for twenty-three chromosomes in a normal gamete to combine with that twenty-four, creating a trisomic individual with forty-seven chromosomes.

Symptoms of Down Syndrome

The slanted appearance of the eyes first reported by Down is caused by a prominent fold of skin called an epicanthic fold (a fold in the upper eyelid near the corner of the eye). This fold of skin is accompanied by excess skin on the back of the neck and abnormal creases in the skin of the palm. In addition, the skull is wide with a flat back and a flat face. The hair on the skull is sparse and straight. The rather benign physical abnormalities are minor compared to the defects in internal organ systems. Almost 40 percent of Down syndrome patients suffer from serious heart defects. They are very prone to cancer of the white blood cells (acute leukemia), the formation of cataracts, and serious recurring respiratory infections. Short of stature with poorly formed joints, they often have poor reflexes, weak muscle tone, and an unstable gait. The furrowed, protruding tongue that often holds the mouth partially open is an external sign of the serious internal digestive blockages frequently present. These blockages must often be surgically repaired before the individual's first birthday. Many suffer from major kidney defects that are often irreparable. Furthermore, a suppressed immune system can easily lead to death from an infectious disease such as influenza or pneumonia.

With all these potential physical problems, it is not surprising that nearly 50 percent of Down syndrome patients die before the age of one. For those who live, there are enormous physical, behavioral, and mental challenges. The mental retardation that always accompanies Down syndrome ranges from quite mild to profound. This mental retardation makes all learning difficult and speech acquisition in particular very slow. Yet most Down syndrome individuals have warm, loving personalities and enjoy art and music very much.

Modern Understanding of Down Syndrome

Although this syndrome was recognized by Down in 1866, true understanding of it dates from the work that Lejeune began in 1953. The seemingly innocuous characteristic of abnormal palm prints and fingerprints fostered an important insight for him. Since those prints are laid down very early in the child's prenatal development, they suggest a profoundly altered embryological course of events. His intuition told him that not one or two altered genes but rather a whole chromosome's genes must be at fault. In 1957, he discovered, by the culturing of cells from Down syndrome children in dishes in the laboratory, that those cells contained forty-seven chromosomes. This work

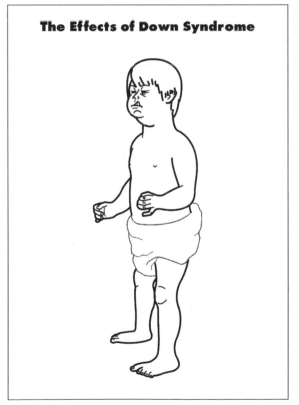

The Effects of Down Syndrome

Down syndrome is characterized not only by impaired mental ability but also by a complex of physical traits that may include short stature, stubby fingers and toes, protruding tongue, a single transverse palm crease, slanting of the eyes, small nose and ears, abnormal finger orientation, congenital heart defects, and other defects that vary from individual to individual.

eventually resulted in his 1959 publication, which was soon followed by the discovery that the extra chromosome present was a third copy of chromosome 21.

The modern development of more sophisticated methods of identifying individual parts of chromosomes has shed much light on the possible mechanisms by which the symptoms are caused. Some affected individuals do not have a whole extra chromosome 21; rather, they possess a third copy of some part of that chromosome. A very tiny strand of DNA, chromosome 21 contains only about fifteen hundred genes. Of these fifteen hundred, only a few hundred are consistently present in those who suffer from Down syndrome, namely the genes in the bottom one-third of the chromosome. Among those genes are several that could very likely cause certain of the symptoms associated with Down syndrome. A leukemia-causing gene and a gene for a protein in the lens of the eye that could trigger cataract formation have both been identified. A gene for the production of the chemicals called purines has been located. The overabundance of purines produced when three copies of this gene are present has been linked to the mental retardation usually seen. Even the fact that Down syndrome individuals have a greatly reduced life expectancy is validated by the presence of an extra gene for the enzyme superoxide dismutase, which seems involved in the normal aging process. Like Alzheimer's disease patients, Down syndrome patients who live past forty years of age have gummy tangles of protein strands called amyloid fibers in their brains. Since one form of inherited Alzheimer's is caused by a gene on chromosome 21, scientists continue to search for links between the impaired mental functioning characteristic of both diseases.

Other modern research has shed light on the long-recognized relationship between the age of the mother and an increased risk of having a Down syndrome child. Using more and more elaborate methods of chromosome banding, geneticists can identify if the extra chromosome 21 came from the mother or the father. In 94 percent of children, the egg brings the extra chromosome. Since the first steps of meiosis to produce her future eggs occur before the mother's own birth, the older the mother, the longer these egg cells have been exposed to potentially harmful chemicals or radiation. On the other hand, paternal age is not a factor because all the steps of meiosis in males occur in cells produced in the few weeks before conception. The continued study of the age factor as well as the symptoms, causes, and ramifications of this already well-studied genetic problem will hopefully lead to a greater understanding for all those affected by Down syndrome.

—*Grace D. Matzen*

See Also: Genetic Counseling; Hereditary Diseases; Meiosis; Nondisjunction and Aneuploidy.

Further Reading: *Down Syndrome: The Facts* (1997), by Mark Selikowitz, presents 192 pages of useful information. "The Causes of Down Syndrome" are well explained in D. Patterson, *Scientific American* 257 (1987). Peter Beighton provides interesting information about John Langdon Down in *The Person Behind the Syndrome* (1997). Helpful advice about the capabilities of affected people is provided by Richard Newton in *The Down's Syndrome Handbook: A Practical Guide for Parents and Carers* (1997).

Drosophila melanogaster

Field of study: Classical transmission genetics

Significance: Drosophila melanogaster *is the scientific name for a species of fruit fly whose study led scientists to discover many of the fundamental principles of the inheritance of traits. The first genetic map that assigned genes to specific chromosomes was developed for* Drosophila. *With advances in molecular technology, continued study of* Drosophila *revealed answers to some essential mysteries of life, such as the role of genetic control over the early embryonic development of a vast array of organisms, including humans.*

Key terms

DEOXYRIBONUCLEIC ACID (DNA): the hereditary material in most organisms, a double-stranded molecule that makes up genes

CHROMOSOME: a long, thread-like structure within a cell composed of DNA and protein; chromosomes occur in pairs called "homologous" chromosomes

GENE: a discrete unit of heredity located along a chromosome and composed of DNA

MITOSIS: the process of cell division in which a cell duplicates its internal contents and splits into two new cells; each new cell contains the exact chromosomal makeup of the original cell

SEX CHROMOSOMES: The X and Y chromosomes, which determine sex in many organisms; in *Drosophila*, a female carries two X chromosomes and a male carries one X and one Y chromosome

Early Studies of *Drosophila*

By the early 1900's, scientists had discovered chromosomes inside of cells and knew that they occurred in pairs, that one partner of each pair was provided by each parent during reproduction, and that fertilization restored the paired condition. This behavior of chromosomes paralleled the observations of Austrian botanist Gregor Mendel, first published in 1866, which showed that traits in pea plants segregated and were assorted independently during reproduction. This led geneticists Walter Sutton, Theodor Boveri, and their colleagues to propose, in 1902, the "chromosome theory of inheritance," which postulated that Mendel's traits, or "genes," existed on the chromosomes. However, this theory was not accepted by all scientists of the time.

Thomas Hunt Morgan's pioneering studies of the genetics of Drosophila *led to many important discoveries.* (Library of Congress)

Thomas Hunt Morgan was an embryologist at Columbia University in New York City, and he chose to study the chromosome theory and inheritance in the common fruit fly, *Drosophila melanogaster*. This organism was an ideal one for genetic studies because a single mating could produce hundreds of offspring, it developed from egg to adult in only ten days, it was inexpensively and easily kept in the laboratory, and it had only four pairs of chromosomes that were easily distinguished with a simple microscope. Morgan was the first scientist to keep large numbers of fly "stocks" (organisms with particular characteristics), and his laboratory became known as the "fly room."

After one year of breeding flies and looking for inherited variations of traits, Morgan found a single male fly with white eyes instead of the usual red. When he bred this white-eyed male with a red-eyed female, he was startled to discover a different inheritance pattern than he expected from Mendel's experiments. In the case of this mating, half of the males and no females had white eyes; Morgan had expected half of all of the males and females to be white-eyed. After many more generations of breeding, Morgan was able to deduce that eye color in a fly was related to its sex, and he located the eye-color gene to the X chromosome of the fruit fly. The X chromosome is one of the sex chromosomes. Because a female fly has two X chromosomes and a male has one X and one Y chromosome, and because the Y chromosome does not carry genes corresponding to those on the X chromosome, any gene on the male's X chromosome is expressed as a trait. This interesting and unusual example of the first mutant gene in flies was called a "sex-linked" trait because the trait was located on the X chromosome.

This important discovery attracted many students to Morgan's laboratory, and before long they found many other unusual inherited traits in flies and determined their inheritance patterns. One of the next major discoveries by members of the "fly lab" was that of genes existing on the same chromosome, information that was used to map the genes to individual chromosomes.

Linked Genes and Chromosome Maps

Many genes are located on a single chromosome. Genes, and the traits they specify, that are situated on the same chromosome tend to be inherited together. Such genes are referred to as "linked" genes. Morgan performed a variety of genetic crosses with linked genes and developed detailed maps of the positions of the genes on the chromosomes based on his results. Morgan did his first experiments with linked genes in *Drosophila* that specified body color and wing type. In fruit flies, a brown body is normal and a blank body is unusual or mu-

tant. Normal wings are very long, while one mutant variant has short, crinkled wings referred to as "vestigial" wings. When Morgan mated brown, normal-winged females with black-bodied, vestigial-winged males, the next generation consisted of all brown, normal-winged flies. When he then mated females from this new generation, which exhibited brown bodies and normal-length wings, with black-bodied, vestigial-winged males, the progeny were one-half brown and normal winged and one-half black and vestigial winged. Because of the equal distribution of these mutant traits between males and females, Morgan knew the genes were not sex linked. Because the traits for body color and wing length seemed to be inherited together, he deduced that they existed on the same chromosome.

As Morgan and his students and colleagues continued their experiments on the inheritance of body color and wing length, they observed that a very small percentage of the offspring did not have a phenotype, or appearance, like that of either parent. In most instances they observed a very small percentage of the offspring that exhibited nonparental phenotypes—some flies would be brown-bodied with vestigial wings, and others would have black bodies and normal-length wings. After repeating these experiments with many different linked genes, Morgan discovered that chromosomes exchange pieces during egg and sperm formation. This exchange of chromosome pieces occurs during a process called "meiosis," which occurs in sexually reproducing organisms and results in the production of gametes, generally eggs and sperm. During meiosis, the homologous chromosomes pair tightly and may exchange pieces; since the homologous chromosomes contain genes for the same trait along their length, this exchange does not present any genetic problems. The eggs or sperm produced through meiosis contain one of each pair of chromosomes.

In some of Morgan's genetic crosses, flies carried one chromosome with genes for black bodies and vestigial wings. The homologous chromosome carried the genes for brown bodies and normal-length wings. During meiosis, a small percentage of the homologous chromosomes exchanged pieces, resulting in some flies receiving chromosomes carrying genes for black bodies and normal wings or brown bodies and vestigial wings. These traits showed up in a small percentage of the progeny. The exchange of chromosome pieces resulting in new combinations of traits in progeny is referred to as "recombination." Morgan's students and colleagues pursued many different traits that showed genetic recombination. In 1917, one of Morgan's students, Alfred Sturtevant, reasoned that the further apart two genes were on a chromosome, the more likely they were to recombine and the more progeny with new combinations of traits would be observed. Over many years of work, Sturtevant and his colleagues were able to collect recombination data and cluster all the then-known mutant genes into four groupings that corresponded to the four chromosomes of *Drosophila*. They generated the first linkage maps that located genes to chromosomes based on their recombination frequencies.

The chromosomes in the salivary glands of the larval stage of the fruit fly are particularly large. Scientists were able to isolate these chromosomes, stain them with dyes, and observe them under microscopes. Each chromosome had an identifying size and shape and highly detailed banding patterns. X rays and chemicals were used to generate new mutations for study in *Drosophila*, and researchers realized that in many cases they could correlate a particular gene with a physical band along a chromosome. Also noted were chromosome abnormalities, including deletions of pieces, inversions of chromosome sections, and the translocation of a portion of one chromosome onto another chromosome. The pioneering techniques of linkage mapping through recombination of traits and physical mapping of genes to chromosome sections provided detailed genetic maps of *Drosophila*. Similar techniques have been used to construct gene maps of other organisms, including humans.

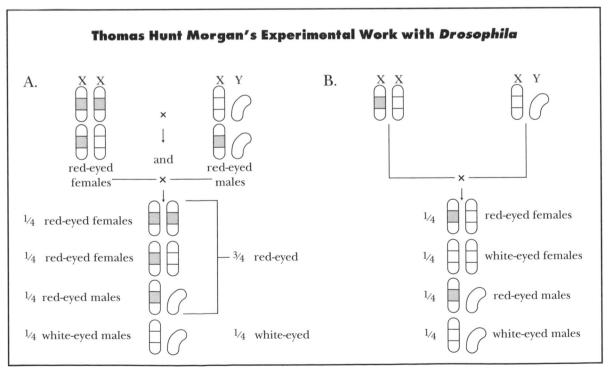

Thomas Hunt Morgan's Experimental Work with *Drosophila*

A.

X X × X Y

↓

and

red-eyed females ——— × ——— red-eyed males

↓

¼ red-eyed females

¼ red-eyed females ——— ¾ red-eyed

¼ red-eyed males

¼ white-eyed males ——— ¼ white-eyed

B.

X X X Y

×

↓

¼ red-eyed females

¼ white-eyed females

¼ red-eyed males

¼ white-eyed males

Morgan's experiments discovered such results as the following: A.: A red-eyed female is crossed with a white-eyed male. The red-eyed progeny interbreed to produce offspring in a ¾-red-to-¼-white ratio. All the white-eyed flies are male. B.: A white-eyed male is crossed with its red-eyed daughter, giving red-eyed and white-eyed males and females in equal proportions.

Control of Genes at the Molecular Level

This seminal genetic work on *Drosophila* was unparalleled in providing insights into the mechanisms of inheritance. Most of the inheritance patterns discovered in the fruit flies were found to be applicable to nearly all organisms. However, the usefulness of *Drosophila* as a research organism did not end with classical transmission genetics; it was found to provide equally valuable insight into the mechanisms of development of all animals at the level of deoxyribonucleic acid (DNA).

Drosophila were discovered to be ideal organisms to use in the study of early development. During its development in the egg, the *Drosophila* embryo orchestrates a cascade of events that results in the embryo having a polarity (a head and a tail), with segments between each end defined to become a particular body part in the adult. For example, the second segment of the thorax will support one pair of wings and one of the three pairs of legs. By studying many types of mutants that showed bizarre appearances as adults (for example, two sets of wings or legs replacing the normal antennae on the head), scientists were able to elucidate some of the mechanisms that control development in nearly all animals.

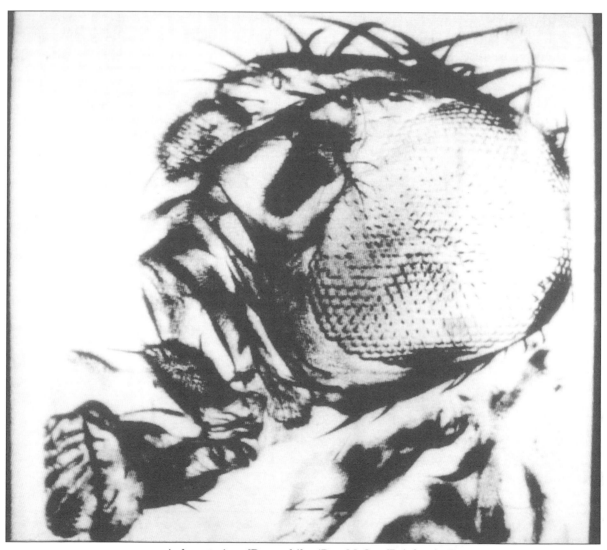

A close-up view of Drosophila. (Dan McCoy/Rainbow)

Developmental instructions from the mother fruit fly are sequestered in the egg. When the egg is fertilized, these instructions begin to "turn on" genes within the fertilized eggs that begin to establish the directionality and segment identity within the embryo. So many genes are involved in this process that a defect in a single one will truncate the rest of development, resulting in a severely mutated fly. It was found that conserved regions of DNA outside of the developmental genes received the signals to "turn on." Such sequences were found to be present in all animals studied. These control regions were termed "homeo boxes" after the homeotic genes that control the overall body plan of an organism in early development.

Many other aspects of *Drosophila* were found to be useful in understanding the structure and function of the DNA of all organisms. It was found that in *Drosophila*, large pieces of DNA will, under certain circumstances, pop out of the chromosome and reinsert themselves at another site. One such element, called a P element, was used by scientists to introduce nonfly DNA into the fruit fly embryo, thus providing information on how DNA is expressed in animals. This work also provided early clues into the successful creation of transgenic animals commonly used in research to study cancer and other diseases.

Impact and Applications

Genetic studies of *Drosophila melanogaster* have provided the world with a fundamental understanding of the mechanisms of inheritance. In addition to the inheritance modes shown by Mendel's studies of pea plants, fruit fly genetics revealed that some genes are sex linked in sexually reproducing animals. The research led to the understanding that while many genes are linked to a single chromosome, the linkage is not necessarily static, and that chromosomes can exchange pieces during recombination. The ease with which mutant fruit flies could be generated led to the development of detailed linkage maps for all the chromosomes and ultimately to the localization of genes to specific regions of chromosomes. With the advent of molecular techniques, it was

discovered that *Drosophila* again provided a wealth of information in terms of mobile genetic elements and developmental studies. All of these breakthroughs were scientifically interesting in terms of the flies themselves; more important, the fundamental principles were found to be consistent among all animals. Most of what is known about human genetics and genetic diseases has come from these pioneering studies with *Drosophila*.

Because of the sheer numbers of offspring from any mating of flies, their very short life cycle, and large numbers of traits that are easily observable, fruit flies have become an ideal system to screen for potential chemical carcinogens (cancer-causing agents) or mutagens (agents that cause mutations in DNA) in humans. Flies are exposed to the chemical in question and mated; then their offspring are analyzed for any abnormal appearances or behaviors, or for low numbers of offspring. Should a test substance cause any variation in the expected outcome of a cross, it is then subjected to more rigorous research in other organisms.

The versatile, easy-to-care-for, inexpensive fruit fly is often a fixture in classrooms around the world. Indeed, many geneticists have traced their passion to their first classroom encounters with fruit flies and the excitement of discovering the inheritance patterns for themselves. Flies are also routinely used in the study of neural pathways, learning patterns, behavior, and population genetics. Because of the ease of study and the volumes of information that have been compiled about its genetics, development, and behavior, *Drosophila* will continue to be an important model organism for biological study.

—*Karen E. Kalumuck*

See Also: Chromosome Theory of Heredity; Classical Transmission Genetics; Homeotic Genes; Linkage Maps; Meiosis.

Further Reading: Neil Campbell's *Biology* (1996) provides a detailed but accessible overview of biology as well as information on Mendelian genetics, mitosis, meiosis, and DNA in general. Another excellent overview written for the nonspecialist is *Essential Cell Biology: An Introduction to the Molecular Biology of the Cell* (1997), edited by Bruce Alberts, which is lavishly illustrated, clear, and concise. A more advanced discussion can be found in *Molecular Biology of the Cell* (1995), by Bruce Alberts et al., a useful resource on the central role of *Drosophila* genetics in the understanding of classical and molecular inheritance. A clear, detailed discussion of homeotic genes can be found in *Molecular Cell Biology* (1995), by Harvey Lodish et al. Peter A. Lawrence, *The Making of a Fly: The Genetics of Animal Design* (1992), delves into the exciting techniques used by imaginative scientists to study fly development.

Epistasis

Field of study: Classical transmission genetics

Significance: *Epistasis is the interaction of genes and their effect on the final performance or external appearance (phenotype) of an organism. Elucidation of the interactions between genes leads to a more complete understanding of the influence of genotype on phenotype, and thus it is important for the improvement of quantitatively inherited characteristics, such as crop yield, and for gaining a greater knowledge of the mechanisms of complexly inherited diseases, such as cancer.*

Key terms

ALLELE: an alternate form of a gene

DIHYBRID CROSS: a cross between parents that differ by two specified genes

F_1: first filial generation, or the progeny resulting from the first cross in a series

F_2: second filial generation, or the progeny resulting from cross of the F_1 generation

Introduction

The term "epistasis" is of Greek and Latin origin, meaning "to stand upon" or "stoppage." The term was originally used by geneticist William Bateson at the beginning of the twentieth century to define genes that mask the expression of other genes. The gene at the initial location (locus) is termed the epistatic gene. The genes at the other loci are "hypostatic" to the initial gene. In its strictest sense, it describes a nonreciprocal interaction between two or more genes, such that one gene modifies, suppresses, or otherwise influences the expression of another gene affecting the same phenotypic (physical) character or process. By this definition, simple additive effects of genes affecting a single phenotypic character or process would not be considered an epistatic interaction. Similarly, interactions between alternative forms (alleles) of a single gene are governed by dominance effects and are not epistatic. Epistatic effects are interlocus interactions. Therefore, in terms of the total genetic contribution to phenotype, three factors are involved: dominance effects, additive effects, and epistatic effects. The analysis of epistatic effects can suggest ways in which the action of genes can control a phenotype and thus supply a more complete understanding of the influence of genotype on phenotype.

A gene can influence the expression of other genes in many different ways. One result of multiple genes is that more phenotypic classes can result than can be explained by the action of a single pair of alleles. The initial evidence for this phenomenon came out of the work of Bateson and British geneticist Reginald C. Punnett during their investigations on the inheritance of comb shape in domesticated chickens. The leghorn breed has a "single" comb, brahmas have "pca" combs, and wyandottes have "rose" combs. Crosses between brahmas and wyandottes have "walnut" combs. Intercrosses among walnut types show four different types of F_2 progeny, in the ratio 9 walnut: 3 rose: 3 pea: 1 single. This ratio of phenotypes is consistent with the classical F_2 ratio for dihybrid inheritance. The corresponding ratio of genotypes, therefore, would be 9 $A\_ B\_$: 3 $A\_ bb$: 3 $aa B\_$: 1 $aa bb$, respectively. (The symbol "_" is used to indicate that the second gene can be either dominant or recessive; for example, $A\_$ means that both AA and Aa will result in the same phenotype.) In this example, one can recognize that two independently assorting genes can affect a single trait. If two gene pairs are acting epistatically, however, the expected 9:3:3:1 ratio of phenotypes is altered in some fashion. Five basic examples of two-gene epistatic interactions can be described: complementary, modifying, inhibiting, masking, and duplicate gene action.

Two-Gene Epistatic Interactions

For complementary gene action, a dominant allele of two genes is required to produce a single effect. An example of this form of epistasis again comes from the observations of Bateson and Punnett of flower color in crosses between two white-flowered varieties of sweet peas. In their investigation, crosses between these two varieties produced an unexpected result: All of the F_1 progeny had purple flowers.

William Bateson (at right), who pioneered the study of epistasis, in a 1924 pose with another genetics pioneer, Wilhelm Johannsen. (California Institute of Technology)

A Punnett Square Showing Flower Pigmentation

White
CCpp
×
↓
White
ccPP

Purple
CcPp
↓

F_1

	CP	Cp	cP	cp
CP	CCPP purple	CCPp purple	CcPP purple	CcPp purple
Cp	CCPp purple	CCpp white	CcPp purple	Ccpp white
cP	CcPP purple	CcPp purple	ccPP white	ccPp white
cp	CcPp purple	Ccpp white	ccPp white	ccpp white

F_2

When white-flowered sweet pea plants were crossed, the first-generation progeny (F_1) all had purple flowers. When these plants were self-fertilized, the second-generation progeny (F_2) revealed a ratio of nine purple to seven white. This result can be explained by the presence of two genes for flower pigmentation, P (dominant) or p (recessive) and C or c. Both dominant forms, P and C, must be present in order to produce purple flowers.

When the F_1 individuals were allowed to self-fertilize and produce the F_2 generation, a phenotypic ratio of nine purple-flowered to seven white-flowered individuals resulted. Their hypothesis for this ratio was that a homozygous recessive genotype for either gene (or both) resulted in the lack of flower pigmentation. A simple model to explain the biochemical basis for this type of flower pigmentation is a two-step process, each step controlled by a separate gene and each gene having a recessive allele that eliminates pigment formation. Given this explanation, each parent must have had complementary genotypes (*AA bb* and *aa BB*), and thus both were white-flowered. Crosses between these two parents would produce double heterozygotes (*Aa Bb*) with purple flowers. In the F_2 generation, $9/16$ would have the genotype *A_ B_* and would have purple flowers. The remaining $7/16$ would be homozygous recessive for at least one of the two genes and, therefore,

would have white flowers. In summary, the phenotypic ratio of the F_2 generation would be 9:7.

The term "modifying gene action" is used to describe a situation whereby one gene produces an effect only in the presence of a dominant allele of a second gene at another locus. An example of this type of epistasis is aleurone color in corn. The aleurone is the outer cell layer of the endosperm (food storage tissue) of the grain. In this system, a dominant gene (*P_*) produces a purple aleurone layer only in the presence of a gene for red aleurone (*R_*) but expresses no effect in the absence of the second gene in its dominant form. Thus, the corresponding F_2 phenotypic ratio is 9 purple: 3 red: 4 colorless. The individuals without aleurone pigmentation would, therefore, be of the genotype *P_ rr* ($3/16$) or *pp rr* ($1/16$). Again, a two-step biochemical pathway for pigmentation can be used to explain this ratio; however, in this example, the product of the second gene (*R*) acts first in the biochemical pathway and allows for the production of red pigmentation and any further modifications to that pigmentation. Thus, the phenotypic ratio of the F_2 generation would be 9:3:4.

Inhibiting action occurs when one gene acts as an inhibitor of the expression of another gene. In this example, the first gene allows the phenotypic expression of a gene, while the other gene inhibits it. Using a previous example (the gene R for red aleurone color in corn seeds), the dominant form of the first gene R does not produce its effect in the presence of the dominant form of the inhibitor gene I. In other words, the genotype $R\_ i\_$ results in a phenotype of red aleurone ($^3/_{16}$), while all other genotypes result in the colorless phenotype ($^{13}/_{16}$). Thus gene R is inhibited in its expression by the expression of gene I. The F_2 phenotypic ratio would be 13:3. This ratio, unlike the previous two examples, includes only two phenotypic classes and highlights a complicating factor in determining whether one or two genes may be influencing a given trait. A 13:3 ratio is close to a 3:1 ratio (the ratio expected for the F_2 generation of a monohybrid cross). Thus it emphasizes the need to look at an F_2 population of sufficient size to discount the possibility of a single gene phenomenon over an inhibiting epistatic gene interaction.

Masking gene action, a form of modifying gene action, results when one gene is the primary determinant of the phenotype of the offspring. An example of this phenomenon is fruit color in summer squash. In this example, the F_2 ratio is 12:3:1, indicating that the first gene in its dominant form results in the first phenotype (white fruit); thus this gene is the primary determinant of the phenotype. If the first gene is in its recessive form and the second gene is in its dominant form, the fruit will be yellow. The fruit will be green at maturity only when both genes are in their recessive form ($^1/_{16}$ of the F_2 population).

Duplicate gene interaction occurs when two different genes have the same final result in terms of their observable influence on phenotype. This situation is different from additive gene action in that either gene may substitute for the other in the expression of the final phenotype of the individual. It may be argued that duplicate gene action is not a form of epistasis, since there may be no interaction between genes (if the two genes code for the same protein product), but this situation may be an example of gene interaction when two genes code for similar protein products involved in the same biochemical pathway and their combined interaction determines the final phenotype of the individual. An example of this type of epistasis is illustrated by seed capsule shape in the herb shepherd's purse. In this example, either gene in its dominant form will contribute to the final phenotype of the individual (triangular shape). If both genes are in their recessive form, the seed capsule has an ovoid shape. Thus, the phenotypic ratio of the F_2 generation is 15:1.

Impact and Applications

Nonallelic gene interactions have considerable influence on the overall functioning of an individual. In other words, the genome (the entire genetic makeup of an organism) determines the final fitness of an individual, not only as a sum total of individual genes (additive effects) or by the interaction between different forms of a gene (dominance effects) but also by the interaction between different genes (intragenomic or epistatic effects). This situation is something akin to a chorus: Great choruses not only have singularly fine voices, but they also perform magnificently as finely tuned and coordinated units. Knowledge of what contributes to a superior genome would, therefore, lead to a fuller understanding of the inheritance of quantitative characters and more directed approaches to genetic improvement. For example, most economically important characteristics of agricultural species (such as yield, pest and disease resistance, and stress tolerance) are quantitatively inherited, the net result of many genes and their interactions. Thus an understanding of the combining ability of genes and their influence on the final appearance of domesticated breeds and crop varieties should lead to more efficient genetic improvement schemes. In addition, it is thought that many important human diseases are inherited as a complex interplay among many genes. Similarly, an understanding of genomic functioning should lead to improved screening or therapies.

—*Henry R. Owen*

See Also: Dihybrid Inheritance; Multiple Alleles; Quantitative Inheritance.

Further Reading: *Practical Genetics* (1991), by Robert N. Jones and G. K. Rickards, presents an applied approach to many topics, including epistasis, and is intended for high school and college students. A reexamination of the whole concept of epistasis, with statistical implications, is provided in Wayne N. Frankel and Nicholas J. Schork, "Who's Afraid of Epistasis?" *Nature Genetics* 14 (December, 1996). *Genetics: A Basic Guide* (1972), by I. J. Pedder and E. G. Wynne, includes diagrams of crosses that illustrate specific examples of epistasis.

Escherichia coli

Field of study: Bacterial genetics

Significance: *Through the study of the genetics of* Escherichia coli (E. coli), *biologists have come to understand the molecular-level regulation of gene expression and how genes direct routine activities of living cells. This understanding and the suitability of* E. coli's *genetics has led to the extensive use of* E. coli *in biotechnology. Such technology permits the introduction of foreign genes into* E. coli *cells, which may result in new bacterial strains capable of solving problems as diverse as environmental pollution, food and energy shortages, and the spread of diseases.*

Key terms

DEOXYRIBONUCLEIC ACID (DNA): the genetic material found in all cells

OPERON: a genetic unit consisting of structural genes coding for amino acid chains; an operator gene controlling the transcriptional (message encoding) activity of the structural genes

REPLICATION: the process of DNA duplication

MUTATION: a process that produces change in DNA or chromosome structure

E. coli: A Suitable Experimental Organism

Escherichia coli (*E. coli*)—discovered in 1885 by Theodor Escherich—is the most intensely studied bacterium in genetics. In fact, of the earth's living organisms, this bacterium is one of the better understood, and its use as a favorite experimental organism dates to the mid-twentieth century. Even before gaining a rich genetic history, the bacterium was selected for genetic research for several reasons: its ease of handling in experiments, its twenty-minute generation time, its single copy of each gene, and its meager genetic material. Results derived from *E. coli* genetic experiments have significantly influenced the thinking of biologists, and the genetics of *E. coli* has provided evidence that explains mechanisms underlying important processes: *E. coli* chromosome organization; regulation of gene expression; deoxyribonucleic acid (DNA) replication, transcription, and translation; mutation and DNA repair; biotechnology; and evolution.

The *E. coli* cell usually contains a single chromosome, although the cell's actual number of chromosomes depends on the bacterium's growth rate. Fast-growing *E. coli* have two to four chromosome copies per cell, while the slow-growing counterparts have one to two copies per cell. These multiple copies, however, are genetically identical, permitting *E. coli* to behave as haploids (cells containing a single chromosome). This chromosome, a dense cellular structure carrying hereditary information from generation to generation, consists of a single molecule of double-stranded DNA. The DNA, in a closed-circle form, is located in the nucleoid, a central region of the *E. coli* cell. *E. coli*'s DNA molecule is probably made up of four million base pairs and carries 2,800 genes. These genes constitute 75 percent of the DNA molecule; the remaining DNA consists of regions between genes, such as the stretch of DNA acting as the unique origin of replication for the *E. coli* DNA molecule.

Packaging this DNA into the nucleoid is an important concept in *E. coli* genetics because the length of the bacterial chromosome containing the DNA is 1,200 times that of the *E. coli* cell. The chromosome, therefore, is packaged in a highly compact form. This compact DNA consists of one hundred independent genetic segments, each having forty thousand base pairs of DNA containing extra twists, with the ends of each genetic segment held, presumably, by proteins. The extra-twisted DNA in one genetic segment is unaffected by events influencing extra twisting of DNA in other genetic segments. Such a structure forms because the DNA, a negatively charged molecule, associates with positively charged structural proteins.

About one-third of E. coli's 2,800 genes have been located on the bacterium's chromosome using gene mapping (which determines the locations of genes along the chromosome) and recombinant DNA techniques such as DNA sequencing (which determines the order of the nucleotides in DNA). Of the located genes, 260 of them are organized into seventy-five operons, with the remaining 740 genes scattered, perhaps randomly, around the rest of the DNA molecule.

Regulation of Gene Expression

E. coli genetics reveals that 26 percent of E. coli's mapped genes are organized in transcriptional units (DNA segments containing message-encoding start and stop signals) called "operons"; these work to regulate gene expression. Operons, coordinately regulated units, often contain genes with related functions. Each regulated unit has a set of adjoining structural genes, a promoter for enzyme binding, and an operator for regulatory protein binding. If the genes encode enzymes involved in an anabolic pathway (in which chemical reactions form larger molecules from smaller ones), they are usually turned off in the presence of the pathway's end product. Alternatively, if the genes encode enzymes involved in a catabolic pathway (in which chemical reactions break down large molecules into smaller ones), they are often expressed in the presence of the enzymes' substrates (molecules whose actions are increased).

The E. coli system was used to unravel the genetic workings of the lactose (lac) and the tryptophan (trp) operons. This earned the bacterium a place in history as being involved in the generation of evidence that explained the regulation of gene expression. In the E. coli system, the lac operon consists of three structural genes—Z, Y, and A—that encode beta-galactosidase, beta-galactoside permease, and beta-galactoside transacetylase, respectively. Other operon components are the promoter and the operator adjoining the Z gene. The regulator gene has its own promoter and adjoins the operon. The E. coli lac operon is an example of an inducible system because the operon's three structural genes are transcribed (put into message code) only in the presence of lactose. In the absence of lactose, the lac repressor (a protein product of the regulator gene) binds to the operator and prevents ribonucleic acid (RNA) polymerase from initiating operon transcription.

In contrast to the lac operon, E. coli's trp operon is a repressible system, in which the production of an enzyme stops with the addition of the end product of the enzyme reaction. Transcription of the operon's five structural genes, which encode enzymes involved in tryptophan production, is repressed in the presence of tryptophan. A second regulatory mechanism also controls the system.

Because of E. coli genetics, biologists know that operon function may change if fused to a new operator. French molecular biologist François Jacob's research team showed this for the structural genes of E. coli's lac operon. The team used particles carrying parts of the lac-pur region, but with an added deletion that eliminated the lac operator and part of the Z gene. These modified particles were inserted into E. coli that were unable to produce the enzymes permease and acetylase. The functional lac enzymes produced by the modified particles were no longer activated by lactose. Such enzymes were instead under control of the deactivated purine operator. As a result, excessive purine caused repression of galactoside permease and acetylase. In the E. coli system, gene expression can be regulated at different levels, but transcriptional regulation is the most common.

DNA Replication, Transcription, and Translation

Early in the study of E. coli, Matthew Meselson and Franklin Stahl determined how DNA duplicates itself in the bacterium. They grew the organism, across several generations, in culture media containing normal or heavy nitrogen, then used cesium chloride density-gradient centrifugation to characterize results. Their findings, verified through autoradiography several years later by John Cairns, showed that in *E. coli*, DNA duplicates itself semiconservatively. This means that the strands of *E. coli*'s DNA double helix separate and form a *Y*-shaped replication fork where DNA duplication begins. Proteins stabilize the unwound helix and assist in relaxing the coiling tension created ahead of the duplication activity. A new, complementary strand of DNA, duplicated in *E. coli* at the rate of thirty thousand nucleotides per minute, is produced on each of the two parental template (guide) strands. The DNA duplication process results in two double-stranded DNA molecules, each having one strand from the parent molecule and one newly

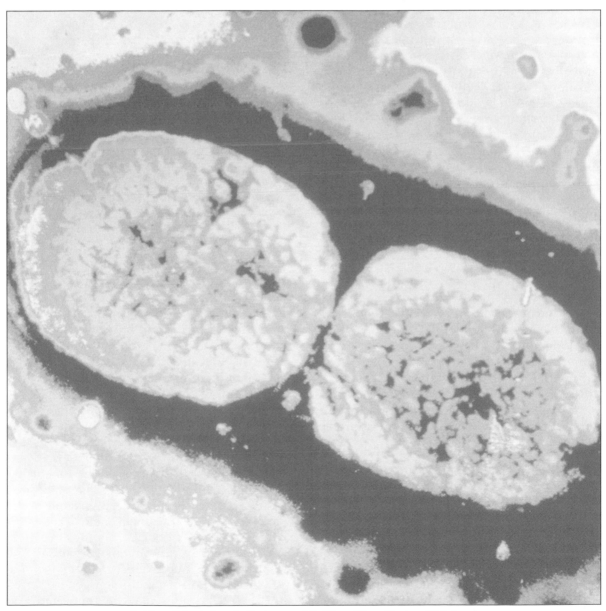

An E. coli *bacterium in the process of dividing.* (Tom Broker/Rainbow)

produced strand. This semiconservative mechanism ensures the faithful copying of the genetic information at each *E. coli* cell division.

During the message-encoding process (transcription), the genic message (RNA transcript) is created step by step, using the DNA template. The template is read in one direction, while RNA is produced in the opposite direction. The process includes initiation, elongation, and termination phases. The transcription initiation site is signalled by the promoter (a short nucleotide sequence recognized by an RNA polymerase). During elongation, RNA polymerase migrates along the DNA molecule, melting and unwinding the double helix as it moves and sequentially attaching ribonucleotides to one end of the growing RNA molecule. Base pairing to the template strand of the gene determines the identity of the ribonucleotide added to each position. By a complex signal, transcription is terminated shortly after the ends of genes. As a result of the process, a gene-complementary, single-stranded RNA molecule (messenger RNA, or mRNA) is created.

Like the message-encoding process, the message-decoding process (translation) consists of initiation, elongation, and termination. In *E. coli*, the small subunit of a ribosome (the cell's interior structure for protein production) attaches to the ribosome-binding site of an mRNA, resulting in an initiation complex. In elongation, the large subunit of the ribosome attaches to the initiation complex, creating two different binding sites for transfer RNA (tRNA), the amino acid transporter. Ribosomes use mRNA-coded information to take amino acids brought by tRNA and assemble them, on ribosomes, into protein.

Mutation and DNA Repair

In the genetics of *E. coli*, phenotypes resulting from changes in the DNA can occur because of either mutation (a change in the nucleotide sequence of a gene) or recombination (a process leading to new combinations of genes on a chromosome). These new combinations can occur following transfer of chromosomal genes from one bacterial cell to another by transformation (in which a recipient cell acquires genes from free DNA in the medium), transduction (in which a virus carries DNA from donor to recipient cell), or conjugation (in which two bacterial cells make contact and exchange DNA). Transposon and insertion elements, both found in *E. coli*, may also change phenotypes in *E. coli*. Transposon, a mobile DNA segment, contains genes for inserting DNA into the chromosome and for mobilizing the element to other chromosome locations. Insertion elements, *E. coli*'s simplest transposable elements, contain only genes for mobilizing the elements and inserting them into chromosomes at new locations.

In *E. coli*, as in all other organisms, many chemical and physical agents cause structural changes in DNA. Consequently, mechanisms are needed for repairing such damaged DNA. Because of the genetics of *E. coli*, biologists know that the needed repair mechanisms exist for the bacterium, although they are complicated and require many different proteins. The three main types of repair mechanisms in *E. coli* are direct repair (the reversal of a structural change), excision repair (in which appropriate enzymes recognize and "label" a damaged nucleotide, excise it, fill in the gap, and seal the strand), and mismatch repair (in which enzymes recognize the mismatch nucleotide and either label it or repair it directly). In *E. coli*, the parental strand is distinguished from the newly created daughter strand by tagging the parental strand with methyl groups attached to adenines occurring within specific sequences. Such modified adenines act as labels for the parent strand, enabling the repair enzymes to recognize which strand should be repaired at a mismatch position.

Biotechnology and Evolution

To test a genetic hypothesis, the genetic history of the organism involved must be well known so that the genetic background of the parents used in the experimental crosses is known. The genetics of *E. coli* provides geneticists with such an experimental organism. As a result, *E. coli* is used extensively in biotechnology. In this industry, a foreign gene inserted into the bacterium may be replicated and sometimes translated in the same manner as the native bacterial DNA, producing a foreign gene product. *E. coli* can accept foreign DNA derived from any organism because the genetic code is nearly universal. As an example, genetic mapping of a free-living, nitrogen-fixing bacterium showed that seventeen genes involved in nitrogen fixation are clustered on one portion of the chromosome. Biologists transferred this gene cluster to a plasmid—double-stranded, independently replicating DNA—and introduced the plasmid into *E. coli* cells, which then produced the enzyme nitrogenase and fixed nitrogen.

A contribution to *E. coli* genetics was made when the synthetic gene coding for somatostatin, an antigrowth hormone important in the treatment of different human growth disorders, was fused with the start of the lacZ' gene contained within an *E. coli* cloning vector (a self-duplicating DNA molecule containing inserted, foreign DNA). After alteration of the bacterium brought about by the incorporation of foreign DNA (transformation), the *E. coli* RNA polymerase transcribed the fused gene, recognizing the lac promoter as its binding site. The mRNA was then translated by ribosomes that recognized the lac ribosome binding sequence. The resulting fused protein was cleaved with cyanogen bromide, which cuts amino acid chains specifically at methionines, resulting in pure-form somatostatin.

The genetics of *E. coli* provides evidence for punctuated evolution caused by the appearance of rare, beneficial mutations. This evidence involved studies that measured changes in cell size over three thousand generations of bacteria in a constant environment. During the studies, periods of stagnancy were interrupted by periods of rapid change. The changes in cell size may be the result of direct selection for a rare, beneficial mutation that caused increased cell size. This mutation swept through the population, producing a change in cell size in one hundred generations or less.

Impact and Applications

Geneticists use experimental organisms for their research. Their favorite organisms, such as *E. coli*, have qualities that make them well suited for genetic experimentation—a rich ge-

netic history, a short life cycle, production of large progeny from a mating, ease in handling, and genetic variation among the individuals in the population. The quantity of genetics involving *E. coli* is a testament to the bacterium's suitability as an experimental organism for testing genetic hypotheses. The hypotheses tested using this experimental organism have contributed in a revolutionary way to the understanding of significant scientific concepts and to the understanding of the genetics of organisms more complex than bacteria, such as humans.

In addition, recombinant DNA technology (techniques for constructing, studying, and using DNA created in a test tube), which uses *E. coli* extensively, is used in all areas of basic genetics research to investigate genetic circumstances. Many biotechnology companies owe their existence to recombinant DNA technology—and to *E. coli*—as they seek to clone and manipulate genes for the production of commercial products, the improvement of plant and animal agriculture, the development of diagnostic tools for genetic diseases, and the development of new or more effective pharmaceuticals.

—*Robert Haynes*

See Also: Biotechnology; Gene Regulation: Bacteria; Gene Regulation: Lac Operon.

Further Reading: Peter Russell, *Genetics* (1996), D. Peter Snustad et al., *Principles of Genetics* (1997), and Terence A. Brown, *Genetics: A Molecular Approach* (1990), give a general treatment of the genetics of bacteria, including *E. coli*.

Eugenics

Field of study: Human genetics
Significance: *The eugenics movement is an infamous chapter in the tale of the power and function of science in society. This movement sought to speed up the process of natural selection through the use of selective breeding and led to the enactment of numerous laws requiring the sterilization of "genetically inferior" individuals and limiting the immigration of supposedly defective groups. Fu-*

eled by economic hardship and racial prejudice, the largest-scale application of eugenics occurred in Nazi Germany, where numerous atrocities, including genocide, were committed in the name of the genetic improvement of the human species. These flawed policies were based on an inadequate understanding of the complexity of human genetics, an underestimation of the role of the environment in gene expression, and the desire of certain groups to claim genetic superiority and the right to control the reproduction of others.

Key terms

NEGATIVE EUGENICS: improving human stocks through the restriction of reproduction

POSITIVE EUGENICS: improving human stocks by encouraging the "naturally superior" to breed extensively with other superior humans

BIOMETRY: the measurement of biological and psychological variables

The Founding of the Eugenics Movement

With the publication of Charles Darwin's *On the Origin of Species* (1859), the concept of evolution came to the forefront and revolutionized the way many aspects of life were viewed. In particular, evolution caused some people to reformulate their thoughts on the human condition. Herbert Spencer and other proponents of Social Darwinism adhered to the belief that social class structure arose through natural selection. Thus, the class stratification in industrial societies, including the existence of a permanently poor underclass, was thought by them to be a reflection of the underlying, innate differences between classes.

During this era there was also a rush to legitimate all sciences with measurement and quantification. There was a blind belief that attaching numbers to a study would ensure its objectivity. Francis Galton, an aristocratic inventor, statistician, and cousin of Darwin, became one of the primary promoters of such quantification. Obsessed with mathematical analysis, Galton measured everything from physiology and reaction times to boredom, the efficacy of prayer, and the beauty of women. He was particularly interested in the differences between human races. Galton eventually founded the field of biometry by applying statistics to biological problems.

A hereditarian, Galton assumed that talent in humans was subject to the laws of heredity. Although Galton did not coin the term "eugenics" until 1883, he published the first discussion of his ideas in 1865, in which he recognized the apparent evolutionary paradox that those of talent often have few, if any, children and that civilization itself diminishes the effects of natural selection on human populations. Fearing that medicine and social aid would lead to the propagation of weak individuals, Galton advocated increased breeding by "better elements" in the population (positive eugenics), while at the same time discouraging breeding of the "poorer elements" (negative eugenics).

Like most in his time, Galton believed in "blending inheritance," whereby hereditary material would mix together like different colors of paint. Trying to reconcile how superior traits would avoid being swamped by such blending, he came up with the statistical concept of the correlation coefficient, and in the process connected Darwinian evolution to the "probability revolution." His work focused on the bell-shaped curve or "normal distribution" demonstrated by many traits and the possibility of shifting the mean by selection pressure at either extreme. His statistical framework deepened the theory of natural selection. Unfortunately, the mathematical predictability he studied has often been misinterpreted as inevitability.

In 1907, Galton founded the Eugenics Education Society of London. He also carefully cataloged eminent families in his *Hereditary Genius* (1869), wherein the Victorian world was assumed to be the ultimate level that society could attain and the cultural transmission of status, knowledge, and social connections were discounted.

Early Eugenics in Britain

Statistician and social theorist Karl Pearson was Galton's disciple and first Galton Professor of Eugenics at the Galton Laboratory at the University of London. His *Grammar of Science* (1892) outlined his belief that eugenic management of society could prevent genetic deterioration and ensure the existence of intelligent rulers, in part by transferring resources from inferior races back into the society. According to philosopher David J. Depew and biochemist Bruce H. Weber, even attorney Thomas Henry Huxley, champion of Darwinism, balked at this "pruning" of the human garden by the administrators of eugenics. For the most part, though, British eugenicists focused on improving the superior rather than eliminating the inferior.

Another of Galton's followers, comparative anatomist Walter Frank Weldon, like Galton before him, set out to measure all manner of things, showing that the distribution of many human traits formed a bell-shaped curve. In a study on crabs, he showed that natural selection can cause the mean of such a curve to shift, adding fuel to the eugenicists' conviction that they could better the human race through artificial selection.

Population geneticist Ronald A. Fisher was Pearson's successor as the Galton Professor of Eugenics. Fisher co-

Francis Galton, the "father of eugenics." (Library of Congress)

founded the Cambridge Eugenics Society and became close to Charles Darwin's sons, Leonard and Horace Darwin. In a speech made to the Eugenics Education Society, Fisher called eugenicists the "agents of a new phase of evolution" and the "new natural nobility," with the view that humans were becoming responsible for their own evolution. The second half of his book *The Genetical Theory of Natural Selection* (1930) deals expressly with eugenics and the power of "good-making traits" to shape society. Like Galton, he believed that those in the higher social strata should be provided with financial subsidies to counteract the "resultant sterility" caused when upper class individuals opt to have fewer children for their own social advantage.

British embryologist William Bateson, who coined the terms "genes" and "genetics," championed the Mendelian genetics that finally unseated the popularity of Galton's ideas in England. In a debate that lasted thirty years, those that believed in Austrian monk Gregor Mendel's particulate inheritance argued against the selection touted by the biometricians, and vice versa. Bateson, who had a deep distrust of eugenics, successfully replicated Mendel's experiments. Not recognizing that the two arguments were not mutually exclusive, Pearson and Weldon rejected genetics, thus setting up the standoff between the two camps.

Fisher, on the other hand, tried to model the trajectory of genes in a population as if they were gas molecules governed by the laws of thermodynamics, with the aim of converting natural selection into a universal law. He used such "genetic atomism" to propose that continuous variation, natural selection, and Mendelian genetics could all coexist. Fisher also mathematically derived Galton's bell-shaped curves based on Mendelian principles. Unfortunately, by emulating physics, Fisher underestimated the degree to which environment dictates which traits are adaptive.

Early Eugenics in the United States

While Mendelians and statisticians were debating in Britain, in the United States, Harvard embryologist Charles Davenport and others embarked on a mission of meshing early genetics with the eugenics movement. In his effort, Davenport created the Laboratory for Experimental Evolution at Cold Springs Harbor, New York. The laboratory was closely linked to his Eugenics Record Office (ERO), which he established in 1910. Davenport raised much of the money for these facilities by appealing to wealthy American families who feared unrestricted immigration and race degeneration. Though their wealth depended on the availability of cheap labor guaranteed by immigration, these American aristocrats feared the cultural impact of a flood of "inferior immigrants."

Unlike the British, U.S. eugenicists thought of selection as a purifying force and thus focused on how to stop the defective from reproducing. Davenport wrongly felt that Mendelian genetics supported eugenics by reinforcing the effects of inheritance over the environment. He launched a hunt to identify human defects and link specific genes (as yet poorly understood entities) to specific traits. His primary tool was the family pedigree chart. Unfortunately, these charts were usually based on highly subjective data, such as questionnaires given to school children to determine the comparative social traits of various races.

The Eugenics Research Association was founded in 1913 to report the latest findings. In 1918, the Galton Society began meeting regularly at the American Museum of Natural History in New York, and in 1923 the American Eugenics Society was formed. These efforts

paid off. By the late 1920's and early 1930's, eugenics was a topic in high school biology texts and college courses across the United States.

Among eugenics supporters was psychologist Lewis M. Terman, developer of the Stanford-Binet intelligence quotient (IQ) test, and Harvard psychologist Robert Yerke, developer of the Army IQ test, who both believed that IQ test performance (and hence intelligence) was hereditary. The administration of such tests to immigrants by eugenicist Henry Goddard represented a supposedly "objective and quantitative tool" for screening immigrants for entry into the United States. Biologist Garland Allen reports that Goddard, in fact, determined that more than 80 percent of the Jewish, Hungarian, Polish, Italian, and Russian immigrants were mentally defective.

Fear that immigrants would take jobs away from hard-working Americans, supported by testimony from ERO's superintendent, Harry Laughlin, and the findings of Goddard's IQ tests resulted in the Johnson Act of 1924, which severely restricted immigration. In the end, legal sterilization and immigration restrictions became more widespread in the United States than in any country other than Nazi Germany. By 1940, more than thirty states in the United States had enacted compulsory sterilization laws. Most were not repealed until after the 1960's.

Eugenics and the Progressive Era

During the Progressive Era, the eugenics movement became a common ground for such diverse groups as biologists, sociologists, psychologists, militarists, pacifists, socialists, communists, liberals, and conservatives. The progressive ideology, exemplified by Theodore Roosevelt's Progressive Party, sought the scientific management of all parts of society. Eugenics attracted the same crowd as preventative medicine, since both were seen as methods of harnessing science to reduce suffering and misfortune. For example, cereal entrepreneur John Harvey Kellogg founded the Race Betterment Foundation, mixing eugenics with hygiene, diet, and exercise. During this period, intellectuals of all stripes were pulled in by the promise of "the improvement of the human race by better breeding." The genetics research of this time focused on improving agriculture, and eugenics was seen as the logical counterpart to plant and animal husbandry.

Davenport did not hesitate to play on their sympathies by making wild claims about the inheritance of "nomadism," "shiftlessness," "love of the sea," and other "traits" as if they were single Mendelian characteristics. Alcoholism, pauperism, prostitution, rebelliousness, criminality, feeblemindedness, chess expertise, and industrial sabotage were all claimed to be determined by one or two pairs of Mendelian genes. In particular, the progressives were lured by the idea of sterilizing the "weak minded," especially after the publication of articles about families in Appalachia and New Jersey that supposedly documented genetic lines cursed by a preponderance of habitual criminal behavior and mental weakness.

Having the allure of a "social vaccination," the enthusiasm to sterilize the "defective" spread rapidly among intellectuals, without regard to political or ideological lines. Sweden's Social Democrats forcibly sterilized some sixty thousand Swedes under a program that lasted from 1935 to 1976 organized by the state-financed Institute for Racial Biology. Grounds for sterilization included not only "feeblemindedness" but also "gypsy features," criminality, and "poor racial quality." The low class or mentally slow were institutionalized in the Institutes for Misled and Morally Neglected Children and released only if they would agree to be sterilized. Involuntary sterilization policies were also adopted in countries ranging from Switzerland and Austria to Belgium and Canada, not to be repealed until the 1970's.

Hermann Müller, a eugenicist who emigrated to the Soviet Union (and later returned to the United States), attacked Davenport's style of eugenics at the International Eugenics Congress in 1932. Müller, a geneticist who won the 1946 Nobel Prize for Physiology or Medicine for his discovery of the mutagenic power of X rays, instead favored the style of eugenics envisioned by English novelist Aldous Huxley's *Brave New World* (1932), with state nurseries, artificial insemination, and the use of other scientific techniques to produce an engineered socialist society.

According to journalist Jonathan Freedland, the British Left, including a large number of socialist intellectuals such as playwright George Bernard Shaw and philosopher Bertrand Russell, was convinced that it knew what was best for society. Concerned with the preservation of their higher intellectual capacities, they joined the fashionable and elitist Eugenics Society in the 1930's, where they advocated the control of reproduction, particularly favoring the idea of impregnating working-class women with sperm of men with high IQs.

The American Movement Spreads to Nazi Germany

The eugenics movement eventually led to grave consequences in Nazi Germany. Negative eugenics reached its peak there, with forced sterilization, "mercy killing," experimentation, and ultimately genocide being used in the name of "racial hygiene." Eugenicists in the United States and Germany formed close and direct alliances, especially after the Nazis came to power in 1933. The ERO's Laughlin gave permission for his article "Eugenical Sterilization" to be reprinted in German in 1928. It soon became the basis of Nazi sterilization policy. Davenport even arranged for a group of German eugenicists to participate in the three hundredth anniversary of Harvard's founding in 1936.

Inspired by the U.S. eugenics movement and spurred by economic hardship that followed World War I, the Nazi Physician's League took a stand that those suffering from incurable disease were a useless waste of medications and posed an economic drain on society, along with the crippled, the feebleminded, the elderly, and the chronic poor. Hereditary defects were considered to be the cause of such maladies, and these people were dubbed "lives not worth living." In 1933, the German Law on Preventing Hereditarily Diseased Progeny made it legal to involuntarily sterilize the blind, deaf, and epileptic, as well as those "useless" people already mentioned. The Nazis set up "eugenics courts" to decide cases of involuntary sterilization. Frederick Osborn, secretary of the American Eugenic Society, wrote a 1937 report summarizing the German sterilization programs, indicative of the fascination American eugenicists had for the Nazi agenda and the Nazi's ability to move this experiment to a scale never possible in the United States.

The Demise of Eugenics

With the Great Depression in 1929, the U.S. eugenics movement lost much of its momentum. Geneticist and evolutionary biologist Sewall Wright, although himself a member of the Eugenics Society of America, found fault with the genetics and the ideology of the movement: "Positive eugenics seems to require . . . the setting up of an ideal of society to aim at, and this is just what people do not agree on." He also wrote several articles in the 1930's challenging the assumptions of Fisher's genetic atomism model. In a speech to the Eugenics Society in New York in 1932, Müller pointed out the economic disincentive for middle and upper classes to reproduce, epitomized by the failure of many eugenicists to have children. Galton himself died childless. This inverse relationship between fertility and social status, coupled with the apparent predatory nature of the upper class, seemed to doom eugenics to failure.

Evolutionary biologist Stephen Jay Gould claims that the demise of the eugenics movement in the United States was more a matter of Adolf Hitler's use of eugenic arguments for sterilization and racial purification than it was of advances in genetic knowledge. Once the Holocaust and other Nazi atrocities became known, eugenicists distanced themselves from the movement. Depew and Weber have written that Catholic conservatives opposed to human intervention in reproduction and progressives, who began to abandon eugenics in favor of behaviorism (nurture rather than nature), were political forces that began to close down the eugenics movement, while Allen points out that the movement had outlived its political usefulness. Russian geneticist Theodosius Dobzhansky had by this time recognized the prime importance of context in genetics and consequently rejected the premise of eugenics, helping to push it into the realm of phony genetics.

Implications

The term "euphenics" is used to describe human genetic research that is aimed at improving the human condition, replacing the tainted term eugenics. Euphenics deals primarily with medical or genetic intervention that is designed to reduce the impact of defective genotypes on individuals (such as gene therapy for those with cystic fibrosis). However, in this age of increasing information about human genetics, it is necessary to keep in mind the important role played by environment and the malleability of human traits.

Allen argues that the eugenics movement may reappear (although probably under a different name) if economic problems again make it attractive to eliminate "unproductive" people. His hope is that a better understanding of genetics, combined with the lessons of Nazi Germany, will deter humans from ever again going down that path that journalist Jonathan Freedland calls "the foulest idea of the 20th century."

—Lee Anne Martínez

See Also: Biological Determinism; Heredity and Environment; Intelligence; Sociobiology; Sterilization Laws.

Further Reading: In *Darwinism Evolving* (1995), David Depew and Bruce Weber discuss the relationship between eugenics and Darwinian evolution and the role played by statistics in the origin of this movement. *The Nazi Connection* (1994), by Stefan Kühl, exposes the ties between the American eugenics movement and the Nazi program of "racial hygiene." Wray Herbert, "The Politics of Biology," *U.S. News and World Report* (April, 1997), considers the fluctuating viewpoint on the validity and social remedy for the genetics of human behavior. Garland E. Allen, "Eugenics and Public Health in American History," *Technology Review* 99 (August/September, 1996), discusses the connection between the eugenics movement and periods of economic or social hardship. Martin S. Pernick, "Science Misapplied: The Eugenics Age Revisited," *The American Journal of Public Health* 87 (November, 1997), is a fascinating exploration of the overlap between the goals, values, and concepts of disease in the public health and the eugenics movements in the early twentieth century.

Eugenics: Nazi Germany

Field of study: Human genetics

Significance: *Under the Nazi regime, Germany undertook the most extreme government-directed eugenics program in history. The German example raised worldwide awareness of the dangers of eugenics and did much to discredit eugenic theory.*

Key terms

ARYAN: the race believed by Nazis to have established the civilizations of Europe and India

NORDIC: the northernmost of the Aryan groups of Europe, believed by the Nazis to be the highest and purest racial group

EUTHANASIA: the killing of suffering people; sometimes referred to as "mercy" killing

Origins of Nazi Eugenic Thought

Nazi eugenic theory and practice grew out of two traditions: the eugenics movement founded by English scientist Francis Galton and racial theories of human nature. Most historians trace the origin of modern racial theories to French diplomat and writer Joseph-Arthur de Gobineau, who maintained that all great civilizations had been products of the Aryan, or Indo-Germanic, race. Through the late nineteenth and early twentieth centuries, German thinkers applied Galton's ideas to the problem of German national progress. The progress of the nation, argued scientists and social thinkers, could be best promoted by improving the German people through government-directed control of human reproduction. This type of eugenic thinking became known as "racial hygiene"; in 1904, eugenicists and biologists formed the Racial Hygiene Society in Berlin.

The Aryan mythology of Gobineau also grew in popularity. In 1899, an English admirer of Germany, Houston Stewart Chamberlain, published a widely read book entitled *The Foundations of the Nineteenth Century*. Chamberlain, heavily influenced by Gobineau, maintained that Europe's accomplishments had been the work of ethnic Germans, members of a healthy and imaginative race. Opposed to the Germans

The racial policies of Nazi Germany brought eugenics into disrepute. (Library of Congress)

were the Jews, who were, according to Chamberlain, impure products of crossbreeding among the peoples of the Middle East.

Basics of Nazi Eugenics

The Law to Prevent Hereditarily Sick Offspring, requiring sterilization of people with hereditary diseases and disabilities, was drafted and decreed in Germany in 1933. Before the Nazis came to power, many segments of German society had supported sterilization as a way to improve future generations, and Adolf Hitler's emergence as a national leader provided the pressure to ensure the passage of the law. Between 1934 and 1945, an estimated 360,000 people (about 1 percent of the German population) who were believed to have hereditary ailments were sterilized. Despite this law, the Nazis did not see eugenics primarily as a matter of discouraging the reproduction of unhealthy individuals and encouraging the reproduction of healthy individuals. Following the theories of Chamberlain, Adolf Hitler and his followers saw race, not individual health or abilities, as the distinguishing characteristic of human beings.

The Schutzstaffel (SS) organization was a key part of Nazi eugenic activities. In January, 1929, Heinrich Himmler was put in charge of the SS, a police force aimed at establishing order among the street fighters who formed a large part of the early Nazi Party. In addition to disciplining rowdy Nazis, the SS quickly emerged as a racial elite, the spearhead of an intended German eugenic movement. Himmler recruited physicians and biologists to help ensure that only those of the purest Nordic heritage could serve in his organization. In 1931, the agriculturalist R. Walther Darre helped Himmler draw up a marriage code for SS men, and Himmler appointed Darre head of an SS Racial Office. Himmler hoped to create the seeds of a German super race by directing the marriages and reproduction of the "racially pure" members of the SS.

Since the Nazis saw Germans as a "master race," a race of inherently superior people, they attempted to improve the human stock by encouraging the birth of as many Germans as possible and by encouraging those seen as racially pure to reproduce. The Nazis declared that women should devote themselves to bearing and caring for children. Hitler's mother's birthday was declared the Day of the German Mother. On this day, public ceremonies awarded medals to women with large numbers of children. The SS set up and maintained an organization of maternity homes for unmarried mothers of acceptable racial background and orphanages for their children; these institutions were known as the Lebensborn ("fountain of life"). There is some evidence that young women with desired racial characteristics who were not pregnant were brought to the Lebensborn to have children by the SS men to create "superior" Nordic children.

Impact

In addition to encouraging the reproduction of those seen as racially pure, the Nazis also sought to eliminate the unhealthy and the racially undesirable. In August, 1939, a committee of physicians and government officials, operating under Hitler's authority, issued a secret decree under which all doctors and midwives would have to register births of malformed or handicapped children. By October of that year, Hitler had issued orders for the "mercy killing" of these children and all those with incurable diseases. This euthanasia movement expanded from sick and handicapped children to those believed to belong to "sick" races. The T4 euthanasia organization, designed for efficient and secret killing, experimented with lethal injections and killing by injection and became a pilot program for the mass murder of the Jews during the Holocaust.

German racial hygienists had long advocated controlling marriages of non-Jewish Germans with Jews in order to avoid "contaminating" the German race. In July, 1941, Nazi leader Hermann Göring appointed SS officer Reinhard Heydrich to carry out the "final solution" of the perceived Jewish problem. At the Wannsee Conference in January, 1942, Hitler and his close associates agreed on a program of extermination. According to conservative estimates, between four million and five million European Jews died in Nazi extermination camps. When the murderous activities of the

Nazis were revealed to the world after the war, eugenics theory and practice fell into disrepute.

—*Carl L. Bankston III*

See Also: Miscegenation and Antimiscegenation Laws; Race; Sterilization Laws.

Further Reading: *Health, Race, and German Politics Between National Unification and Nazism, 1870-1945* (1989), by Paul Weindling, offers a definitive history of German racial eugenics. *Of Pure Blood* (1976), by Clarissa Henry and Marc Hillel, investigates the Lebensborn organization. In *Hitler's Willing Executioners: Ordinary Germans and the Holocaust* (1996), Daniel J. Goldhagen argues that the German people participated in the mass murder of Jews because Germans had come to see Jews as a racial disease.

Evolutionary Biology

Field of study: Population genetics

Significance: *The principles of evolution are the framework upon which the understanding of the rest of biology is built; a complete understanding of biological data comes only when this information is placed within the historical context of evolutionary biology. While the existence of evolution is firmly established, many questions remain about the specific causes of evolutionary change in particular groups of organisms. The science of evolutionary biology focuses on reconstructing the actual history of life and on understanding how evolutionary mechanisms operate in nature.*

Key terms

ADAPTATION: a feature that increases the ability of its bearer to survive and reproduce under prevailing environmental conditions

EVOLUTION: the process of change in the genetic structure of a population over time; descent with modification

FITNESS: the relative reproductive contribution of one individual to the next generation as compared to that of others in the population

GENETIC DRIFT: chance fluctuations in allele frequencies within a population resulting from random variation in the number of offspring produced by different individuals

GENOTYPE: the genetic makeup of an individual or group

HYBRIDIZATION: the exchange of genetic material between populations or species through mating between unlike forms

MUTATION: a change in the sequence of nucleotides along a strand of DNA

NATURAL SELECTION: the phenomenon of differing survival and reproduction rates among various genotypes; the frequency of the favored genotypes increases in succeeding generations

PHENOTYPE: the properties of an organism produced by the interaction of its genotype and its environment

PHYLOGENY: the history of descent of a group of species from a common ancestor

RECOMBINATION: the production of new combinations of alleles as a result of "crossing-over" between DNA strands during meiosis

An Evolutionary Context

Life is self-perpetuating, with each generation connected to previous ones by the thread of DNA passed from ancestors to descendants. Life on Earth thus has a single history much like the genealogy of a family, the shape and characteristics of which have been determined by internal and external forces. The effort to uncover that history and describe the forces that shape it constitutes the field of evolutionary biology. Moreover, the recognition of the historical nature of biological systems establishes a common and powerful perspective from which all biologists interpret their particular observations.

As an example of the need for this perspective, consider three vertebrates of different species, two aquatic (a whale and a fish) and one terrestrial (a deer). The two aquatic species share a torpedolike shape and oarlike appendages. These two species differ, however, in that one lays eggs and obtains oxygen from the water using gills, while the other produces live young and must breathe air at the surface. The terrestrial species has a different, less streamlined, shape and appendages for walking, but it too breathes air using lungs and produces live young. All three species are the same in having a bony skeleton typical of vertebrates. In order to understand why the various organisms display the features they do, it is necessary to consider what forces or historical constraints influence their genotype and subsequent phenotype.

It is logical to hypothesize that a streamlined shape is beneficial to the swimming creatures, as is the structure of their appendages. This statement is itself an evolutionary hypothesis; it implies that streamlined individuals will be more successful than less streamlined ones and so will become prevalent. However, it may initially be difficult to reconcile the differences between the two aquatic forms swimming side-by-side with the similarities between one of them and the terrestrial species walking around on dry land. However, if it is understood that the whale is more closely related to the terrestrial deer than it is to the fish, much of the confusion disappears. Using this comparative approach, it is unnecessary, and scientifically unjustified, to construct an elaborate scenario whereby breathing air at the surface is more advantageous to a whale than gills would be; the simpler explanation is that the whale breathes air because it (like the deer) is a mammal, and both species inherited this trait from a common ancestor sometime in the past.

Organisms are thus a mixture of two kinds of traits. Ecological traits are those the particular form of which reflects long-term adaptation to the species' habitat. Two species living in the same habitat might then be expected to be similar in such features, and different from species in other habitats. Evolutionary characteristics, on the other hand, indicate common ancestry rather than common ecology. Here, similarity between two species indicates that they are related to each other, just as familial similarity can be used to identify siblings in a crowd of people. In reality, all traits are somewhere along a continuum between these two extremes, but this distinction highlights the importance of understanding the evolutionary history of organisms and traits being studied. The value of an evolutionary perspective comes from its comparative and historical basis, which allows biologists to place their snapshot-in-time observations within a broader context of the continuous history of life.

Along with the principles of Mendelian genetics, the theory of evolution represents one of two truly universal sets of concepts in biology. These core concepts have applications to every area of biology, and all biologists must be familiar with them at some level. Both Mendelian genetics and the theory of evolution are at first glance (and in retrospect) remarkably simple and even obvious. The theory of evolution, however, is paradoxical in that it leads to extremely complex predictions and thus is often misunderstood, misinterpreted, and misapplied.

Early Evolutionary Thought

It is important to distinguish between the phenomenon of evolution and the various processes or mechanisms that may lead to evolution. The idea that species might be mutable, or subject to change over generations, dates back to at least the mid-eighteenth century, when the French naturalist Comte de Buffon, the Swiss naturalist Charles Bonnet, and even the Swedish botanist Carolus Linnaeus suggested that species (or at least "varieties") might be modified over time by intrinsic biological or extrinsic environmental factors. Other biologists after that time also promoted the idea that populations and species could evolve. Nevertheless, with the publication of *On the Origin of Species* in 1859, Charles Darwin became the first to propose that all species had descended from a common ancestor and that there was a single "tree of life." These claims regarding the historical fact of evolution, however, are distinct from the problem of how, or through what mechanisms, evolution could come about.

In the first decade of the nineteenth century, Jean Baptiste de Lamarck promoted the theory of inheritance of acquired characteristics to explain how species could adapt over time to their environments. His famous giraffe example illustrates the Lamarckian view: Individual giraffes acquire longer necks as a result of reaching for leaves high on trees, then pass that modified characteristic to their offspring. According to Lamarck's theories, a result of such adaptation, the species (and, in fact, each individual member of the species) is modified over time. While completely in line with early nineteenth century views of inheritance, this view of the mechanism of evolution has since been shown to be wholly incorrect.

In the mid-nineteenth century, Darwin and Alfred Russel Wallace independently developed the theory of evolution via natural selection, a theory that has proven perfectly consistent with the subsequent understanding of the genetics of inheritance first described by Gregor Mendel. Both Darwin and Wallace's arguments center on four observations of nature and a logical conclusion derived from those observations (presented here in standard genetics terminology, although Darwin and Wallace used different terms). First, variation exists in the phenotypes of different individuals in a population. Second, some portion of that variation is heritable, or capable of being passed from parents to offspring. Third, more individuals are produced in a population than will survive. Fourth, some individuals are, because of their particular phenotype, better able to survive and reproduce than others. From this, Darwin and Wallace deduced that those individuals whose phenotypes conferred on them greater fitness for survival would produce more offspring (genetically and phenotypically similar to themselves) than would less fit individuals; as such, the frequency of individuals with the favored genotype would increase in the next generation, though each individual would be unchanged throughout its lifetime. This process would continue as long as new genetic variants continued to arise and selection favored some over others. The theory of natural selection provided a workable and independently testable natural mechanism by which evolution of complex and sometimes very different adaptations could occur within and among species.

Evolutionary Biology After Darwin

Despite their theoretical insight, Darwin and Wallace had no knowledge of the genetic basis of inheritance. Mendel published his work describing the particulate theory of inheritance in 1865, but Darwin and Wallace appear to have been unaware throughout their lives that this vexing problem had been solved. In fact, Mendel's work went unnoticed by the entire scientific community for nearly fifty years; it was rediscovered, and its significance appreciated, in the first decade of the twentieth century. Over the next three decades, theoreticians integrated Darwin's theory of natural selection with the principles of Mendelian genetics. Simultaneously, Ernst Mayr, G. Ledyard Stebbins, George Gaylord Simpson, and Julian Huxley demonstrated that evolution of species and the patterns in the fossil record could be readily explained using Darwinian principles. This effort culminated in the 1930's and 1940's in the "modern synthesis," a fusion of thought that resulted in the development of the field of population genetics, a discipline via which biologists seek to describe and predict, quantitatively, evolutionary changes in populations and higher groups of organisms.

Since the modern synthesis, biologists have concentrated their efforts on applying the theories of population genetics to understanding the evolutionary dynamics of particular groups of organisms. More recently, techniques of phylogenetic systematics have been developed to provide a means of reconstructing ancestor-descendant, or genealogical, relationships among species. This effort has emphasized the need for a comparative and evolutionary approach to biology, which is essential to correct interpretation of data. In the 1960's, Motoo Kimura proposed the neutral theory of evolution, which challenged the "selectionist" view that patterns of genetic and phenotypic variation in most traits are determined by natural selection. The "neutralist" view maintains that much genetic variation, especially that seen in the numerous alleles of enzyme-coding genes, has little effect on fitness and therefore must be controlled by mechanisms other than selection. The last remaining frontier in the quest for a unified model of evolution is the integration of evolutionary theory with the understanding of the processes of development.

Evolutionary Mechanisms

Natural selection as described by Darwin and Wallace leads to the evolution of adaptations. However, many traits (perhaps the majority) are not adaptations; that is, differences in the particular form of those traits from one member of the species to the next do not lead to differences in fitness among those individuals. Such traits cannot evolve through natural selection, yet they can and do evolve. Thus, there must be additional mechanisms that lead to changes in the genetic structure of biological systems over time.

Evolutionary mechanisms are usually envisioned to act on individual organisms within a population. For example, natural selection may eliminate some individuals while others survive and produce a large number of offspring similar to themselves. As a result, evolution occurs within those populations. A key tenet of Darwinian evolution (which distinguishes it from Lamarckian evolution) is that populations evolve, but the individual organisms that constitute that population do not. While evolution of populations is certainly the most familiar scenario, this is not the only level at which evolution occurs.

Ernst Mayr (left) accepting a 1946 award from the Academy of Natural Sciences. (AP/Wide World Photos)

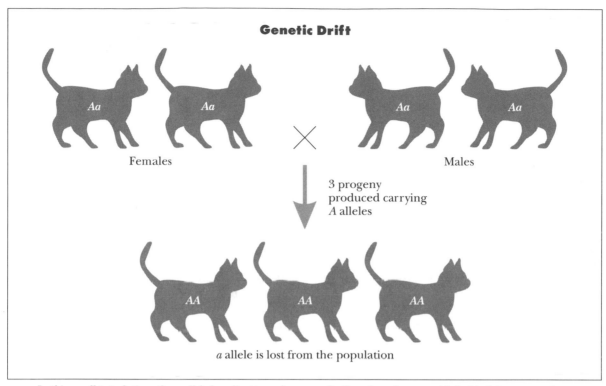

Genetic Drift

Aa   *Aa*   ×   *Aa*   *Aa*

Females   Males

3 progeny
produced carrying
A alleles

AA   *AA*   *AA*

a allele is lost from the population

In this small population, the a *allele has disappeared as a result of random chance and is lost to future generations.*

Richard Dawkins energized the scientific discussion of evolution with his book *The Selfish Gene,* first published in 1976. Dawkins argued that natural selection could operate on any type of "replicator," or unit of biological organization that displayed a faithful but imperfect mechanism of copying itself and that had differing rates of survival and reproduction among the variant copies. Under this definition, it is possible to view individual genes or strands of DNA as focal points for evolutionary mechanisms such as selection. Dawkins used this framework to consider how the existence of DNA selected to maximize its chances of replication (or "selfish DNA") would influence the evolution of social behavior, communication, and even multicellularity.

Recognizing that biological systems are arranged in a hierarchical fashion from genes to genomes (or cells) to individuals through populations, species, and communities, Elisabeth Vrba and Niles Eldredge in 1984 proposed that evolutionary changes could occur in any collection of entities (such as populations) as a result of mechanisms acting on the entities that make up that collection (individuals). Because each level in the biological hierarchy (at least above that of genes) has as its building blocks the elements of the preceding one, evolution may occur within any of them. Vrba and Eldredge further argued that evolution could be viewed as resulting from two general kinds of mechanisms: those that introduce genetic variation and those that sort whatever variation is available. At each level, there are processes that introduce and sort variation, though they may have different names depending on the level being discussed.

Natural selection is a sorting process. Other mechanisms that sort genetic variation include sexual selection, whereby certain variants are favored based on their ability to enhance reproductive success (though not necessarily survival), and genetic drift, which is especially important in small populations. While these forces are potentially strong engines for driving changes in genetic structure, their action—and, therefore, the direction and magnitude of evolutionary changes that they can cause—is constrained by the types of variation available and the extent to which that variation is genetically controlled.

Processes such as mutation, recombination, development, migration, and hybridization introduce variation at one or more levels in the biological hierarchy. Of these, mutation is ultimately the most important, as changes in DNA sequences constitute the raw material for evolution at all levels. Without mutation, there would be no variation and thus no evolution. Nevertheless, mutation alone is a relatively weak evolutionary force, only really significant in driving evolutionary changes when coupled with processes of selection or genetic drift that can quickly change allele frequencies. Recombination, development, migration, and hybridization introduce new patterns of genetic variation (initially derived from mutation of individual genes) at the genome, multicellular-organism, population, and species levels, respectively.

The Reality of Evolution

It is impossible to prove that descent with modification from a common ancestor is responsible for the diversity of life on Earth. In fact, this dilemma of absolute proof exists for all scientific theories; as a result, science proceeds by constructing and testing potential explanations, gradually accepting those best supported by new observations until they are either clearly disproved or replaced by another theory even more consistent with the data.

Darwin's concept of a single tree of life is supported by vast amounts of scientific evidence. In fact, the theory of evolution is among the most thoroughly tested and best-supported theories in all of science. As such, the view that evolution has and continues to occur is not debated by biologists; there is too much evidence to support its existence, across every biological discipline.

On a small scale, it is possible to demonstrate evolutionary changes experimentally or through observation. Spontaneous mutations that introduce genetic variation are well-documented; the origination and spread of drug-resistant forms of viruses and other patho-gens is clear evidence of this potential. Agricultural breeding programs and other types of artificial selection illustrate that the genetic structure of lineages containing heritable variation can be changed over time. For example, work by John Doebley begun in the late 1980's has suggested that the evolution of corn from its wild ancestor teosinte may have involved changes in as few as five major genes and that this transition likely occurred as a result of domestication processes established in Mexico between seven thousand and ten thousand years ago. The effects of natural selection can likewise be observed in operation: Peter Grant and his colleagues have demonstrated that during drought periods when seed availability is limited, deep-billed individuals of the Galapagos Island finch *Geospiza fortis* increase in proportion to the general population of the species, as only the deep-billed birds can crack the large seeds remaining after the supply of smaller seeds is exhausted. These and similar examples demonstrate that the evolutionary mechanisms put forward by Darwin and others do occur and lead to micro-evolution, or evolutionary change within single species.

Attempts to account for larger-scale macro-evolutionary patterns, such as speciation and origin of major groups of organisms, rely on indirect tests using morphological and genetic comparisons among different species, observed geographic distributions of species, and the fossil record. Such comparative studies rely on the concept of homology, the presence of corresponding and similarly constructed features among species.

At the most basic level, organization of the genetic code is remarkably similar across species; only minor variations exist among organisms as diverse as archaea (bacteria found in extreme environments such as hot springs, salt lakes, and habitats lacking in oxygen), bacteria, and eucarya (organisms with a true nucleus, including plants, animals, fungi, and their unicellular counterparts). This genetic homology extends as well to the presence of shared and similarly functioning gene sequences across biological taxa, such as homeotic genes common among all eukaryotes. Morphological homologies are also widespread; the limbs of mammals, birds, amphibians, and reptiles, for example, are all built out of the same arrangement of bones (although the particular shape of these bones can vary greatly among groups). None of these patterns would be predicted from theories of special creation or immutability of species. Rather, the conclusion that emerges from this weight of independent evidence is that structural homologies reflect an underlying evolutionary homology, or descent from a common ancestor.

The fossil record also contains patterns strongly consistent with the idea of evolution. Remains of vastly different organisms no longer in existence and the absence of many modern-day forms in the fossil record confirms that the diversity of life on Earth has changed over time. Further, many fossils exist that seem to document the evolutionary transition of one group to the next. *Archaeopteryx* was a feathered and probably flying reptile, while a large sequence of fossils illustrates the transition of bones in the skull of mammal-like reptiles (including *Diarthrognathus*, which had two functioning jaw joints). The order of appearance of fossils also supports an evolutionary view of life; for example, the expected sequence from non-vascular mosses and seedless ferns to seed and then flowering plants is perfectly laid out in the layers of rock.

The Practice of Evolutionary Biology

Contemporary evolutionary biology builds upon the theoretical foundations established by Darwin, Wallace, and the framers of the modern synthesis. While the reality of evolution is no longer in doubt, considerable debate remains about the importance of the various mechanisms in the history of particular groups of organisms. Much effort continues to be directed at reconstructing the particular historical path that life on Earth has taken and that has led to the enormous diversity of species. Likewise, scientists seek a fuller understanding of how new species arise, as the process of splitting lineages represents a watershed event separating microevolution and macroevolution.

Unlike in many other fields of biology, in evolutionary biology it is difficult to design simple cause-and-effect tests of hypotheses. Much of what evolutionary biologists are interested in understanding occurred in the past and over vast periods of time. In addition, the evolutionary outcomes observed in nature depend on such a large number of environmental, biological, and random factors that re-creating and studying the circumstances that could have led to a particular outcome is virtually impossible. Finally, organisms are complex creatures exposed to conflicting evolutionary pressures, such as the need to attract mates while simultaneously attempting to remain hidden from predators; such compromise-type situations are hard to simulate under experimental conditions.

Many evolutionary studies rely on making predictions about the patterns one would expect to observe in nature if evolution in one form or another were to have occurred, and such studies often involve synthesis of data derived from field work, theoretical modeling, and laboratory analysis. While such indirect tests of evolutionary hypotheses are not based on the sort of controlled data that are used in direct experiments, if employed appropriately, the indirect tests can be equally valid and powerful. Their strength comes from the ability to formulate predictions based on one species or type of data that may then be supported or refuted by examining additional species or data from another area of biology. In this way, evolutionary biologists are able to use the history of life on Earth as a natural experiment, and, like forensic scientists, to piece together clues to solve the greatest biological mystery of all.

—*Doug McElroy*

See Also: Classical Transmission Genetics; Hardy-Weinberg Law; Lamarckianism; Mutation and Mutagenesis; Natural Selection; Population Genetics; Speciation.

Further Reading: *The Selfish Gene*, 2d ed. (1989), by Richard Dawkins, provides a fascinating view of the driving force behind evolution; Dawkins has also written *The Blind Watchmaker* (1987) and *Climbing Mount Improbable* (1996), clear discussions of how natural selection can produce complex adaptations without the need for a "designer." Jonathan Weiner's *The Beak of the Finch* (1995) describes the work of Peter and Rosemary Grant on the evolution of Darwin's finches in the Galapagos Islands. In *Flight of the Dodo* (1997), David Quammen chronicles the rich experiences of Alfred Russel Wallace, so different from yet paralleled by the life of Charles Darwin. *Evolutionary Analysis* (1998), by Scott Freeman and Jon C. Herron, is an excellent textbook that presents evolutionary biology as a dynamic field of scientific inquiry. Stephen Jay Gould has authored numerous collections of essays on evolution, including *Eight Little Piggies* (1993); each provides thought-provoking case histories and insight into the pattern and process of evolution. Finally, Charles Darwin's *On the Origin of Species* (1859), while difficult, is an enormously thorough and visionary treatise on evolution and natural selection.

Extrachromosomal Inheritance

Field of study: Classical transmission genetics

Significance: *Extrachromosomal inheritance refers to the transmission of traits that are controlled by genetic factors located in chloroplasts or mitochondria, which are located in the cytoplasm. Nuclear or chromosomal traits are determined equally by both parents, but cytoplasm contribution is almost always by the female parent. The understanding of extrachromosomal inheritance is crucial since many important traits in plants and animals display this type of transmission. Mitochondrial DNA mutations have also been implicated in human disease and aging.*

Key terms

CHLOROPLAST: a self-replicating, chlorophyll-containing plant organelle (plastid)

GENOME: hereditary material in the nucleus of a cell

PLASMON: genetic factors in the cytoplasm of a cell (plasmagenes or cytogenes); a plastid plasmon is referred to as a "plastome"

MITOCHONDRIA: self-replicating, semiautonomous organelles in cells of most plants and animals that produce energy for the cell

PLASMAGENE: a self-replicating gene in a cytoplasmic organelle

The Discovery of Extrachromosomal Inheritance

Carl Correns, one of the three geneticists who rediscovered Austrian botanist Gregor Mendel's laws of inheritance in 1900, and Erwin Baur first described, independently, extrachromosomal inheritance of plastid color in 1909. However, they did not know then that they were observing the transmission patterns of organelle genes. Correns studied the inheritance of plastid color in the albomaculata strain of four-o'clock plants (*Mirabilis jalapa*), whereas Baur investigated garden geraniums (*Pelargonium zonate*). Correns observed that seedlings resembled the maternal parent regardless of the color of the male parent (uniparental-maternal inheritance). Seeds obtained from plants with three types of branches—with green leaves, white leaves, and variegated (a mixture of green and white) leaves—provided interesting results. Seeds from green-leaved branches produced only green-leaved seedlings, and seeds from white-leaved branches produced only white-leaved seedlings. However, seeds from branches with variegated leaves resulted in varying ratios of green-leaved, white-leaved, and variegated-leaved offspring. The explanation is that plastids in egg cells of the green-leaved branches and white-leaved branches were only of one type (homoplasmic or homoplastidic)—that is, normal chloroplasts in the green-leaved cells and white plastids (leukoplasts) in the white-leaved cells. The cells of the variegated branches, on the other hand, contained both chloroplasts and leukoplasts (heteroplasmic or heteroplastidic) in varying proportions. Some descendants of the heteroplastidic cells received only chloroplasts, some received only leukoplasts, and some received a mixture of the two types of plastids in varying proportions in the next generation, hence variegation.

Baur observed similar progeny from reciprocal crosses between normal green and white *Pelargonium* plants. Progeny in both cases were of three types: green, white, and variegated, in varying ratios. This indicated that cytoplasm was inherited from the male as well as the female parent; however, the transmission of plastids was cytoplasmic. Male transmission of plastids has also been observed in oenothera, snapdragons, beans (*Phaseolus*), potatoes, and rye. Rye is the only member of the grass family that exhibits both maternal and paternal inheritance of plastids.

The investigations on plastid inheritance also clearly established that in plants exhibiting uniparental-maternal inheritance, a variegated maternal parent always produces green, white, and variegated progeny in varying proportions because of its heteroplastidic nature. Crosses between green and white plants always yield green or white progeny, depending upon the maternal parent, when the parental plants are homoplasmic for plastids.

Extrachromosomal Inheritance Versus Nuclear Inheritance

Extrachromosomal inheritance has been found in many plants, including barley, maize, and rice. Traits are inherited through chloroplasts, mitochondria, or plasmids (small, self-replicating structures). Inheritance of traits that are controlled by organelle genomes (plasmons) can be called "nonnuclear" or "cytoplasmic." The cytoplasm contains, among other organelles, mitochondria in all higher organisms and chloroplasts in plants. Because cytoplasm is almost always totally contributed by the female parent, this type of transmission may also be called "maternal" or "uniparental" inheritance.

Most chromosomally inherited traits obey Mendel's law of segregation, which states that a pair of alleles or different forms of a gene separate from each other during meiosis (the process that halves the chromosome number in gamete formation). They also follow the law of independent assortment, in which two alleles of a gene assort and combine independently with two alleles of another gene. Such traits may be called Mendelian traits. Extrachromosomal inheritance is one of the exceptions to Mendelian inheritance. Thus, it can be called non-Mendelian inheritance. (Mendel only studied and reported on traits controlled by nuclear genes.) Mendelian heredity is characterized by regular ratios in segregating generations for qualitative trait differences and identical results from reciprocal crosses. On the contrary, non-Mendelian inheritance is characterized by a lack of regular segregation ratio and nonidentical results from reciprocal crosses.

The mitochondria are the sites of aerobic respiration (the breaking down of organic substances to release energy in the presence of oxygen) in both plants and animals. They are, like plastids, self-replicating entities and exhibit genetic continuity. The mitochondrial genes do not exhibit the Mendelian segregation pattern either. Mitochondrial genetics began around 1950 with the discovery of "petite" mutations in baker's yeast (*Saccharomyces cerevisiae*). Researchers observed that one or two out of every one thousand colonies grown on culture medium were smaller than normal colonies. The petite colonies bred true (produced only petite colonies). The petite mutants were respiration deficient under aerobic conditions. The slow growth of the petite colonies was related to the loss of a number of respiratory (cytochrome) enzymes that occur in mitochondria. These mitochondrial mutants, termed "vegetative petites," can be induced with acriflavine and other related dyes. Another type of mutation, called a "suppressive petite," was found to be caused by defective, rapidly replicating mitochondrial deoxyribonucleic acid (DNA). Petite mutants that are strictly under nuclear gene control have also been reported and are called "segregational petite" mutants. Most respiratory enzymes are under both nuclear and mitochondrial control, which is indicative of collaboration between the two genetic systems.

In the fungus neurospora, mitochondrial inheritance has been demonstrated for mutants referred to as "poky" (a slow-growth characteristic). The mutation resulted from an impaired mitochondrial function related to cytochromes involved in electron transport. The mating between poky female and normal male yields only poky progeny, but when the cross is reversed, the progeny are all normal, confirming maternal inheritance for this mutation.

According to a 1970 study, cytoplasmic male sterility is found in about eighty plant species. The molecular basis of cytoplasmic male sterility in maize through electrophoretic separation of restriction-endonuclease-created fragments of DNA has been traced to mitochondrial DNA. Cytoplasmic male sterility can be overcome by nuclear genes. Plasmids, residing in mitochondria, are also important extrachromosomal DNA molecules that are especially important in antibiotic resistance. Plasmids have been found to be extremely useful in genetic engineering.

Mutator Genes

Plastome mutations can be induced by nuclear genes. A gene that increases the mutation rate of another gene is called a "mutator." One such gene is the recessive, nuclear iojap (*ij*) mutation in maize. In the homozygous (*ij ij*) condition, it induces a plastid mutation. The name "iojap" has been derived from "Iowa" (the maize strain in which the mutation is found) and "japonica" (a type of striped variety that the mutation resembles). Once the plastid gene mutation caused by the *ij* gene has been initiated, the inheritance is non-Mendelian, and it no longer depends on the nuclear *ij* gene. As long as the iojap plants are used as female parents, the inheritance of the trait is similar to that for plastids in the albomaculata variety of four o'clock plants.

The *chm* mutator gene causes plastid mutations in the plant arabidopsis, and mutator "striata" in barley causes mutations in both plastids and mitochondria. Cases of mutator-induced mutations in the plastome have also been reported in rice and catnip.

Chloroplast and Mitochondrial DNA

Plastids contain DNA, have their own DNA polymerase (the enzyme responsible for DNA replication), and undergo mutation. The chloroplast DNA (ctDNA) is a circular, self-replicating system that carries genetic information that is transcribed (DNA to ribonucleic acid, or RNA) and translated (RNA to protein) in the plastid. It replicates in a semiconservative manner—that is, an original strand of DNA is conserved and serves as the template for a new strand in a manner similar to nuclear DNA replication.

The soluble enzyme ribulose biphosphate carboxylase/oxygenase (rubisco) is involved in photosynthetic carbon dioxide fixation. In land plants and green algae, its large subunit is a ctDNA product, while its small subunit is controlled by a nuclear gene family. Thus, the rubisco protein is, as are chloroplast ribosomes, a product of the cooperation between the nuclear and chloroplast genes. In all other algae, both the large and small subunits of rubisco are encoded in ctDNA.

Mitochondrial DNA (mtDNA) molecules are also circular and self-replicating. Human, yeast, and higher plant mtDNA are the major systems that have been studied. The human mtDNA has a total of 16,569 base pairs. The yeast mtDNA is five times larger than that (84 kilobases), and maize mtDNA is much larger than the yeast mtDNA. Every base pair of human mtDNA may be involved in coding for a mitochondrial messenger RNA for a protein, a mitochondrial ribosomal RNA, or a mitochondrial transfer RNA. It is compact, showing no intervening, noncoding base sequences between genes. It has only one major promoter (a DNA region to which an RNA polymerase binds and initiates transcription) on each strand. Most codons—triplets of nucleotides (bases) in messenger RNA carrying specific instructions from DNA—have the same meaning as in the universal genetic code, except the following differences: UGA represents a "stop" signal (universal), but represents tryptophan in yeast and human mtDNA; AUA represents isoleucine (universal), but methionine in human mtDNA; CUA represents leucine (universal), but threonine in yeast mtDNA; and CGG represents arginine (universal), but tryptophan in plant mtDNA.

The mtDNA carries the genetic code (plasmagene names in parentheses) for proteins, such as cytochrome oxidase subunits I (coxl), II (cox2), and III (cox3); cytochrome B (cytb); and ATPase subunits 6 (atp6), 8 (atp8), and 9 (atp9). It also contains the genetic codes for several ribosomal RNAs, such as mtrRNA 16s and 12s in mouse; mtrRNA 9s, 15s, and 21s in yeast; and mtrRNA 5s, 18s, and 26s in maize. In

The Structure of Eukaryotic Cells

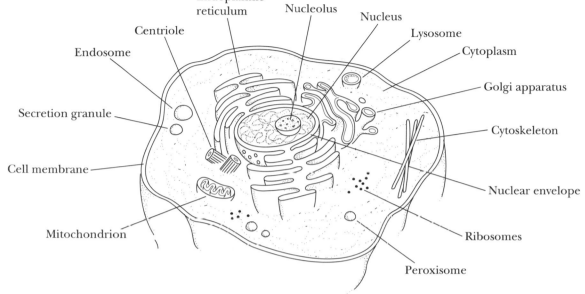

Endoplasmic reticulum

Nucleolus

Nucleus

Centriole

Lysosome

Endosome

Cytoplasm

Golgi apparatus

Secretion granule

Cytoskeleton

Cell membrane

Nuclear envelope

Mitochondrion

Ribosomes

Peroxisome

Eukaryotic Animal Cell

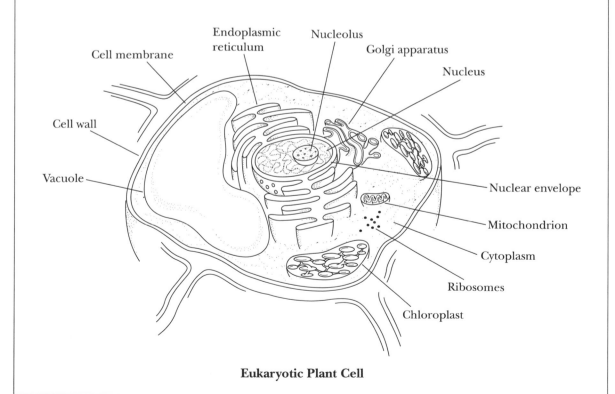

Cell membrane

Endoplasmic reticulum

Nucleolus

Golgi apparatus

Cell wall

Nucleus

Vacuole

Nuclear envelope

Mitochondrion

Cytoplasm

Ribosomes

Chloroplast

Eukaryotic Plant Cell

addition, twenty-two transfer RNAs in mice, twenty-four in yeast, and three in maize are encoded in mtDNA.

Chlamydomonas Reinhardtii

Chlamydomonas reinhardtii (*C. reinhardtii*) is a unicellular green algae in which chloroplast and mitochondrial genes show uniparental transmission. In 1954, Ruth Sager discovered the chloroplast genetic system. Resistance to high levels of streptomycin (a trait controlled by chloroplast genes) has been shown to be transmitted uniparentally by the mt^+ mating type parent. The mt^- mating type transmits the mitochondrial genes uniparentally. Mutants in chloroplasts have been identified for antibiotic and herbicide resistance. Genetic recombination is common in *C. reinhardtii*, which occurs in zygotes (resulting from the fusion of gametes of opposite sexes) when biparental cytogenes are in a heterozygous (union of unlike genes) state. This is an ideal system among plants for recombination studies since there is only one large plastid per cell. In higher plants, study of genetic recombination is difficult because of a large number of plastids in cells and a lack of genetic markers.

Mutations in mitochondria of *C. reinhardtii* can be induced with acriflavine or ethidium bromide dyes. Point mutations for myxothiazol resistance mapping in the *cytb* gene have been isolated. The mitochondrial genome of this algae has been completely sequenced. It encodes five of more than twenty-five subunits of the reduced nicotinamide-adenine dinucleotide (NADH) dehydrogenase of complex I (nad1, nad2, nad4, nad5, and nad6), the COX I subunit of cytochrome oxidase (cox1), and the apocytochrome b (cob) subunit of complex III. All of these proteins have a respiratory function.

Origin of Plastid and Mitochondrial DNA

According to the endosymbiont theory, plastids and mitochondria in eukaryotes are the descendants of prokaryotic organisms that invaded primitive eukaryotes. Subsequently, they developed a symbiotic relationship and became dependent upon each other. There is much support for this theory. Researchers in 1972 showed homology (genetic similarity) be-

tween ribosomal RNA from cyanobacteria and DNA from the chloroplasts of *Euglena gracilis*. This provided support for chloroplasts as the descendants of cyanobacteria. Mitochondria are believed to have come from primitive bacteria and plastids from blue-green algae. Molecular evidence strongly supports the endosymbiotic origin of mitochondria from alpha purple bacteria.

In 1981, Lynn Margulis summarized evidence for this theory. There are many similarities between prokaryotes and organelles: Both have circular DNA and the same size ribosomes, both lack histories and a nuclear membrane, and both show similar response to antibiotics that inhibit protein synthesis. Both also show a primitive mode of translation that begins with formulated methionine. The discovery of promiscuous DNA (DNA segments that have been transferred between organelles or from a mitochondrial genome to the nuclear genome) in eukaryotic cells also lends support to this theory.

Impact and Applications

Genetic investigations have helped tremendously in constructing a genetic map of maize ctDNA. Important features of the map, including two large, inverted, repeat segments containing several ribosomal RNA and transfer RNA genes, are now known. Detection and quantification of mutant mtDNA are essential for the diagnosis of diseases and for providing insights into the molecular basis of pathogenesis, etiology, and ultimately the treatment of diseases. This should help enhance the knowledge of mitochondrial biogenesis. Mitochondrial dysfunction, resulting partly from mutations in mtDNA, may play a central role in organismal aging.

A number of human diseases associated with defects in mitochondrial function have been identified. Large-scale deletions and transfer RNA point mutations (base changes) in mtDNA are associated with clinical mitochondrial encephalomyopathies. Heteroplasmy (the coexistence of more than two types of mtDNA) has provided experimental systems in which the transmission of mtDNA in animals can be studied. Numerous deleterious point mutations of mtDNA are associated with various types of human disorders involving deficiencies in the mitochondrial oxidative phosphorylation (respiration) apparatus. Leigh disease is caused by a point mutation in mtDNA. Deletions of mtDNA have been associated with diseases such as isolated ocular myopathy, chronic progressive external ophthalmoplegia, Kearn-Sayre's syndrome, and Pearson's syndrome.

The influence of the mitochondrial genome and mitochondrial function on nuclear gene expression is poorly understood, but progress is being made toward understanding why a few genes are still sequestered in the mitochondria and toward developing new tools to manipulate mitochondrial genes.

—Manjit S. Kang

See Also: Chloroplast Genes; Hereditary Diseases; Mitochondrial Genes; Mutation and Mutagenesis; Protein Synthesis.

Further Reading: *Methods in Enzymology: Mitochondrial Biogenesis and Genetics,* volumes 260 (1995) and 264 (1996), edited by Giuseppe M. Attardi and Anne Chomyn, provide reviews of mitochondrial genetics. Nicholas W. Gillham, *Organelle Genes and Genomes* (1994), contains a comprehensive review of the genetic-molecular aspects of chloroplasts and mitochondria.

Forensic Genetics

Field of study: Human genetics
Significance: *Forensic genetics uses deoxyribonucleic acid (DNA) or the inherited traits derived from DNA to identify individuals involved in criminal cases. Blood tests and DNA testing are used to determine the source of evidence, such as blood stains or semen, left at a crime scene.*

Key terms

ALLELES: alternative versions of genes that determine an individual's traits

DEOXYRIBONUCLEIC ACID (DNA): the biological structure that stores the genetic information in most living organisms; two linear molecules made of nucleotides entwine to form the DNA double helix

DNA FINGERPRINTING: a DNA test used by forensic scientists to aid in the identification of criminals or to resolve a parentage dispute

ENZYMES: proteins that assist or accelerate specific biological reactions without themselves being altered by the reactions

FORENSIC SCIENCE: the application of scientific knowledge to questions of civil and criminal law, especially in court proceedings

Forensic Science and DNA Analysis

Forensic scientists use genetics for two major legal applications: identifying the source of a sample of blood, semen, or other tissue, and establishing the biological relationship between two people in paternity or maternity suits. This evidence can be used in the solution of criminal cases; furthermore, forensic scientists are frequently called upon to testify as expert witnesses in criminal trials. One of the most useful sources of inherited traits for forensic science purposes is blood. Such traits include blood type, proteins found in the plasma, and enzymes found in blood cells. The genes in people that determine such inherited traits have many different forms (alleles), and the specific combination of alleles for many of the inherited blood traits can be used to identify an individual. The number of useful blood group systems is small, however, which means that a number of individuals might have blood groups identical to those of the subject being tested.

The ultimate source of genetic information for identification of individuals is the DNA found in the chromosomes in the nucleus of a cell. Using a class of enzymes known as restriction enzymes, DNA strands are cut into segments, forming bands similar to a supermarket bar code that vary with people's family lines. The pattern, termed a DNA "fingerprint" or profile, is inherited much like the alleles for blood traits. DNA fingerprinting can be used to establish biological relationships (including paternity) with great reliability because a child cannot have a variation that is not present in one of the parents. Since DNA is stable and can be reliably tested in dried blood or semen even years after a crime has been committed, DNA fingerprinting has revolutionized the solution of criminal cases in which biological materials are the primary evidence. The likelihood of false matches ranges from one per million to one per billion. These numbers, however, do not include the possibility of mishandling of evidence, laboratory errors, or planting of evidence.

Criminal Cases Involving DNA Evidence

On November 6, 1987, serial rapist Tommy Lee Andrews became the first American ever convicted in a case involving DNA evidence. Samples of semen left at the crime scene by the rapist and blood taken from Andrews were sent to a New York laboratory for testing. Using the techniques of DNA fingerprinting, the laboratory isolated DNA from each sample, compared the patterns, and found a DNA match between the semen and the blood. Andrews was sentenced to twenty-two years in prison for rape, aggravated battery, and burglary.

The 1990-1991 *United States v. Yee* homicide trial in Cincinnati, Ohio, was the first major case that challenged the soundness of DNA testing methods. DNA analysis by the Federal Bureau of Investigation (FBI) showed a match between blood from the victim's van and from Steven Yee's car. The defense claimed that the matching DNA data were ambiguous or inconsistent, citing what they claimed to be errors, omissions, lack of controls, and faulty analysis. However, after a fifteen-week hearing, the judge accepted the DNA testing as valid.

In 1994, O. J. Simpson was arrested and charged with the murders of his ex-wife, Nicole Brown, and her friend, Ronald Goldman. Blood with DNA that matched Simpson's was found at Brown's home, and blood spots in Simpson's car contained DNA matching Brown's, Goldman's, and Simpson's. Furthermore, blood at Simpson's home contained DNA that matched Brown's and Goldman's. For the most part, the defense admitted the accuracy of the DNA tests and did not scientifically challenge the results of the DNA fingerprinting. Instead, they argued that the biological evidence had been contaminated by shoddy

laboratory work and by planting of evidence; the jury found Simpson not guilty of the charges against him.

Impact and Applications

Between 1994 and 1997, forensic genetics were used as evidence in approximately 10,000 criminal investigations and over 130,000 paternity testing cases per year in the United States. In addition, there have been numerous cases in which forensic DNA testing has been used to free previously convicted and incarcerated individuals, some of whom have been in prison for over twelve years. Furthermore, most states now have data banks containing DNA profiles of people already convicted of sexual or related offenses. When police begin investigating a

O. J. Simpson and attorneys discuss strategy for cross-examining a forensic scientist during Simpson's 1995 murder trial. (AP/Wide World Photos)

new crime, they can test DNA collected at the scene to see if it matches anybody in the data bank with a history of a similar offense.

—*Alvin K. Benson*

See Also: Behavior; DNA Fingerprinting; Human Genetics; Paternity Tests; Restriction Enzymes.

Further Reading: *An Introduction to Forensic DNA Analysis* (1997), by Keith Jarman and Norah Rudin, discusses applications of DNA fingerprinting to forensic science. *Convicted by Juries, Exonerated by Science* (1996), by Edward Connors et al., provides case studies in the use of DNA evidence to establish innocence after conviction in a trial. *The Evaluation of Forensic DNA Evidence* (1996), by the United States National Research Council, presents an evaluation of the power and validity of forensic DNA methods. *DNA in the Courtroom* (1994), by Howard Coleman and Eric Swenson, gives a good overview of DNA fingerprinting, expert evidence in court, and applications of forensic genetics.

Fragile X Syndrome

Field of study: Human genetics
Significance: *There are more than fifty mental retardation disorders associated with the X chromosome, but their frequencies are rare. Fragile X syndrome represents a distinctive type of mental retardation. It is the most common inherited form of mental retardation, affecting an estimated one in fifteen hundred males and one in twenty-five hundred females.*

Key terms

DEOXYRIBONUCLEIC ACID (DNA): the molecules composing genes and chromosomes; four types of molecules make up DNA: adenine (A), thymine (T), cytosine (C), and guanine (G)

SEX CHROMOSOMES: gender is determined by the presence of two X chromosomes for females and an X chromosome and a Y chromosome for males; these chromosomes are received from an individual's parents, each of whom contributes one sex chromosome to their offspring

GENE: a linear organization of molecules in DNA serving as the blueprint for a protein; chromosomes are composed of genes

INHERITANCE: the genes that are received from one's parents; there are two genes for every trait, one contributed by each parent

History of Fragile X Syndrome

In 1969, a family of four mentally retarded brothers was described in which the tip of the X chromosome appeared to be detached from the rest of the chromosome. It is now recognized that this fragile site occurs in the vicinity of a gene that has been named for the fragile X syndrome associated with it: the *FMR1* gene.

Males affected with fragile X syndrome have moderate to severe mental retardation and show distinctive facial features, including a long and narrow face, large and protruding ears, and a prominent jaw. Additional features include velvet-like skin, hyperextensible finger joints, and double-jointed thumbs. These features are generally not observed until maturity. Prior to puberty, the only symptoms a child may have are delayed developmental milestones, such as sitting, walking, and talking. Fragile X children may also display an abnormal temperament marked by tantrums, hyperactivity, or autism. A striking feature of most adult fragile X males is an enlarged testicular volume (macroorchidism). This enlargement is not a result of testosterone levels, which are normal in these men. Fragile X men are fertile, and offspring have been documented, but those with significant mental retardation rarely reproduce.

The intelligence quotient (IQ) of the majority of affected males is in the moderate to severely retarded range. Only a few affected males have IQs above seventy-five. Fragile X males frequently show delayed speech development and a paucity of language. Repetitive speech patterns may also be present. Hyperactivity among these males appears common and may be the only symptom observable in a young boy.

Mode of Inheritance

The pattern of inheritance of a gene on the X chromosome is known as X-linked inheritance. In males, any abnormal gene on the X

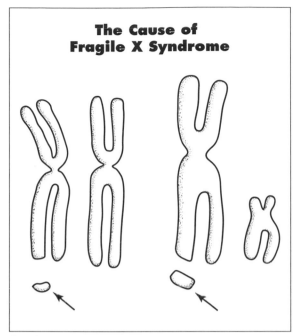

The Cause of Fragile X Syndrome

Fragile X syndrome in females (left) and males (right). Note the apparently detached tips of the X chromosomes.

chromosome is expressed because males have only one X chromosome. In females, two copies of the X chromosome are present in each cell, but only one of the X chromosomes is active in each cell. Which X chromosome is active is a random event. Therefore, approximately half of a person's cells have an active maternal X chromosome, whereas the others have an active paternal X chromosome. For this reason, females with fragile X syndrome will often have half of their cells with a normal X chromosome and will not appear as affected as males.

The pattern of inheritance for fragile X is unusual. Fragile X syndrome increases in severity through successive generations. This is explained by an increase in the defect of the *FMR1* gene as it is passed from mothers to sons. Since males contribute the Y chromosome to their sons, fathers do not pass the fragile X

gene to their sons. They will, however, contribute their X chromosome to their daughters. Because these daughters also receive an X chromosome from their mothers, they generally appear normal or only mildly affected. It is only when these daughters have a son that the condition is expressed. Therefore, the increasing severity of the syndrome results while the gene is a part of the son's mother.

An explanation for this increasing severity through generations was discovered by analyzing the DNA sequence of the *FMR1* gene. The molecules composing DNA are adenine (A), thymine (T), cytosine (C), and guanine (G). In fragile X syndrome, a region of three molecules, CGG, was found to have been repetitively increased hundreds of times. This caused the *FMR1* gene not to be expressed in males with the fragile X syndrome. Individuals not having the fragile X syndrome have a working *FMR1* gene. From this evidence, researchers have concluded that the *FMR1* gene actually prevents mental retardation.

—*Linda R. Adkison*

See Also: Aggression; Classical Transmission Genetics; Criminality; Genetic Testing; X Chromosome Inactivation.

Further Reading: Stephen T. Warren, "Trinucleotide Repetition and Fragile X Syndrome," *Hospital Practice* 32 (April 15, 1997), provides more detail about CGG repeats in fragile X syndrome. Michael A. Schmidt, "Fragile X Syndrome: Diagnosis, Treatment, and Research," *The Journal of the American Medical Association* 277 (April 9, 1997) provides a more detailed discussion of the fragile X syndrome. David Hirsch, "Fragile X Syndrome: Medications for Aggressive Behavior?" *The Exceptional Parent* 26 (October, 1996), answers parental concerns about medications for males displaying aggressive behavior. David Hirsch, "Fragile X Syndrome," *The Exceptional Parent* 25 (June, 1995), answers parental concerns about fragile X syndrome genetic testing.

Gel Electrophoresis

Field of study: Molecular genetics

Significance: *Gel electrophoresis is a laboratory technique involving the movement of charged molecules in a buffer solution when an electric field is applied to the solution. The technique allows scientists to separate DNA, RNA, and proteins according to their size. The method is the most widely used way to determine the molecular weight of these molecules and can be used to determine the approximate size of most DNA molecules and proteins.*

Key terms

DEOXYRIBONUCLEIC ACID (DNA): the genetic material found in all cells

GEL: a support matrix formed by interconnecting long, multiply repeated unit chains into a porous, solid material that retards the movement of molecules entrapped in it

RIBONUCLEIC ACID (RNA): a form of nucleic acid in the cell used primarily for genetic expression through transcription and translation

STAINING DYE: a chemical with a high affinity for DNA, RNA, or proteins that cause a visible color signal that allows the detection of these molecules in the gel

Basic Theory of Electrophoresis

As part of their molecular experiments, scientists often need to determine the approximate size of deoxyribonucleic acid (DNA) fragments, ribonucleic acid (RNA), or proteins. Although a short stretch of DNA can contain amazing amounts of information for the production of proteins capable of widely varied functions, both the gene and gene product are much too small to be visualized with normal microscopes. The size of a piece of DNA capable of carrying all the information needed for a single gene may be only 2 microns long and 20 angstroms wide, while the protein encoded by this gene might form into a globular ball only 2.5 to 10 nanometers in diameter. Therefore, some indirect method of "seeing" the length of these molecules must be used to determine their sizes. The easiest and by far most common way to do this is gel electropho-

resis. The basic idea is very simple. If molecules can be induced to move in the same direction through a tangled web of material, smaller molecules will be less retarded and will be able to move further through the matrix. This will cause a separation by size, and the distance a molecule moves will be related to its size in a way that will allow an approximation of its relative molecular weight.

As an analogy, imagine a family with two children picnicking by a thick, brushy forest. Their small dog runs into the brush, and the whole family runs in after it. The dog, being the smallest, penetrates into the center of the forest. The six-year-old can duck through many of the branches and manages to get two-thirds of the way in; the twelve-year-old makes it halfway; the mother gets tangled up and must stop after only a short distance; the father, too large to fit in anywhere, cannot enter at all. This is what happens to molecules moving through a gel: Some travel through with no retardation, others are separated into easily visualized size groups, and others cannot even enter the matrix.

The Electrophoresis Setup

The electrophoresis setup usually involves two glass plates held a short distance apart (usually from 0.1 to 3 millimeters) by thin plastic strips (spacers) on either edge. A matrix solution is poured between the plates in a liquid form that solidifies into a gel. A piece of plastic with alternating indentations like an oversized comb is pushed into the top. When the gel has solidified, the "comb" is removed, leaving small depressions in the matrix (wells) into which the DNA, RNA, or protein sample is applied. The plates are then attached to an apparatus that submerges the top and bottom in a salt buffer attached to an electric power supply. The buffer allows an even application of the electric field.

Since the molecules of interest are so small, matrices with small pore size must be created. It is important to find a matrix that will properly separate the molecules being studied. To return to the previous analogy, if the family is next to a forest of tall trees and the only blockage is from widely spaced trunks, they will move at roughly the same rate and there will be no size separation. On the other hand, if the brush is a thick hedge, no one will be able to enter, leaving everyone stuck outside the brush. The key is to find a material that creates pores large enough to let DNA or proteins enter but small enough to create significant slowing of larger molecules. The two most common components used are agarose and acrylamide. By using different concentrations of these materials, scientists can create gels that will separate anything from very short pieces of DNA that differ by only a single nucleotide to whole chromosomes.

Agarose is composed of long, linear chains of multiple saccharides (sugars). At high temperatures (95 degrees Celsius), the agarose will melt in a water/salt buffer solution. This can be poured into the gel plates. As the gel cools to around 50 degrees Celsius, the long chains begin to wrap around each other and solidify into a gel. The concentration of agarose determines the pore size since a larger concentration will create more of a tangle. Agarose is usually used with large DNA or RNA molecules.

Acrylamide is a short molecule made up of a core of two carbons connected through a double bond with a short side-chain with a carboxyl and amino group. When the reactive chemicals ammonium persulfate and TEMED are added, the carbon ends fuse together to create long chains of polyacrylamide ("many acrylamides"). If this were the only reaction, the end result would be much like agarose. However, a small amount (usually 5 percent or less) of the acrylamides are a related molecule called bis-acrylamide, a two-headed version of the acrylamide molecule. This allows the formation of interconnecting branch points every twenty to fifty acrylamides on the chain, which creates a pattern more like a net than the tangled strands of agarose. This results in a narrower pore size than agarose, which allows

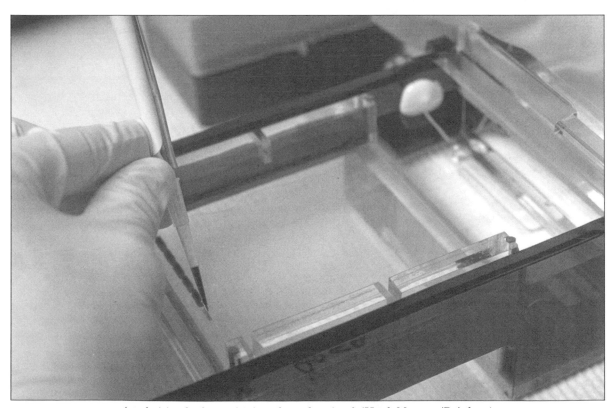

A technician loads proteins into electrophoresis gel. (Hank Morgan/Rainbow)

the separation of much smaller fragments. Acrylamide is used to separate proteins and small DNA fragments and for sequencing gels in which DNA fragments differing in size by only a single nucleotide must be clearly separated.

Why Nucleic Acids and Proteins Move in a Gel

DNA and RNA will migrate in an electric field since every base is connected by a negatively charged phosphodiester bond. This means that the DNA is negatively charged and will migrate toward the positive pole if placed in an electric field. In fact, since each base contributes the same level of charge, the amount of negative charge is directly proportional to the length of the DNA. This means that the electromotive force on any piece of DNA or RNA is directly proportional to its length (and therefore its mass) and that the rate of movement (unless the gel matrix retards it) should be identical for any piece of DNA.

The charge on different amino acids varies considerably, and the number of each amino acid in a protein varies drastically. Therefore, the charge on a protein has nothing to do with its length and can vary widely from one protein to another. To correct for this, proteins are mixed with the detergent sodium dodecyl sulfate, or SDS (the same material that gives most shampoos their suds), before being loaded onto the gel. The detergent coats the protein evenly. This has two important effects. The first is that the globular structure of the protein is disrupted, and the polypeptide chain will largely exist as a long strand. This is important because a tightly balled protein would be less retarded in moving through the polyacrylamide matrix than a linear molecule, and proteins with the same molecular weight might appear to be different sizes. More important, each SDS molecule has a slight negative charge, so the even coating of the protein results in a negative charge that is directly proportional to the size of the protein.

Once the molecules have been subjected to the electric field for a length of time sufficient to separate them on the gel, they must be visualized so that they can be identified. This is done by soaking the gel in a solution that contains a dye that stains the electrophoresed molecules. For DNA and RNA, this dye is usually ethidium bromide, a compound whose structure resembles a base pair. This creates an affinity for nucleic acids, and the dye slides in or intercalates into the helix. The dye, when exposed to ultraviolet light, glows orange, revealing the position of the nucleic acid to which it is bound. For proteins, the dye Coomassie Blue is usually used.

—J. Aaron Cassill

See Also: DNA Fingerprinting; DNA Structure and Function; Protein Structure; Restriction Enzymes.

Further Reading: *Gel Electrophoresis of Nucleic Acids: A Practical Approach* (1982) and *Gel Electrophoresis of Proteins: A Practical Approach* (1981), both edited by B. D. Hames and D. Rickwood, review the basic theories of electrophoresis and discuss many methods and applications; the books include many diagrams of the equipment used and the results typically achieved. *Recombinant DNA* (1992), by James D. Watson et al., covers the use of many molecular genetics techniques in elucidating the function of genes.

Gender Identity

Field of study: Human genetics
Significance: *Researchers have long sought an understanding of the basis of human gender identity. Discoveries in the field of human genetics have opened the way to examine how genes affect sexual behavior and sexual identity.*

Key terms

HERMAPHRODITE: an individual who has both male and female sex organs

RESTRICTION FRAGMENT LENGTH POLYMORPHISM (RFLP): a technique involving the cutting of DNA that allows researchers to compare genetic sequences from various sources

SEX DETERMINATION: the chromosomal sex of an individual; normal human females have two X chromosomes; normal human males have one X and one Y chromosome

SEXUAL ORIENTATION: the actual sexual behavior exhibited by an individual

Boy or Girl?

The question of what is "male" and what is "female" can have a variety of answers, depending on whether one is thinking of chromosomal (genetic) sex, gonadal sex, phenotypic sex, or self-identified gender. Chromosomal sex is determined at the time of conception. The fertilized human egg has a total of forty-six chromosomes, including one pair of sex chromosomes. If the fertilized egg has a pair of X chromosomes, its chromosomal, or genetic, sex is female. If it has one X chromosome and one Y chromosome, its genetic sex is male. Toward the end of the second month of prenatal development, processes are initiated that lead to the development of the gonadal sex of the individual; the embryo develops testes if male, ovaries if female. Although the chromosomal sex may be XX, it does not always happen that the sexual phenotype will be female; likewise, if the chromosomal sex is XY, the sexual phenotype does not always turn out to be male. Naturally occurring chromosomal variations or single-gene mutations may interfere with normal development and differentiation, leading to sexual phenotypes that do not correspond to the chromosomal sex.

One such case is that of hermaphrodites, individuals who possess both ovaries and testes. They usually carry both male and female tissue. Some of their cells may be of the female chromosomal sex (XX), and some may be of the male chromosomal sex (XY). Such individuals are called "sex chromosome mosaics," and their resulting phenotype may be related to the number and location of cells that are XX and those that are XY. Another example is testicular feminization syndrome, in which a single gene affects sexual differentiation. Individuals with this syndrome have the chromosomal sex of a normal male but have a female phenotype. XY males with this gene, located on the X chromosome, exhibit initial development of the testes and normal production of male hormones. However, the mutant gene prevents the hormones from binding to receptor cells; as a result, female characteristics develop.

Gender Identity

Gender identity disorder, or transsexualism, is defined by researchers as a persistent feeling of discomfort or inappropriateness concerning one's anatomic sex. The disorder typically begins in childhood and is manifested in adolescence or adulthood as cross-dressing. About one in eleven thousand men and one in thirty thousand women are estimated to display transsexual behavior. Hormonal and surgical sex reassignment are two forms of available treatment for those wanting to take on the physical characteristics of their self-identified gender. Little is known about the causes of gender identity disorder. In some cases, research shows a strong correlation between children who exhibit cross-gender behavior and adult homosexual orientation. Adults with gender identity disorder and adult homosexuals often recall feelings of alienation beginning as early as preschool.

Although some clinical aspects are shared, however, gender identity disorder is different from homosexuality. One definition for homosexuality proposed by Paul Gebhard is "the physical contact between two individuals of the same gender which both recognize as being sexual in nature and which ordinarily results in sexual arousal." Other researchers have underscored the difficulty in defining and measuring sexual orientation. Whatever measure is used, homosexuality is far more common than transsexualism.

Impact and Applications

Biological and genetic links to gender identity have been sought for more than a century. Studies on twins indicate a strong genetic component to sexual orientation. There appears to be a greater chance for an identical twin of a gay person to be gay than for a fraternal twin. Heritability averages about 50 percent in the combined twin studies. The fact that heritability is 50 percent rather than 100 percent, however, may indicate that other biological and environmental factors play a role. One study using restriction fragment length polymorphisms (RFLPs) to locate a gene on the X chromosome associated with male homosexual behavior showed a trend of maternal inheritance. However, not all homosexual brothers had the gene, and some heterosexual brothers shared the gene, indicating that other factors, whether genetic or nongenetic, influence sexual orientation.

Although some genetic factors have been found to influence sexual orientation, most researchers believe that no single gene causes homosexuality. It is also apparent that gender identity and homosexuality are influenced by complexes of factors dictated by biology, environment, and culture. Geneticists and social scientists alike continue to design studies to define how these complexes are interrelated.

—*Donald J. Nash*

See Also: Hermaphrodites; Human Genetics; Pseudohermaphrodites; Testicular Feminization Syndrome.

Further Reading: *The Psychology of Sexual Orientation, Behavior, and Identity: A Handbook* (1995), edited by L. Diamant and R. McAnuity, provides a good overview of matters related to gender identity and homosexuality. Simon Le Vay and Dean Hamer describe the early studies linking a single gene to male homosexuality in "Evidence for a Biological Influence in Male Homosexuality," *Scientific American* (January, 1994). J. Money describes numerous gender variations in *Sex Errors of the Body and Related Syndromes: A Guide to Counseling Children, Adolescents, and Their Families* (1994).

Gene Regulation: Bacteria

Field of study: Bacterial genetics

Significance: *Gene regulation is the process by which the synthesis of gene products is controlled. The study of gene regulation in bacteria has led to an understanding of how cells respond to their external and internal environments.*

Key terms

RIBONUCLEIC ACID (RNA): a form of nucleic acid in the cell used primarily for genetic expression through transcription and translation

TRANSCRIPTION: the use of deoxyribonucleic acid (DNA) as the template in the synthesis of RNA

TRANSLATION: the use of an RNA molecule as the guide in the synthesis of a protein

GENE: a sequence of base pairs that specifies a product (either RNA or protein); the average gene in bacteria is one thousand base pairs long

ALLELE: an alternate form of a gene; for example, $lacI^+$, $lacI^-$, and $lacI^S$ are alleles of the $lacI$ gene

CONTROLLING SITE: a sequence of base pairs to which regulatory proteins bind to affect the expression of neighboring genes

OPERON: one or more genes plus one or more controlling sites that regulate the expression of the genes

The Discovery of Gene Regulation

In 1961, a landmark paper by French researchers François Jacob and Jacques Monod outlined what was known about genes involved in the breakdown of sugars, the synthesis of amino acids, and the reproduction of a bacterial virus called "lambda." Jacob and Monod described in detail the induction of enzymes that break down the sugar lactose. These enzymes were induced by adding the sugar or, in some cases, structurally related molecules to the media. If these inducer molecules were removed, the enzymes altering lactose were no longer synthesized. Bacteria without the $lacI$ gene ($lacI^-$) produced the enzymes for metabolizing lactose whether the inducer was present or not. Bacteria heterozygous for the $lacI$ gene ($lacI^-/lacI^+$) functioned like normal bacteria ($lacI^+$), indicating that the $lacI^+$ allele was dominant to the $lacI^-$ allele. Certain alleles of the operator site, $lacO^C$, result in the synthesis of lactose-altering enzymes whether or not the inducer was present and even when $lacI^+$ was present. These observations suggested that the $lacI^+$ gene specified a repressor that might bind to $lacO^+$ and block transcription of the genes involved in lactose metabolism. Jacob and Monod concluded that inducers interfered with the repressor's ability to bind to $lacO^+$. This allowed transcription and translation of the lactose operon. In their model, the repressor protein is unable to bind to the altered operator site, $lacO^C$. This explained how certain mutations in the operator caused the

Jacques Monod (left) and François Jacob pose with Andre Lwoff, with whom they shared a 1965 Nobel Prize for their work on gene regulation. (Archive Photos/AFP)

enzymes for lactose metabolism to be continuously expressed.

Seeing a similarity between the expression of the genes for lactose metabolism, the genes for amino acid synthesis, and the genes for lambda proliferation, Jacob and Monod proposed that all genes might be under the control of operator sites that are bound by repressor proteins. An operon consists of the genes that the operator controls. Although the vast majority of operons have operators and are regulated by a repressor, there are some operons without operator sites that are not controlled by a repressor. Generally, these operons are regulated by an inefficient promoter or by transposition of the promoter site, whereas some are inhibited by attenuation. The only controlling site absolutely necessary for gene expression is the

promoter site, where ribonucleic acid (RNA) polymerase binds.

Lactose Operon: Negatively Controlled Genes

The lactose operon (*lacZYA*) consists of three controlling sites (*lacCRP*, *lacP$_{ZYA}$*, and *lacO*) and three structural genes (*lacZ*, *lacY*, and *lacA*). The lactose operon is controlled by a neighboring operon, the lactose regulatory operon, consisting of a single controlling site (*lacP$_I$*) and a single structural gene (*lacI*). The order of the controlling sites and structural genes in the bacterial chromosome is *lacP$_I$*, *lacI*, *lacCRP*, *lacP$_{ZYA}$*, *lacO*, *lacZ*, *lacY*, *lacA*. Transcription of the regulatory operon proceeds to the right from the promoter site, *lacP$_I$*. Similarly, transcription of the L-arabinose operon occurs rightward from $lacP_{ZYA}$. A cyclic-adenosine monophosphate receptor (CRP) bound by cyclic-adenosine monophosphate (cAMP), referred to as a CRP-cAMP complex, attaches to the $lacCRP$ site.

The $lacI$ gene specifies the protein subunit of the lactose repressor, a tetrameric protein that binds to the operator site, $lacO$, and blocks transcription of the operon. The $lacZ$ gene codes for beta-galactosidase, the enzyme that cleaves lactose into galactose plus glucose. This enzyme also converts lactose into the effector molecule allolactose, which actually binds to the repressor inactivating it. The $lacY$ gene specifies the enzyme, known as the "lactose permease," that transports lactose across the plasma membrane and concentrates it within the cell. The $lacA$ gene codes for an enzyme called "transacetylase," which adds acetyl groups to lactose.

In the absence of lactose, the repressor occasionally diffuses from the operator, allowing RNA polymerase to attach to $lacP_{ZYA}$ and make a single RNA transcript. This results in extremely low levels of enzymes called the "basal" level. With the addition of lactose, a small amount of allolactose binding to the repressor induces a conformational change in the repressor so that it no longer binds to $lacO$. The levels of permease and beta-galactosidase quickly increase, and within an hour the enzyme levels may be one thousand times greater than they were before lactose was added.

Normally, cells do not produce levels of lactose messenger RNA (mRNA) or enzymes that are more than one thousand times greater than basal level because the lactose operon is regulated by catabolite repression. As cells synthesize cellular material at a high rate, lactose entrance and cAMP synthesis is inhibited, whereas cAMP secretion into the environment is increased. This causes most of the CRP-cAMP complex to become CRP. CRP is unable to bind to $lacCRP$ and promote transcription from $lacP_{BAD}$.

If lactose is removed from the fully induced operon, repressor quickly binds again to $lacO$ and blocks transcription. Within a few hours, lactose mRNA and proteins return to their basal levels. Since the lactose operon is induced

and negatively regulated by a repressor protein, the operon is classified as an inducible, negatively controlled operon.

Arabinose Operon: Positively Controlled Genes

The L-arabinose operon ($araBAD$) has been extensively characterized since the early 1960's by American researchers Ellis Englesberg, Nancy Lee, and Robert Schleif. This operon is under the control of a linked regulatory operon consisting of ($araC$, $araO2$) and ($araP_C$, $araO1$). The parentheses indicate that the regions overlap: $araO2$ is an operator site in the middle of $araC$, whereas $araP_C$ and $araO1$ represent a promoter site and an operator site respectively, which overlap. The order of the controlling sites and genes for the regulatory operon and the L-arabinose operon is as follows: ($araC$, $araO2$), ($araP_C$, $araO1$), $araCRP$, $araI1$, $araI2$, $araP_{BAD}$, $araB$, $araA$, $araD$. RNA polymerase binding to $araP_C$ transcribes $araC$ leftward, whereas RNA polymerase binding to $araP_{BAD}$ transcribes $araBAD$ rightward.

The $araA$ gene specifies an isomerase that converts L-arabinose to L-ribulose, the $araB$ gene codes for a kinase that changes L-ribulose to L-ribulose-5-phosphate, and the $araD$ gene contains the information for an epimerase that turns L-ribulose-5-phosphate into D-xylulose-5-phosphate. Further metabolism of D-xylulose-5-phosphate is carried out by enzymes specified by genes in other operons.

The $araC$ product is in equilibrium between two conformations, one having repressor activity and the other having activator activity. The conformation that functions as an activator is stabilized by the binding of L-arabinose or by certain mutations ($araC^C$). In the absence of L-arabinose, almost all the $araC$ product is in the repressor conformation; however, in the presence of L-arabinose, nearly all the $araC$ product is in the activator conformation.

In the absence of L-arabinose, bacteria will synthesize only basal levels of the lactose regulatory protein and the enzymes involved in the breakdown of L-arabinose. The repressor binding to $araO2$ prevents $araC$ transcription beginning at $araP_C$ from being completed, whereas repressor binding to $araI1$ prevents $araBAD$

transcription beginning at *araP_{BAD}*.

The addition of L-arabinose causes repressor to be converted into activator. Activator binds to *araI1* and *araI2* and stimulates *araBAD* transcription. Activator is absolutely required for the metabolism of L-arabinose since bacterial cells with a defective or missing L-arabinose regulatory protein, *araC⁻*, only produce basal levels of the L-arabinose enzymes. This is in contrast to what happens to the lactose enzymes when there is a missing lactose regulatory protein, *lacI⁻*. Because of the absolute requirement for an activator, the L-arabinose operon is considered an example of a positively controlled, inducible operon.

Transcription of the *araBAD* operon is also dependent upon the cyclic-adenosine monophosphate receptor protein (CRP), which exists in two conformations. When excessive adenosine triphosphate (ATP) and cellular constituents are being synthesized from L-arabinose, cAMP levels drop very low in the cell. This results in CRP-cAMP acquiring the CRP conformation and dissociating from *araCRP*. When this occurs, the *araBAD* operon is no longer transcribed. The L-arabinose operon is controlled by catabolite repression very much like the lactose operon.

Tryptophan Operon: Genes Controlled by Attenuation

The tryptophan operon (*trpLEDCBA*) consists of the controlling sites and the genes that are involved in the synthesis of the amino acid tryptophan. The order of the controlling sites and genes in the tryptophan operon are as follows: (*trpP*, *trpO*), *trpL*, *trpE*, *trpD*, *trpC*, *trpB*, *trpA*. RNA polymerase binds to *trpP* and initiates transcription at the beginning of *trpL*.

An inactive protein is specified by an unlinked regulatory gene (*trpR*). The regulatory protein is in equilibrium between its inactive and its repressor conformation, which is stabilized by tryptophan. Thus, if there is a high concentration of tryptophan, the repressor binds to *trpO* and shuts off the tryptophan operon. This operon is an example of an operon that is repressible and negatively regulated.

The tryptophan operon is also controlled by a process called "attenuation," which involves the mRNA transcribed from the leader region, *trpL*. The significance of leader region mRNA is that it hydrogen-bonds with itself to form a number of hairpinlike structures. Hairpin-III interacts with the RNA polymerase, causing it to fall off the deoxyribonucleic acid (DNA). Any one of several hairpins can form, depending upon the level of tryptophan in the environment and the cell. When there is no tryptophan in the environment, the operon is fully expressed so that tryptophan is synthesized. This is accomplished by translation of the leader region right behind the RNA polymerase up to a couple of tryptophan codons, where the ribosomes stall. The stalled ribosomes cover the beginning of the leader mRNA in such a way that only hairpin-II forms. This hairpin does not interfere with transcription of the rest of the operon and so the entire operon is transcribed.

When there is too much tryptophan, the operon is turned off to prevent further synthesis of tryptophan. This is accomplished by translation of the leader region up to the end of the leader peptide. Ribosomes synthesizing the leader peptide cover the leader mRNA in such a way that only hairpin-III forms. This hairpin causes attenuation of transcription.

In some cases, the lack of amino acids other than tryptophan can result in attenuation of the tryptophan operon. In fact, cells starved for the first four amino acids (N-formylmethionine, lysine, alanine, and isoleucine) of the leader peptide result in attenuation. When these amino acids are missing, hairpins-I and III both form, resulting in attenuation because of hairpin-III.

Flagellin Operons: Operons Controlled by Transposition

Some pathogenic bacteria change their flagella to avoid being recognized and destroyed by the host's immune system. This change in flagella occurs by switching to the synthesis of another flagellar protein. The phenomenon is known as "phase variation." The genes for flagellin are in different operons. The first operon consists of a promoter site, an operator site, and the structural gene for the first flagellin (*flgP_{H1}*, *flgO1*, *flgH1*). The first operon is under the negative control of a repressor specified by the second operon. The second operon also specifies the second flagellin and a transposase that causes part of the second operon to reverse itself. This portion of the operon that "flips" is called a "transposon." The promotor sites for the transposase gene (*flgT2*), flagellin gene (*flgH2*), and repressor gene (*flgR2*) are located on either side of the transposase gene in sequences called "inverted repeats." Transcription from both promoters in the second operon occurs from left to right: *flgP_{T2}*, *flgT2*, *flgP_{H2R2}*, *flgH2*, *flgR2*.

When the second operon is active, the repressor binds to *flgO1*, blocking the synthesis of the first flagellin (*flgH1*). Consequently, all bacterial flagella will be made of the second flagellin (*flgH2*). Occasionally, the transposase will catalyze a recombination event between the inverted repeats, which leads to the transposon being reversed. When this occurs, neither *flgH2* nor *flgR2* are transcribed. Consequently, the first operon is no longer repressed by *flgR2*, and *flgH1* is synthesized. All the new flagella will consist of *flgH1* rather than *flgH2*.

Impact and Applications

Many of the genetic procedures developed to study gene regulation in bacteria have contributed to the development of genetic engineering and the production of biosynthetic consumer goods. One of the first products to be manufactured in bacteria was human insulin. The genes for the two insulin subunits were spliced to the lactose operon in different populations of bacteria. When induced, each population produced one of the subunits. The cells were cracked open, and the subunits were purified and mixed together to produce functional human insulin. Many other products have been made in bacteria, yeast, and even plants and animals.

Considerable progress has been made toward introducing genes into plants and animals to change them permanently. In most cases, this is difficult to do because the controlling sites and gene regulation are much more complicated in higher organisms than in bacteria. Nevertheless, many different species of plants have been altered to make them resistant to desiccation, herbicides, insects, and various plant pathogens. Although curing genetic defects by introducing good genes into animals and humans has not been very successful, animals have been transformed so that they produce a number of medically important proteins in their milk. Goats have been genetically engineered to release tissue plasminogen activator, a valuable enzyme used in the treatment of heart attack and stroke victims, into their milk. Similarly, sheep have been engineered to secrete human alpha-1 antitrypsin, useful in treating emphysema. Cattle that produce more than ten times the milk that sheep or goats produce may potentially function as factories for the synthesis of all types of valuable proteins specified by artfully regulated genes.

—*Jaime S. Colomé*

See Also: Bacterial Genetics and Bacteria Structure; Central Dogma of Molecular Biology; *Escherichia coli*; Gene Regulation: Lac Operon; Molecular Genetics; Transposable Elements.

Further Reading: Avraham Rasooly and Rebekah Sarah Rasooly, "How Rolling Circle Plasmids Control Their Copy Number," *Trends in Microbiology* 5 (November, 1997), illustrates how regulatory genes control the rate of synthesis of plasmids in bacteria. Toshifumi Inada et al., "Mechanism Responsible for Glucose-Lactose Diauxie in *Escherichia coli*: Challenge to the cAMP Model," *Genes to Cells* 1 (March, 1996) provides an understandable discussion of catabolite repression with numerous diagrams. Stephen M. Soisson et al., "Structural Basis of Ligand-Regulated Oligomerization of AraC," *Science* 276 (April 18, 1997), explains how two molecules of *AraC* protein interact with the inducer, with each other, and with controlling sites to regulate the expression of the L-arabinose operon.

Gene Regulation: Lac Operon

Field of study: Bacterial genetics
Significance: *Studies of the regulation of the lactose (lac) operon in* Escherichia coli *have led to an understanding of how the expression of a gene is*

turned on and off through the binding of regulator proteins to the DNA. This has served as the groundwork for understanding not only how bacterial genes work but also how genes of higher organisms are regulated.

Key terms

ACTIVATOR: a protein that binds to deoxyribonucleic acid (DNA) to enhance a gene's conversion into a product that can function within the cell

OPERATOR: a sequence of DNA adjacent to (and usually overlapping) the promoter of an operon; binding of a repressor to this DNA prevents transcription of the genes that are controlled by the operator

OPERON: a group of genes that all work together to carry out a single function for a cell

PROMOTER: a sequence of DNA to which the gene expression enzyme (RNA polymerase) attaches to begin transcription of the genes of an operon

REPRESSOR: a protein that prevents a gene from being made into a functional product when it binds to the operator

Inducible Genes and Repressible Genes

In order for genes or genetic information stored in deoxyribonucleic acid (DNA) to be used, the information must first be transcribed into messenger ribonucleic acid (mRNA); mRNA is synthesized by an enzyme, RNA polymerase, which uses the DNA as a template for making a single strand of RNA that can be translated into proteins. The proteins are the functional gene products that act as enzymes or structural elements for the cell. The process by which DNA is transcribed and then translated is referred to as expression of the genes.

All cells of an organism contain the same DNA, but it is clear that not all of the DNA is expressed all of the time. For example, in humans, only some cells produce the antibodies that protect against disease, while other cells produce the proteins that become hair fibers. Some genes are always expressed in cells; that

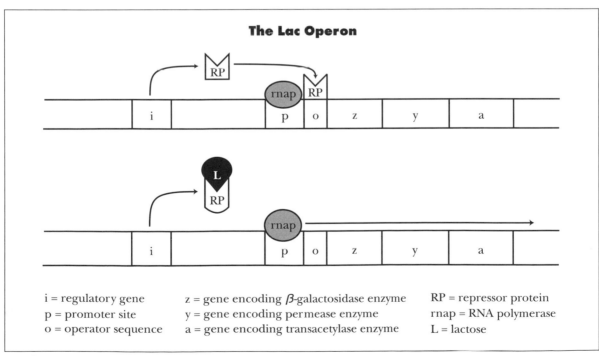

The Lac Operon

i = regulatory gene
p = promoter site
o = operator sequence

z = gene encoding β-galactosidase enzyme
y = gene encoding permease enzyme
a = gene encoding transacetylase enzyme

RP = repressor protein
rnap = RNA polymerase
L = lactose

1. In the absence of lactose, the repressor protein binds to the operator, blocking the movement of RNA polymerase. The genes are turned off.

2. When lactose is present, it preferentially binds the repressor protein, freeing up the operator and allowing RNA polymerase to move through the operon. The genes are turned on.

is, they are continually being transcribed into mRNA, which is translated into functional proteins (gene products) of the cell. The genes involved in using glucose as an energy source are included in this group. Other genes are inducible (expressed only under certain circumstances). The genes for using lactose as an energy source are included in this group. β-galactosidase, the enzyme that converts lactose into glucose and galactose so that the sugar can be easily metabolized by a cell, is only made in cells when lactose is present. Synthesizing proteins uses a large amount of energy. In order for the cell to conserve energy, it produces proteins only when they are needed. There is no need to make β-galactosidase if there is no lactose in the cell, so it is synthesized only when lactose is present.

As early as the 1940's Francois Jacob, Jacques Monod, and their associates were studying the mechanisms by which β-galactosidase was induced in *Escherichia coli*. They discovered that when there is no lactose in the cell, a repressor protein binds to the DNA of the operon at the operator site. Under these conditions, transcription of genes in the operon cannot occur since the RNA polymerase is physically prevented from binding to the promoter when the repressor is in place. This occurs because the promoter and operator sequences are overlapping. The lactose (lac) operon is, therefore, an example of a repressible operon. When lactose is present, an altered form of the lactose attaches to the repressor in such a way that the repressor can no longer bind to the operator. With the operator sequence vacant, it is possible for the RNA polymerase to begin transcription of the operon genes at the promoter. Lactose (or its metabolite) serves as an inducer for transcription. Only if it is present are the lactose operon genes expressed. The lactose operon is, therefore, also an inducible operon. Inducible and repressible control of gene expression follow from the arrangement of components in the operon; this is often referred to as negative control since binding of the repressor prevents expression of the genes. In 1965, Jacob and Monod were awarded the Nobel Prize in Physiology or Medicine in recognition of their discoveries concerning the genetic control of enzyme synthesis.

Lac Operon Expression in the Presence of Glucose

When a culture of *Escherichia coli* is given equal amounts of glucose and lactose for growth and is compared with cultures given either glucose alone or lactose alone, the cells given two sugars do not grow to twice the concentration in a single growth spurt but rather show two distinct growth cycles. β-galactosidase is not synthesized; therefore, lactose is not used until all the glucose has been metabolized. Laboratory observations show that having lactose present is necessary but not a sufficient condition for the lactose operon to be expressed. An activator protein must bind at the promoter in order to unravel the DNA double helix so that the RNA polymerase can bind more efficiently. The activator protein binds only when there is little or no glucose in the cell. If a cell has glucose available, it is used preferentially to other sugars because it is most easily metabolized to make energy in the form of adenosine triphosphate (ATP). ATP is made through a series of reactions from an intermediate molecule, cyclic adenosine monophosphate (cAMP). The cAMP concentration decreases when ATP is being made but builds up when no ATP synthesis occurs. When the glucose has been used, the concentration of cAMP rises. The cAMP binds to the activator protein to enable it to bind at the operon's promoter. With the activator bound, transcription of the genes, including the β-galactosidase, occurs.

The activation of a DNA-binding protein by cAMP is a global control mechanism. The lactose operon is only one of many that are controlled in this way. Global control allows bacteria to prevent or turn on transcription of a group of genes in response to a single signal. It ensures that the bacteria always utilize the most efficient energy source if more than one is available. This type of global control only occurs, however, when the operon is also under the control of another DNA-binding protein (the lac repressor in the case of the lac operon), which makes the operon inducible or repressible or both. Control of transcription through the binding of an activator protein is an example of positive control since binding of the activator turns on gene expression.

Impact and Applications

Jacob and Monod developed the concept of an operon as a functional unit of gene expression in bacteria. What they learned from studying the lac operon has led to a more general understanding of gene transcription and the use of mRNA as an information-bearing intermediate in the process of gene expression. The operon concept has proven to be a universal mechanism by which bacteria organize their genes. Although genes of higher cells are not usually organized in operons and although negative control of expression is rare in them, similar positive control mechanisms occur in both bacterial and higher cell types. Studies of the lactose operon have made possible the understanding of how DNA-binding proteins can attach to a promoter to enhance transcription.

—Linda E. Fisher

See Also: Bacterial Genetics and Bacteria Structure; *Escherichia coli*; Gene Regulation: Bacteria.

Further Reading: Horace Freeland Judson, *The Eighth Day of Creation: Makers of the Revolution in Biology* (1996), provides an interesting account of the discoveries and the personalities behind the discoveries that form the basis of modern molecular biology. "Molecular Machines That Control Genes," *Scientific American* 271 (February, 1995), by Robert Tijan, discusses regulatory proteins that direct transcription of DNA and what happens when they malfunction. "Genetic Repressors," *Scientific American* 222 (June, 1970), by Mark Ptashne and Walter Gilbert, summarizes repression mechanisms that turn genes on and off, using the lac operon and the λ bacterial virus as models.

Gene Therapy

Field of study: Genetic engineering and biotechnology

Significance: *Gene therapy offers the hope of treating, curing, and preventing many human diseases that cannot be ameliorated by traditional medical approaches. Gene therapy utilizes the techniques of recombinant deoxyribonucleic acid (DNA) technology to insert genes into human cells to correct genetic dysfunction at the molecular level.*

Key terms

PLASMID: DNA of bacterial origin that contains regulatory sequences permitting the production of many copies of the DNA; plasmids can be combined with specific genes using recombinant DNA technology to permit their amplification in host systems

VECTOR: a viral or nonviral vehicle for the transmission of recombinant plasmids containing therapeutic genes to recipient cells

TRANSFECTION: the process of inserting genes into cells during which the DNA is taken up by recipient cell and enters the nucleus

GENE EXPRESSION: the production of protein gene products from a specific therapeutic gene after its uptake by a recipient cell

RECESSIVE DISORDER: a condition in which both copies of a gene are dysfunctional; therefore, no normal gene products are made by the cell

MULTIGENE DISORDER: a condition in which more than one gene defect is involved in the cause of a specific genetic disorder

CLINICAL RESEARCH: research involving treatment strategies in human patients

Significant Discoveries in Recombinant DNA Technology

The tools of recombinant deoxyribonucleic acid (DNA) technology have been used successfully to provide information on the role of specific genes in the cause of many inherited single-gene disorders as well as complex, multigene disorders such as cancer. The most exciting implication of these discoveries is that they open the possibility of treating genetically based diseases at the molecular level by gene therapy. Gene therapy involves the insertion of genes into cells in order to correct genetic dysfunction. These approaches are rooted in fundamental discoveries in the area of recombinant DNA technology. In the mid-1950's, John Holland and colleagues at the University of California at San Diego observed that cells could take up ribonucleic acid (RNA) or DNA from viruses and express the viral genes to make proteins. Subsequent developments in the field of recombinant DNA technology permitted researchers to produce large amounts of specific genes by inserting them into plasmid vectors to form recombinant molecules. Plasmid vectors are small, circular DNAs of bacterial origin that contain regulatory sequences permitting the amplification of specific genes to high copy number. Additional discoveries by Paul Berg and colleagues at Stanford University in the late 1970's showed that recombinant plasmid DNA could be inserted into cultured mammalian cells by a process called transfection and that specific gene products could be expressed in the "transfected" cells.

Research in the area of virology has also contributed significantly to the development of gene therapy. In many ways, viruses represent ideal vectors for the insertion of genetic material into cells since they are essentially packets of genetic information that enter specific cells and amplify their genetic material within the host system. Recombinant DNA technology approaches involve replacing nonessential or harmful viral genes with therapeutic human genes and then packaging the recombinant viral genome into virus particles that can deliver the human gene to specific cells by viral infection.

Nonviral gene therapy approaches include the development of "artificial viruses" that use synthetic materials to transfer genes to recipient cells. DNA complexes containing lipids, proteins, or synthetic polymers have been used as gene delivery systems to lung, liver, endothelium, epithelium, and tumor cells. It has also been observed that "naked" DNA can be directly injected into patients and can express functional gene products in cells that take up the DNA; these injected plasmids may be used to generate a new class of DNA vaccines.

Taken together, the critical discoveries that have opened the door to gene therapy approaches in the treatment of genetic disease have involved strategies for identifying and amplifying human genes as well as gene transfer and expression within mammalian systems by infection with recombinant viruses or by transfection using mammalian expression vectors.

In 1989, the National Institutes of Health (NIH) approved the first clinical trials involving the transfer of therapeutic genes into humans. By 1997, the Recombinant DNA Advisory Committee had approved more than one hundred gene therapy studies. Numerous clinical trials have demonstrated that human genes can be successfully transferred to recipient cells outside the body (ex vivo) or directly into the body (in vivo) and that the transferred genes can be activated to produce therapeutic gene products.

Types of Gene Therapy

Gene therapy approaches can be distinguished on the basis of type of target cell, desired therapeutic effect, and mode of gene transfer. Two general types of gene therapy can be defined based on type of target cell: germline gene therapy and somatic cell gene therapy. Germ-line gene therapy involves the insertion of genes into the reproductive cells or "germ" cells, which results in the permanent alteration of the genetic content of the individual and generates an altered genome that is transmitted to offspring in successive generations. Germ-line genetic insertions have been carried out in mice to generate transgenic animals that carry foreign genes in the germline. Transgenic mice have been used to study the functions of specific genes in development and physiology and also to create mouse models of human diseases. In contrast, germline gene therapy is not contemplated for human use because of important physiological, technical, and ethical considerations.

Somatic cell gene therapy involves the insertion of genes into specific tissues of the body; the genetic changes are not heritable. Somatic cell gene therapy could theoretically be applied to any nonreproductive tissue of the body. In practice, some types of target tissue are more amenable to gene therapy than others, based on accessibility and available vectors. The first somatic cell target for gene therapy was hematopoietic (blood-forming) tissue; additional targets under experimentation include lung, liver, muscle, skin, and neural tissue.

The nature of the desired therapeutic outcome of gene therapy depends on the type of genetic dysfunction. In genetic disorders resulting from the absence of a single gene product caused by gene inactivation, such as adenosine deaminase deficiency or cystic fibrosis, gene replacement therapy is used to provide the missing gene function directly to the malfunctioning cell. Another type of gene therapy involves the inactivation of abnormally functioning genes that interfere with cell physiology or the destruction of abnormal cells such as cancer cells. In this case, the product of the transferred gene blocks gene function or destroys the abnormal cell. Additional modes of gene therapy involve the implantation of genetically engineered cells or recombinant plasmid DNA whose gene products affect the function of other tissues within the body. Examples include growth factors that stimulate blood vessel formation (angiogenesis), injected muscle cells that secrete circulatory proteins, and DNA vaccines that trigger immune responses to specific gene products from infectious disease microorganisms or abnormal cells within the body. Each of these therapeutic protocols involves the use of different types of vectors and strategies for achieving targeted gene expression.

Types of Vectors Used in Gene Therapy

Modes of gene transfer include the use of viral or nonviral vectors or even the transfer of "naked" DNA. The mode of gene transfer utilized depends on the nature of the target cell as well as the desired clinical effect. For example, the first clinical trials of human gene therapy involved the use of hematopoietic tissue and a recombinant retroviral vector to treat an inherited disorder called severe combined immunodeficiency disorder (SCID), which results from adenosine deaminase (ADA) deficiency. In this disorder, the immune system fails to function normally because of the absence of the enzyme adenosine deaminase. The goal of gene therapy is to introduce a normal copy of the ADA gene into hematopoietic precursor cells using a recombinant retrovirus to infect the cells and then to insert the therapeutic gene into the genome of the recipient cells. Expression of the normal ADA gene results in the production of active enzymes and the res-

toration of immune system function. This type of gene therapy can be performed ex vivo, although most types of corrective therapy require gene transfer to cells in vivo. In addition, this type of genetic correction requires the long-term expression of the desired gene product in a significant fraction of recipient cells. Use of a retroviral vector for this type of therapy is useful since the virus inserts its genetic material directly into the host genome so that the new gene is transmitted to daughter cells in cell division; however, since the gene is inserted randomly into the recipient cell genome, disruption of host cell's gene function may occur. Retrovirus infection is limited to dividing cells only; since most adult tissues consist primarily of nondividing cells, their use as vectors is mainly limited to ex vivo transfer to proliferating cells. For use in gene therapy, the structural genes of the virus are removed and replaced with a human therapeutic gene; however, the regulatory components of the virus are left intact since they activate gene product formation.

Recombinant adenoviruses are also used in gene therapy; clinical trials have included genetic correction in alpha-1 antitrypsin deficiency and cystic fibrosis by administering recombinant adenovirus containing the therapeutic gene to the lungs in the form of an inhalant. The advantages of this system include the fact that these viruses are capable of injecting most tissues of the body, including nondividing cells, and that the therapeutic gene is not inserted into the host cell genome so that its integrity is preserved. Limitations of adenovirus vectors include the possibility of germline infection, the possibility of recombinations to generate infectious virus, and the instability of the transferred gene since it remains extrachromosomal (not inserted into the genome). In addition, adenoviruses provoke host immune responses that preclude their long-term use in patients. Adeno-associated viruses have also been used in gene therapy; these small viruses can accommodate only small genes, but they produce stable gene transfer by genomic insertion of the therapeutic gene and cannot recombine to form infectious virus. Herpes viruses have been modified to permit insertion of therapeutic genes into neural tissue.

Nonviral vectors are also used to effect gene transfer to target tissues. Use of nonviral vectors removes any possibility of disease from altered viral vectors. The most common type of synthetic vectors are liposomes (small lipid spheres that can combine with recombinant plasmid DNA containing therapeutic genes to form packages of DNA that can be incorporated into target cells). Problems associated with the use of liposomes in gene therapy include a low rate of uptake into target cells and a failure to effect the insertion of target genes into recipient cell genomes in vivo. "Lipoplexes" represent modified liposomes that contain positively charged cationic lipids; lipoplexes have been used therapeutically to transfer the immune system gene *HLA-B7*, which encodes a major histocompatibility antigen that triggers immune cell responses, into tumor cells of patients with malignant melanoma by intratumor injection. Nonlipid synthetic vectors and even "naked" recombinant plasmid DNA can be used in gene transfer. The overall efficiency of these processes in unmodified form is low; however, even transient, low-level gene expression from injected DNA may be sufficient to trigger significant host immune responses, making this a useful strategy in the development of vaccines.

Treatment of Single-Gene Disorders

The most fundamental application of gene therapy involves its use in the treatment of inherited genetic disorders in which the genetic dysfunction resides within a single abnormal gene. Many single-gene disorders are recessive, which means that the genetic disorder results from inactivation of both copies of a gene as a consequence of mutation; therefore, no functional gene product is made. Genetic disorders of this type result from a loss of function of a single gene product, usually enzymatic in nature. The loss of enzyme function disrupts metabolic pathways within the cell, resulting in specific clinical consequences. Hundreds of single-gene recessive disorders have been identified, and many of the responsible genes have been cloned. The recessive disorder severe combined immunodeficiency disorder (SCID)

was the first to enter gene therapy clinical trials; additional diseases under clinical trials include cystic fibrosis, alpha-1 antitrypsin deficiency, familial hypercholesterolemia, Gaucher's disease, and Hunter's disease. These genetic disorders represent good candidates for corrective gene therapy since the diseases result from loss of enzyme function, which can be restored by the low-level insertion of therapeutic genes, and since the primary affected tissues are readily accessible to gene transfer approaches. Problems encountered in gene therapy clinical trials include failure to achieve long-term, stable levels of gene expression in targeted cells and difficulties with in vivo gene transfer methods, resulting in low-efficiency uptake of therapeutic genes.

Other types of inherited single-gene disorders are more refractory to gene therapy approaches. These include genetic disorders resulting from defects in structural gene products that are needed in large amounts by the cell or in amounts that must be precisely regulated. Sickle-cell anemia is in this category, since it results from a defect in the B-hemoglobin gene whose gene product is made in large, highly regulated amounts by the cell in order to form the adult hemoglobin molecule. To achieve effective corrective therapy, it is necessary to insert a functional copy of the B-hemoglobin gene along with appropriate regulatory components to permit regulated, high-level gene expression. Other types of genetic disorders result from gene defects that result in the production of abnormal proteins that may act to disrupt the structure and function of their normal counterparts or accumulate to highly toxic levels within the body. Corrective gene therapy involving the simple addition of a normally functioning gene to the target cell cannot ameliorate these clinical conditions; rather, at the genetic level, these types of disorders require the replacement of the abnormal gene with a normal copy of the gene. Gene replacement requires an exact and precise genetic exchange mechanism involving targeted replacement of the defective gene with a therapeutic gene. Targeted genetic exchange occurs in yeast but represents a major challenge to human gene therapy research.

Other approaches aimed at blocking the production of dysfunctional gene products involve "antisense" gene therapy, in which a synthetic gene capable of blocking the production of the abnormal gene by binding to the gene itself or to intermediates required for protein synthesis is inserted into the cell.

Treatment of Multigene or Multifactorial Disorders

Multigene disorders are conditions in which more than one dysfunctional gene is involved in the genesis of a clinical syndrome. Many types of cancer involve dysfunctions in multiple genes required to maintain proliferative cell cycle control. Multifactorial disorders have both genetic and environmental causes. Many types of coronary heart disease result from combinations of genetic and environmental factors. Simple corrective gene therapy involving single therapeutic genes is unlikely to achieve clinical cures in these highly complex diseases; however, gene therapy may play a critical role in treatment strategies for these groups of diseases, which are responsible for the highest rate of mortality in adults.

In cancer treatment, novel therapeutic strategies include the use of gene therapy to block cell proliferation, inhibit metastasis, reverse multidrug resistance, and promote cell death mechanisms. Clinical studies have shown that gene therapy involving the insertion of functional copies of the *p53* tumor suppressor gene into tumor cells lacking this gene can significantly inhibit tumor cell growth and may sensitize cells to chemotherapeutic destruction. Other therapeutic genes for cancer treatment in clinical trials include tumor necrosis factor (TNF), which destroys malignant cells, and "antisense" genes, which block the activity of abnormal growth-promoting gene products in tumor cells. In addition, cancer immunotherapy involving intratumor injection or vector-mediated transfer of cytokine genes, such as interleukin-2, -4, and -7 and granulocyte-macrophage colony stimulating factor (GM-CSF), to elicit immune responses aimed at selective tumor cell destruction represents an active area of research. One approach to tumor immunotherapy in melanoma and sarcoma involves the use of lethally irradiated tumor cells transfected with a GM-CSF gene in a plasmid expression vector by particle-mediated gene transfer (PMGT), a procedure in which the therapeutic gene is coated with gold particles and delivered to tumor cells ex vivo using helium pressure. These treated tumor cells, which express biologically active therapeutic gene products, are then used to vaccinate the patient to elicit an immune response against the malignant tissue. In the case of transfected genes aimed at directly destroying tumor cells, it is not possible to deliver the therapeutic gene to all cells of a tumor; however, researchers rely on bystander effects (involving the spread of expressed gene products from transfected genes to nearby nontransfected cells by means of cell junctions) to achieve widespread therapeutic effects.

Developments in the use of gene therapy to treat coronary heart disease include the injection of genes that trigger blood vessel formation to bypass blocked arteries by stimulating the growth of new blood vessels. Other gene therapy approaches include the grafting of genetically manipulated cells into brain tissue to produce specific gene products to promote nerve cell survival in the treatment of neurodegenerative disorders such as Parkinson's disease.

Impact and Applications

Gene therapy represents the ultimate achievement of the genetic revolution that began with the identification of DNA as the genetic material by Oswald Avery and colleagues in the 1940's at the Rockefeller Institute in New York. The methods of recombinant DNA technology have permitted insight into the molecular basis of genetic disease, involving the identification of many affected genes and a determination of the role of the dysfunctional genes in generating the clinical syndromes associated with specific disorders. This knowledge has been combined with methodologies of gene transfer in mammalian systems to correct genetic dysfunction, which has led to promising clinical trials in the treatment of recessive single-gene disorders such as adenosine deaminase deficiency. Clinical trials involving the use of gene therapy in the treatment of cancer have also yielded promising results. In addition, gene therapy may prove useful in the development of novel vaccines consisting of genes that elicit systemic immune responses against viruses and other infectious disease microorganisms. Other possibilities include the treatment of neurological disorders by the insertion of genetically engineered nerve cells to produce neurotransmitters and other factors that stabilize nerve cell function. Similar approaches could restore function to other organ systems of the body. As the understanding of the roles of specific genes in immune system function increases, this knowledge will be used to devise gene therapy strategies effective against acquired immunodeficiency syndrome (AIDS) and cancer.

While the possibilities seem endless, significant roadblocks to the successful application of gene therapy remain. Critical areas of research involve the design of vectors capable of recognizing specific receptors to permit target-cell specificity for gene insertion in vivo and to increase the efficiency of DNA uptake by recipient cells. Another area of investigation involves strategies to achieve stable, high-level gene expression by transfected cells. Finally, breakthroughs in the area of targeted gene replacement aimed at precisely directing the insertion site of transfected genes in recipient cells would represent a major advance toward gene replacement therapy and lead to many new gene therapy approaches.

—Sarah Crawford Martinelli

See Also: Biotechnology; Gene Therapy: Ethical and Economic Issues; Genetic Medicine; Genetic Testing; Human Genetics; Human Genome Project; Molecular Genetics.

Further Reading: R. Michael Blaese, "Gene Therapy for Cancer and AIDS," *Scientific American* 276 (June, 1997), explains how gene therapy approaches can be exploited to elicit immune responses that result in the destruction of human immunodeficiency virus (HIV)-infected cells and cancer cells. Theodore Friedmann, "Overcoming the Obstacle to Gene Therapy," *Scientific American* 276 (June, 1997), presents the major approaches to gene therapy and the technical challenges to achieving desired results. Philip Felgner, "Nonviral Strategies for Gene Therapy," *Scientific American* 276 (June, 1997), describes gene therapy approaches involving nonviral vectors such as liposomes, lipoplexes, and naked DNA. Glenn Dranoff and Richard Mulligan, "Gene Transfer as Cancer Therapy," *Advances in Immunology* 58 (1995), discusses immunologic approaches to cancer therapy based on intratumor gene therapy. J. Gossen and J. Vigg, "Transgenic Mice as Model Systems for Studying Gene Mutations In Vivo," *Trends in Genetics* 9 (1993), explains how germline gene transfer in mice can be used to determine the roles of specific genes in development. Richard Mulligan, "The Basic Science of Gene Therapy," *Science* 260 (1993), reviews the major principles of gene therapy approaches and the methods for designing gene therapy vectors. Theodore Friedmann, "A Brief History of Gene Therapy," *Nature Genetics* 2 (1992), gives a historical perspective of the major scientific advances leading to the development of gene therapy approaches.

Gene Therapy: Ethical and Economic Issues

Field of study: Genetic engineering and biotechnology

Significance: *Gene therapy has the potential to cure many diseases once viewed as untreatable, such as cystic fibrosis. At the same time, gene therapy presents ethical dilemmas ranging from who decides who will benefit from new therapies to whether humans should attempt to manipulate natural evolutionary processes.*

Key terms

GERM CELLS: reproductive cells such as eggs and sperm

GERM-LINE THERAPY: alteration of germ cells resulting in a permanent genetic change in the organism and succeeding generations

INSULIN: a pancreatic hormone that is essential to metabolize carbohydrates and that is used in the control of diabetes mellitus

RECOMBINANT DNA: genetically engineered DNA prepared by cutting up DNA molecules and splicing together specific DNA fragments, often from more than one species of organism

SOMATIC CELL THERAPY: treatment of specific tissue with therapeutic genes

Gene Therapy

Advances in molecular biology and genetics in the 1990's presented researchers in a variety of fields with the tantalizing possibility of new therapeutic tools for treating medical conditions once viewed as incurable. Strong economic incentives exist to pressure researchers into pursuing genetic engineering and genetic therapies in multiple problem areas. At the same time, each area of economic promise or possible medical advance raises new ethical questions as scientists face the possibility of unintended consequences of gene therapies.

In agriculture, for example, genetic engineers have succeeded in developing strains of grain that are resistant to the effects of herbicides and in altering pigs so that hormones used in treating human disorders are produced in the animals' milk. The latter application holds out considerable economic promise to pharmaceutical companies, as many hormones necessary for treatment of various medical conditions cannot be synthesized and are, consequently, extremely expensive and in short supply.

While the ready availability of rare hormones would seem to be an advance for human medicine, altering pigs to serve as a source of hormones raises questions about the ethical treatment of animals. Do humans have the right to manipulate the genetic makeup of animals in this manner? Supporters of research in this area generally respond by noting that humans have been altering the heredity of livestock for millennia through the process of selective breeding for desirable characteristics and that genetic engineering is simply another form of that process. Scientists further note that many of the hormones that they hope to produce through living animals are only available through harvesting organs from human cadavers; therefore, creating pigs that produce hormones in their milk will make hormone therapy available to a much wider patient population.

Somatic Cell Therapy

While genetic engineering of plants and animals raises troubling questions for ethicists concerned about whether humans should interfere with natural evolutionary processes, it is the application of gene therapy to humans that generates the most heated debates. Somatic cell therapy, for example, could possibly be used to free insulin-dependent diabetics from reliance on external sources of insulin by restoring the ability of the patient's own body to manufacture insulin. Scientists have already succeeded in genetically engineering bacteria to grow recombinant insulin, eliminating the need to harvest insulin from animal pancreatic tissue obtained from slaughterhouses. Having achieved cultivating recombinant bacteria to manufacture insulin in a laboratory setting, it is not surprising that some researchers hope that the next step will be the use of somatic cell therapy to treat individual diabetics.

While diabetes mellitus appears to be a fairly clear-cut case in which any somatic cell therapy developed could be viewed in a positive light, ethicists wonder about other conditions, such as the insufficient production of growth hormone. A shortage of human growth hormone can result in a child suffering from pituitary dwarfism, and the use of somatic cell therapy to correct the condition would be beneficial. Even without the availability of gene therapy, however, human growth is an area that is subject to abuses. In a society in which height is associated with success, wealthy parents have been known to pressure doctors to prescribe human growth hormone to their children who are only slightly smaller than average and not truly suffering from a pituitary gland disorder. If somatic cell gene therapy became widely available for human growth, how many parents would succumb to the temptation to give their children a boost in height?

Germ-Line Therapy

While somatic cell gene therapy raises some ethical questions, including the usual economic issues of who will benefit and how widely available such therapies will be, germ-line therapy presents the most troubling questions. Somatic cell therapy does not lead to a permanent genetic change in the organism being treated. It may, for example, be used to fight cancer but does not allow patients to pass the cancer resistance on to their offspring. On the other hand, germ-line therapy, because it is performed on germ cells, changes the characteristics an organism passes on to its offspring. Humans suffer from a variety of inherited diseases, including hemophilia, Huntington's chorea, and cystic fibrosis, and physicians have long recognized that certain conditions, such as coronary artery disease and diabetes, seem to run in some families. Some researchers have suggested that rather than pursuing somatic cell therapies to treat each individual sufferer, scientists should simply eliminate these medical conditions through germ-line therapies. That is, if a person with a family history of Huntington's chorea wanted to have children, germ-line therapy would be performed in vitro to ensure that any fetus carried in a pregnancy would be free of the Huntington's chorea gene. Eventually, Huntington's chorea would be eliminated from the general population.

Tempting though it is to see this as a good thing, ethicists believe that such an approach could be extremely susceptible to abuse. They view discussions of human germ-line therapy as an attempt to resurrect the failed agenda of the eugenics movement of the 1920's and 1930's. If scientists are allowed to manipulate human heredity to eliminate certain characteristics, what is to prevent those same scientists from manipulating the human genome to enhance other characteristics? Would parents be able to request custom-tailored offspring, children who would be tall with predetermined hair color and eyes? Questions concerning class divisions and racial biases have also been raised. Would therapies be equally available to all people who requested them, or would such technology lead to a future in which the wealthy custom-tailor their offspring while the poor must rely on conventional biology? Would those poor people whose parents had been unable to afford germ-line therapy then find themselves denied access to medical care or employment based on their "inferior" or "unhealthy" genetic profiles?

In addition, many ethicists and scientists raise cautionary notes about putting too much faith in new genetic engineering technologies too soon. Most scientists concede that not enough is known about the interdependency of various genes and the role they play in overall health and human evolution to begin a program to eliminate so-called bad genes. Genes that in one combination may result in a disabling or life-threatening illness may in another have beneficial effects that are not yet known. Germ-line therapy could eliminate one problem while at the same time opening the door to a new and possibly worse condition. Thus, while the economic benefits of genetic engineering and gene therapies can be quite tempting, ethicists remind everyone that many questions remain unanswered. Some areas of genetic research, particularly germ-line therapy, may simply be best left unexplored until a clearer understanding of both the potential social and biological cost emerges.

—*Nancy Farm Mannikko*

See Also: Biotechnology; Eugenics; Eugenics: Nazi Germany; Genetic Engineering; Genetic Screening.

Further Reading: Geneticist Doris Zallen's *Does It Run in the Family?* (1998) provides a concise and accessible guide for the average person to use in evaluating gene therapies. Jeremy Rifkin, in *The Biotech Century: Harnessing the Gene and Remaking the World* (1998), discusses a variety of economic and ethical concerns regarding biotechnology, as does J. P. Harpignies in *Double Helix Hubris: Against Designer Genes* (1997). An anthology edited by Gerhold K. Becker and James P. Buchanan, *Changing Nature's Course: The Ethical Challenge of Biotechnology* (1996), offers a variety of perspectives, while Munawar A. Anees, *Islam and Biological Futures: Ethics, Gender, and Technology* (1989), provides insight into reproductive biotechnologies.

Genetic Clocks

Field of study: Human genetics
Significance: *Genetic clocks control those periodic behaviors of living systems that are a part of their*

normal function. The rhythms may be of a daily, monthly, yearly, or even longer periodicity. In some cases, the clocks may be "programmed" to regulate processes that may occur at some point in the lifetime of the individual, such as those processes related to aging. Altered or disturbed rhythms may result in disease.

Key terms

ALZHEIMER'S DISEASE: a disorder characterized by brain lesions leading to loss of memory, personality changes, and deterioration of higher mental functions

CIRCADIAN RHYTHM: a cycle of behavior, approximately twenty-four hours long, that is expressed independent of environmental changes

FREE-RUNNING CYCLE: the rhythmic activity of an individual that operates in a constant environment

HUNTINGTON'S CHOREA: an autosomal dominant genetic disorder characterized by loss of mental and motor functions in which symptoms do not appear until after age thirty

SUPRACHIASMATIC NUCLEUS (SCN): a cluster of several thousand nerve cells that contains a central clock mechanism that is active in the maintenance of circadian rhythms

Genetic Clocks and Biological Rhythms

Biological clocks control a number of physiological functions, including sexual behavior and reproduction, hormonal levels, periods of activity and rest, body temperature, and other activities. In humans, phenomena such as jet lag and shift-work disorders are thought to result from disturbances to the innate biological clock.

The most widely studied cycles are circadian rhythms. These rhythms have been observed in a variety of animals, plants, and microorganisms and are involved in regulating both complex and simple behaviors. Typically, circadian rhythms are innate, self-sustaining, and have a cyclicity of nearly, but not quite, twenty-four hours. Normal temperature ranges do not alter them, but bursts of light or temperature can change the rhythms to periods of more or less than twenty-four hours. Circadian rhythms are apparent in the activities of many species, in-cluding humans, flying squirrels, and rattlesnakes. They are also seen to control feeding behavior in honeybees, song calling in crickets, and hatching of lizard eggs.

What is known about the nature of the biological clock? The suprachiasmatic nucleus (SCN) consists of a few thousand neurons or specialized nerve cells that are found at the base of the hypothalamus, the part of the brain that controls the nervous and endocrine systems. The SCN appears to play a major role in the regulation of circadian rhythms in mammals and affects cycles of sleep, activity, and reproduction. The seasonal rhythm in the SCN appears to be related to the development of seasonal depression and bulimia nervosa. Light therapy is effective in these disorders. Blind people, whose clocks may lack the entraining effects of light, often show free-running rhythms.

Genetic control of circadian rhythms is indicated by the findings of single-gene mutations that alter or abolish circadian rhythms in several organisms, including the fruit fly (*Drosophila*) and the mouse. A mutation in *Drosophila* affects the normal twenty-four-hour activity pattern so that there is no activity pattern at all. Other mutations produce shortened (nineteen-hour) or lengthened (twenty-nine-hour) cycles. The molecular genetics of each of these mutations is known. Minor alterations in the deoxyribonucleic acid (DNA) molecule result in the production of different proteins, and the small change causes major differences in the rhythms of the flies.

A semidominant autosomal mutation, CLOCK, in the mouse produces a circadian rhythm one hour longer than normal. Mice that are homozygous (have two copies) for the CLOCK mutation develop twenty-seven- to twenty-eight-hour rhythms when initially placed in darkness and lose circadian rhythmicity completely after being in darkness for two weeks. No anatomical defects have been seen in association with the CLOCK mutation.

Genetic Clocks and Aging

Genes present in the fertilized egg direct and organize life processes from conception until death. There are genes whose first effects may not be evident until middle age or later. Huntington's chorea is such a disorder. An individual who inherits this autosomal dominant gene is "programmed" around midlife to develop involuntary muscle movement and signs of mental deterioration. Progressive deterioration of body functions leads to death, usually within fifteen years. It is possible to test individuals early in life before symptoms appear and provide reproductive choices for them.

Alzheimer's disease (AD) is another disorder in which genes seem to program processes to occur after middle age. AD is a progressive, degenerative disease that results in a loss of intellectual function. Symptoms worsen until a person is no longer able to care for himself or herself, and death occurs on an average of eight years after the onset of symptoms. AD may appear as early as forty years of age, although most people are sixty-five or older when they are diagnosed. Age and a family history of AD are clear-cut risk factors. Gene mutations associated with AD have been found on human chromosomes 1, 14, 19, and 21. Although these genes, especially the apolipoprotein *E4* gene, increase the likelihood of a person getting AD, the complex nature of the disorder is underscored when it is seen that the mutations account for less than half of the cases of AD and that some individuals with the mutation never get AD.

Impact and Applications

Evidence has accumulated that human activities are regulated by genetic clocks. It has also become evident that many disorders and diseases, and even processes that are associated with aging, may be affected by abnormal clocks. As understanding of how genes control genetic clocks develops, possibilities for improved therapy and prevention should emerge. It may even become possible to slow some of the harmful processes associated with normal aging.

—*Donald J. Nash*

See Also: Aging; Alzheimer's Disease; Huntington's Chorea.

Further Reading: *Living with Our Genes* (1998), by Dean Hamer and Peter Copeland, contains a good chapter on genetic clocks and aging. A review of Alzheimer's disease is provided in *Alzheimer's: Answers to Hard Questions for Families* (1996), by James Lindemann Nelson and Hilde Lindemann Nelson. Doris Teichler Zallen gives useful information on genetic disorders, including Huntington's chorea and AD, in *Does It Run in the Family?* (1997).

Genetic Code

Field of study: Molecular genetics
Significance: *The molecules of life are made directly or indirectly from instructions contained in DNA. The instructions are interpreted according to the genetic code, which describes the relationship used in the synthesis of proteins from nucleic acid information.*

Key terms

CODON: a three-nucleotide unit of ribonucleic acid (RNA) that determines which amino acid will be added to a growing protein chain

READING FRAME: the phasing of reading RNA codons, determined by starting at the first, second, or third position

TRANSFER RNA (tRNA): molecules that carry amino acids to messenger RNA (mRNA) codons, allowing amino acid polymerization into proteins

TRANSLATION: the process of forming proteins according to instructions contained in an RNA molecule

Elements of the Genetic Code

Every time a cell divides, each daughter cell receives a full set of instructions that allows it to grow and divide. The instructions are contained within deoxyribonucleic acid (DNA). These long nucleic acid molecules are made of nucleotides linked end to end. Four kinds of nucleotides are commonly found in the DNA of all organisms. These are designated A, G, T, and C for the variable component of the nucleotide (adenine, guanine, thymine, and cytosine, respectively). The sequence of the nucleotides in the DNA chain provides the information necessary for life. DNA does not contain the molecules and structures characteristic of a cell or an organism; rather, its information must be decoded to build other molecules. Many of these serve to build other molecules and to obtain the energy needed for construction. The rules by which the sequences are decoded are called the genetic code.

DNA is written in a variety of codes. There are encoded signals for where to start and where to stop reading the code to make ribonucleic acid (RNA) molecules in transcription. RNA molecules are also strings of nucleotides linked end to end in the order specified by the DNA. In RNA, the nucleotide uracil (U) is used in place of T. An RNA molecule has the same information as the section of DNA from which it was made. Some RNAs function directly in the structure and activity of cells, but most serve to bring DNA information to ribosomes. This latter type is known as messenger RNA (mRNA). The ribosome machinery scans the RNA nucleotide sequence to find signals to start the synthesis of polypeptides, the mole-cules of which proteins are made. When the start signals are found, the machinery reads the code in the RNA to convert it into a sequence of amino acids in the polypeptide, a process called translation. Translation stops at termination signals. The term "genetic code" is sometimes reserved for the rules for converting a sequence of nucleotides into a sequence of amino acids.

The Protein Genetic Code: General Characteristics

Experiments in the laboratories of Gobind Khorana, Heinrich Matthaei, Marshall Nirenberg, and others led to the deciphering of the protein genetic code. They knew that the code was more complicated than a simple one-to-one correspondence between nucleotides and amino acids, since there were twenty or more different amino acids in proteins and only four nucleotides in RNA. They found that three

The Full Genetic Code

second row → first row ↓	U	C	A	G	third row ↓
U	Phenylalanine Phenylalanine Leucine Leucine	Serine Serine Serine Serine	Tyrosine Tyrosine END CHAIN END CHAIN	Cysteine Cysteine END CHAIN Tryptophan	U C A G
C	Leucine Leucine Leucine Leucine	Proline Proline Proline Proline	Histidine Histidine Glutamine Glutamine	Arginine Arginine Arginine Arginine	U C A G
A	Isoleucine Isoleucine Isoleucine Methionine	Threonine Threonine Threonine Threonine	Asparagine Asparagine Lysine Lysine	Serine Serine Arginine Arginine	U C A G
G	Valine Valine Valine Valine	Alanine Alanine Alanine Alanine	Aspartic Acid Aspartic Acid Glutamic Acid Glutamic Acid	Glycine Glycine Glycine Glycine	U C A G

The amino acid specified by any codon can be found by looking for the wide row designated by the first base letter of the codon shown on the left, then the column designated by the second base letter along the top, and finally the narrow row marked on the right, in the appropriate wide row, by the third letter of the codon. Many amino acids are represented by more than one codon. The codons UAA, UAG, and UGA do not specify an amino acid but instead signal where a protein chain ends.

adjacent nucleotides are required for each amino acid. Since each of the three nucleotide positions can be occupied by any one of four different nucleotides, sixty-four different sets are possible. Each set of three nucleotides is called a codon. Each codon leads to the insertion of one kind of amino acid in the growing polypeptide chain.

Two of the twenty amino acids (tryptophan and methionine) have only a single codon for them. Nine amino acids are each represented by a pair of codons. These nine include all whose second codon nucleotide is an A as well as phenylalanine and cysteine. The third position of these codons is occupied by A or G for some amino acids and by C or T for the others. Because of the reduced stringency on the identity of the nucleotide in the third position, it is often called the wobble position. For six amino acids, any one of the four nucleotides occupies the wobble position. The three codons for isoleucine can be considered as belonging to this class, with the exception that AUG is reserved for methionine. Three amino acids (leucine, arginine, and serine) are unusual in that they can each be specified by any one of six codons.

Punctuation

The protein genetic code is often said to be "commaless." The bond connecting two codons cannot be distinguished from bonds connecting nucleotides within codons. There are no spaces or commas to identify which three nucleotides constitute a codon. As a result, the choice of which three nucleotides are to be read as the first codon during translation is very important. For example, if "EMA" is chosen as the first set of meaningful letters in the following string of letters, the result is gibberish:

TH EMA NHI TTH EBA TAN DTH EBA TBI THI M.

On the other hand, if "THE" is chosen as the first set of three letters, the message becomes clear:

THE MAN HIT THE BAT AND THE BAT BIT HIM.

The commaless nature of the code also means that one sequence of nucleotides can be read three different ways, starting at the first, second, or third letter. These ways of reading are called "reading frames." A frame is said to be open if none of its codons are stop codons. In most mRNA, only one reading frame is open for any appreciable length. However, in some mRNA, more than one reading frame is open. Some mRNAs produces two, rarely three, polypeptide sequences.

Exceptions to the Universal Code

The universal genetic code was discovered primarily through experiments with extracts from the *Escherichia coli* bacterium and from rabbit cells. Further work suggested that the code was the same in other organisms. It came to be known as the universal genetic code. The code was deciphered before scientists knew how to determine the sequence of nucleotides in DNA efficiently. After nucleotide sequences began to be determined, scientists could, using the universal genetic code, predict the sequence of amino acids. Comparison with the actual amino acid sequence revealed excellent overall agreement. Yet the universal genetic code assignments of codons to amino acids had apparent exceptions. Some turned out to be caused by programmed changes in the mRNA information. In selected codons of some mRNA, a C is changed to a U. In others, an A is changed so that it acts like a G. Editing of mRNA does not change the code used by the ribosomal machinery, but it does mean that the use of DNA sequences to predict protein sequences has pitfalls.

Some exceptions to the universal genetic code are true variations in the code. For example, the UGA universal stop codon codes for tryptophan in mollicutes (bacteria without walls) and in fungal, insect, and vertebrate subcellular organelles (mitochondria). Ciliated protozoans use UAA and UAG universal stop codons for the insertion of glutamine residues. Methionine, which only has one codon in the universal genetic code (AUG), is also encoded by AUA in vertebrate and insect mitochondria and in some, but not all, fungal mitochondria. The vertebrate mitochondria also use the universal arginine codons AGA and AGG as termination codons. AGA and AGG are serine rather than arginine codons in insect mitochondria.

Interpreting the Code

How is the code interpreted? The mRNA codons organize small RNA molecules called transfer RNA (tRNA). There is at least one tRNA for each of the twenty amino acids. They are *L*-shaped molecules. At the top of the *L*, tRNA has a set of three nucleotides (the anticodon) that can pair with the three nucleotides of the mRNA codon. Some tRNA can pair with several of the codons for the same amino acid. They do not pair with codons for other amino acids. At the bottom end of the *L*, tRNA has a site for the attachment of an amino acid.

The role of interpreter in deciphering the genetic code falls to enzymes called aminoacyl tRNA synthetases (RS enzymes), which attach amino acids to the tRNA. There is one RS enzyme for each of the twenty amino acids. Interpretation is possible because each RS enzyme can bind only one kind of amino acid and only to tRNA that pairs with the codons for that amino acid. After binding the amino acid and the tRNA, the RS enzyme makes a covalent bond between the amino acid and the tRNA, attaching the amino acid to the bottom end of the *L*. These charged tRNAs are ready to participate in protein synthesis directed by the codons of the mRNA. Information is stored in the RNA in forms other than the triplet code. A special tRNA for methionine exists to initiate all peptide chains. It responds to AUG. However, proteins also have methionines in the main part of the polypeptide chain. Those methionines are carried by a different tRNA. It also responds to AUG. The ribosome and associated factors must distinguish an initiating AUG from one for an internal methionine.

Distinction occurs differently in eukaryotes and bacteria. In bacteria, AUG will serve as an initiation codon only if it is near a sequence that can pair with a section of the RNA in the ribosome. Two things are required of eukaryotic initiator AUG codons: First, they must be in a proper context of surrounding nucleotides, both downstream and upstream of the codon; second, they must be the first AUG from the mRNA beginning that is in such a context. Context is also important for the incorporation of the unusual amino acid selenocysteine into several proteins. In a limited number of genes, a special UGA stop codon is used as a codon for selenocysteine. Sequences additional to UGA are needed for selenocysteine incorporation. Surrounding nucleotide residues also allow certain termination codons to be bypassed. For example, the mRNA from tobacco mosaic virus encodes two polypeptides, both starting at the same place; however, one is longer than the other. The extension is caused by the reading of a UAG stop codon by tRNA charged with tyrosine.

The production of two proteins with identical beginnings but different ends can also occur by frameshifting. In this mechanism, signals in the mRNA direct the ribosome machinery to advance or backtrack one nucleotide in its reading of the mRNA codons. Frameshifting occurs at a specific sequence in the RNA. Often the code for the frameshift includes a string of seven or more identical nucleotides and a complex RNA structure (a "pseudoknot").

Further codes are embedded in DNA. The linear sequence of amino acids, derived from DNA, has a code for folding in three-dimensional space, a code for its delivery to the proper location, a code for its modification by the addition of other chemical groups, and a code for its degradation. The production of mRNA requires nucleotide codes for beginning RNA synthesis, for stopping its synthesis, and for stitching together codon-containing regions (exons) should these be separated by noncoding regions (introns). RNA also contains signals that can tag them for rapid degradation. DNA has a code recognized by protein complexes for the initiation of DNA replication and signals recognized by enzymes that catalyze DNA rearrangements.

Impact and Applications

A major consequence of the near universality of the protein genetic code is that biotechnologists can move genes from one species into another and have those genes expressed in the second species. Since the code is the same in both organisms, the same protein is produced. This has resulted in the large-scale production of specific proteins in bacteria, yeast, plants, and domestic animals. These proteins are of immense pharmaceutical, industrial, and research value.

Scientists developed rapid methods for sequencing nucleotides in DNA in the 1970's. Since the protein genetic code was known, it suddenly became easier to predict the amino acid sequence of a protein from the nucleotide sequence of its gene than it did to determine the amino acid sequence of the protein by chemical methods. The instant knowledge of the amino acid sequence of a particular protein greatly simplified the proposal of hypotheses about the function of that protein and thus of its gene. This resulted in the molecular understanding of a wide variety of inherited human diseases and the development of rational therapies based on this new knowledge.

—*Ulrich Melcher*

See Also: Central Dogma of Molecular Biology; Genetic Code, Cracking of; Molecular Genetics; Protein Synthesis; RNA Structure and Function.

Further Reading: *Molecular Biology Made Simple and Fun* (1997), by David Clark and Lonnie Russell, is a detailed and entertaining account of molecular genetics. Horace Freeland Judson, *The Eighth Day of Creation* (1979), provides a fascinating history that includes the deciphering of the genetic code. *The Genetic Code* (1984), by Brian F. C. Clark, consists of a brief description of the genetic code.

Genetic Code, Cracking of

Field of study: Molecular genetics

Significance: *The deciphering of the genetic code was a significant accomplishment for molecular biologists. The identification of the "words" used in the code explained how the information carried in DNA can be interpreted, via an RNA intermediate, to direct the specific sequence of amino acids found in proteins.*

Key terms

AMINO ACID: a nitrogen-containing compound that serves as a building block of proteins

ANTICODON: a sequence of three nucleotide bases on the transfer RNA (tRNA) that recognizes the codon

CODON: a sequence of three nucleotide bases on the messenger RNA (mRNA) that specifies a particular amino acid

RIBONUCLEIC ACID (RNA): a molecule composed of ribonucleotides that functions as an intermediate between genes and the synthesis of proteins directed by the genes

The Nature of the Puzzle

Soon after deoxyribonucleic acid (DNA) was discovered to be the genetic material, scientists began to examine the relationship between DNA and the proteins that are specified by the DNA. DNA is composed of four deoxyribonucleotides containing the bases adenine (A), thymine (T), guanine (G), and cytosine (C). Proteins are composed of twenty different building blocks known as amino acids. The dilemma that confronted scientists was to explain the mechanism by which the four bases in DNA could be responsible for the specific arrangement of the twenty amino acids during the synthesis of proteins.

The solution to the problem arose as a result of both theoretical considerations and laboratory evidence. Experiments done in the laboratories of Charles Yanofsky and Sidney Brenner provided evidence that the order, or sequence, of the bases in DNA was important in determining the sequence of amino acids in proteins. Francis Crick proposed that the bases formed triplet "code words." He reasoned that if a single base specified a single amino acid, it would only be possible to have a protein made up of four amino acids. If two bases at a time specified amino acids, it would only be possible to code for sixteen amino acids. If the four bases were used three at a time, Crick proposed, it would be possible to produce sixty-four combinations, more than enough to specify the twenty amino acids. Crick also proposed that since there would be more than twenty possible triplets, some of the amino acids might have more than one code word. The eventual assignment of multiple code words for individual amino acids was termed "degeneracy." The triplet code words came to be known as "codons."

Identifying the Molecules Involved

Since DNA is found in the nucleus of most cells, there was much speculation as to how the

Marshall Nirenberg shared a 1968 Nobel Prize for his role in cracking the genetic code. (Jim Willier/Stokes Imaging)

codons of DNA could direct the synthesis of proteins, a process that was known to take place an another cellular compartment, the cytosol. Several classes of molecules related to DNA known as ribonucleic acid (RNA) were shown to be involved in this process. These molecules consist of ribonucleotides containing the bases A, C, and G (as in DNA) but uracil (U) rather than thymine (T). Ribosomal RNA was found to be contained in structures known as ribosomes, the sites where protein synthesis occurs.

Messenger RNA (mRNA) was shown to be another important intermediate. It is synthesized in the nucleus from a DNA template in a process known as transcription, and it carries an imprint of the information contained in DNA. For every A found in DNA, the mRNA carries the base U. For every T in DNA, the mRNA carries an A. The Gs in DNA become Cs in mRNA, and the Cs in DNA become Gs in mRNA. The information in mRNA is found in a form that is complementary to the nucleotide sequence in DNA. The mRNA is transported to the ribosomes and takes the place of DNA in directing the synthesis of a protein.

Deciphering the Code

The actual assignment of codons to specific amino acids resulted from a series of elegant experiments that began with the work of Marshall Nirenberg and Heinrich Matthaei in 1961. They obtained a synthetic mRNA consisting of polyuridylic acid, or poly (U), made up of a string of Us. They added poly (U) to a cell-free system that contained ribosomes and all other ingredients necessary to make proteins in vitro. When the twenty amino acids were added to the system, the protein that was produced contained a string of a single amino acid, phenylalanine. Since the only base in the synthetic mRNA was U, Nirenberg and Matthaei had discovered the code for phenylalanine: UUU. Because UUU in mRNA is complementary to AAA in DNA, the actual DNA bases that direct the synthesis of phenylalanine are AAA. By convention, the term "codon" is used to designate the mRNA bases that code for specific amino acids. Therefore UUU, the first code word to be discovered, was the codon for phenylalanine.

Using cell-free systems, other codons were soon discovered by employing other synthetic mRNAs. AAA was shown to code for lysine, and CCC was shown to code for proline. Scientists working in the laboratory of Severo Ochoa began to synthesize artificial mRNAs using more than one base. These artificial messengers produced proteins with various proportions of amino acids. Using this technique, it was shown that a synthetic codon with twice as many Us as Gs specified valine. It was not clear, however, if the codon was UUG, UGU, or GUU. Gobind Khorana and his colleagues began to synthesize artificial mRNA with predictable nucleotide sequences, and the use of this type of mRNA contributed to the assignment of additional codons to specific amino acids.

In 1964, Philip Leder and Marshall Nirenberg developed a cell-free protein-synthesizing system in which they could add triplet codons of known sequence. Using this new system, as well as Khorana's synthetic messengers, scientists could assign GUU to valine and eventually were able to assign all but three of the possible codons to specific amino acids. These three codons, UAA, UAG, and UGA, were referred to as "nonsense" codons because they did not code for any of the twenty amino acids. The nonsense codons were later found to be a type of genetic punctuation mark; they act as stop signals to specify the end of a protein.

There is no direct interaction between the mRNA codon and the amino acid for which it codes. Yet another type of RNA molecule was found to act as a bridge or, in Crick's terminology, an "adaptor" between the mRNA codon and the amino acid. This type of RNA is a small molecule known as transfer RNA (tRNA). Specific enzymes connect the amino acids to their corresponding tRNA; the tRNA then carries the amino acid to the appropriate protein assembly location specified by the codon. The tRNA molecules contain recognition triplets known as anticodons, which are complementary to the codons on the mRNA. Thus, the tRNA that carries phenylalanine and recognizes UUU contains an AAA anticodon.

By 1996, all the codons had been discovered. Since some codons had been identified as "stop" codons, scientists had begun searching for one or more possible "start" codons. Since all proteins were shown to begin with the amino acid methionine or a modified form of methionine (which is later removed), the methionine codon, AUG, was identified as the start codon for most proteins. It is interesting that AUG also codes for methionine when this amino acid occurs at other sites within the protein.

The cracking of the genetic code has given scientists a valuable genetic tool. If the amino acid sequence is known for a protein, or for even a small portion of a protein, knowledge of the genetic code now allows scientists to search for the gene that codes for the protein or, in some cases, to design and construct the gene itself. It is also possible to predict the sequence of amino acids in a protein if the sequence of nucleotide bases in a gene is known. Knowledge of the genetic code has also been invaluable in understanding the genetic basis of mutation and in attempts to correct these mutations by gene therapy techniques.

—*Barbara Brennessel*

See Also: Biochemical Mutations; Genetic Code; Mutation and Mutagenesis; Protein Synthesis; RNA Structure and Function.

Further Reading: *A Century of DNA: A History of the Discovery of the Structure and Function of the Genetic Substance* (1977), by Franklin H. Portugal and Jack S. Cohn, provides a comprehensive historical background and identifies many of the scientists who worked to solve the genetic code. *Blueprints: Solving the Mystery of Evolution* (1989), by Maitland A. Edey and Donald C. Johnson, focuses on evolution from the molecular genetic perspective and is written to emphasize the process of scientific discovery; three chapters are devoted to the genetic code. "The Genetic Code—Yesterday, Today, and Tomorrow," by Francis H. C. Crick, *Cold Spring Harbor Symposia on Quantitative Biology* 31 (1966), summarizes how the genetic code was solved and serves as an introduction to papers presented during a symposium on the genetic code. Another summary of the elucidation of the genetic code, "The Genetic Code III," also by Crick, can be found in *Scientific American* 215 (October, 1966) and was reprinted in *The Chemical Basis of Life: An Introduction to Molecular and Cell Biology* (1973).

Genetic Counseling

Field of study: Human genetics
Significance: *Genetic counseling is a health-care profession that deals with situations in which individuals or families are concerned about a genetic abnormality occurring in a member of the family. Genetic counselors provide information regarding the occurrence or risk of occurrence of genetic disorders, discuss available options for dealing with those risks, and help families determine their best course of action.*

Key terms

GENETIC SCREENING: the process of investigating a specific population of people to detect the presence of disease

PEDIGREE ANALYSIS: analysis of a family's history by listing characteristics such as age, sex, and state of health of family members; it is used to determine the risk of passing a genetic abnormality on to future generations

PRENATAL DIAGNOSIS: the process of detecting a variety of birth defects and inherited disorders before a baby is born by testing maternal blood, fetal tissue, and fetal visualization

NONDIRECTIVE COUNSELING: a practice that values patient autonomy and encourages patients to reach a decision that is right for them based upon their personal beliefs and values

The Establishment of Genetic Counseling

Historically, people understood that some physical characteristics were hereditary and that particular defects were often common among relatives. This vague belief was widely accepted by expectant parents and influenced the thinking of many scientists who experimented with heredity in plants and animals. Many efforts were made to understand, predict, and control the outcome of reproduction in humans and other organisms. Gregor Mendel's experiments with garden peas in the mid-1800's led to the understanding of the relationship between traits in parents and those in their offspring. During the early twentieth century, Walter Sutton proposed that newly discovered hereditary factors were physically located on complex structures within the cells of living organisms. This led to the chromosome theory of inheritance, which explains mechanically how genetic information is transmitted from parent to offspring in a regular, orderly manner. In 1953, James Watson and Francis Crick (along with Maurice Wilkins and Rosalind Franklin) discovered the double-helix structure of deoxyribonucleic acid (DNA), the molecule that carries the genetic information in the cells of most living organisms. Three years later, human cells were found to contain forty-six chromosomes each.

These discoveries, along with other developments in genetics, periodically generated efforts (often misguided) to control the existence of "inferior" genes, a concept known as eugenics. Charles F. Dight, a physician influenced by the eugenics movement, left his estate in 1927 "To Promote Biological Race Betterment—betterment in Human Brain Structure and Mental Endowment and therefor[e] in Behavior." The Dight Institute for Human Genetics began to function in 1941 with a practice directed less toward eugenics and more toward the future studies of individual families. In 1947, Sheldon Reed began working at the Dight Institute as a genetic consultant to individual families. Reed believed that his profession should put the clients' needs before all other considerations and that it should be separated from the concept of eugenics. He rejected the older names for his work, such as "genetic hygiene," and substituted "genetic counseling" to describe the type of social work contributing to the benefit of the family. As a result, the field of genetic counseling was born and separated itself from the direct concern of its effect upon the state or politics. In fact, Reed predicted that genetic counseling would have been rejected had it been presented as a form of eugenics.

Genetic counseling developed as a preventive tool and became more diagnostic in nature as it moved from academic centers to the major medical centers. In 1951, there were ten genetic counseling centers in the United States employing academically affiliated geneticists. Melissa Richter and Joan Marks were instrumental in the development of the first graduate program in genetic counseling at Sarah Lawrence College in New York in 1969. By the early 1970's, there were nearly nine hundred genetic counseling centers worldwide. In 1998, there were more than fifteen hundred genetic counselors in the United States not only working with individual families concerning genetic conditions but also involved in teaching, research, screening programs, public health, and the coordination of support groups. In 1990, the Human Genome Project began as a fifteen-year effort coordinated by the U.S. Department of Energy and the National Institutes of Health to map and sequence the entire human genome, prepare a model of the mouse genome, expand medical technologies, and study the ethical, legal, and social implications of genetic research.

The Training of the Genetic Counselor

Most genetic counseling students have undergraduate degrees in genetics, nursing, psychology, biology, social work, or public health. Training programs for genetic counselors are typically two-year masters-level programs and include field training in medical genetics and counseling in addition to a variety of courses focusing on genetics, psychosocial theory, and counseling techniques. During the two-year program, students obtain an in-depth background in human genetics and counseling through coursework and field training at genetic centers. Coursework incorporates information on specific aspects of diseases, including the prognoses, consequences, treatments, risks of occurrence, and prevention as they relate to individuals or families. Field training at genetic centers enables students to develop research, analytical, and communication skills necessary to meet the needs of individuals at risk for a genetic disease.

Many genetic counselors work with M.D. or Ph.D. geneticists and may also be a part of a health-care team that may include pediatricians, cardiologists, psychologists, endocrinologists, cytologists, nurses, and social workers. Other genetic counselors are in private practice or are engaged in research activities related to the field of medical genetics and genetic counseling. Genetic counseling most commonly takes place in medical centers, where specialists work together in a clinical genetics units with access to diagnostic facilities such as genetic laboratories and equipment for prenatal screening.

The Role of the Genetic Counselor

Prior to the 1960's, most genetic counselors were individuals with genetic training who consulted with patients or physicians about specific risks of occurrence of genetic diseases. It was not until 1959, when French geneticist Jérôme Lejeune discovered that children with Down syndrome have an extra chromosome 21, that human genetics was finally brought to the attention of ordinary physicians. Rapid growth in knowledge of inheritance patterns, improvements in the ability to detect chromosomal abnormalities, and the advent of screening programs for certain diseases in high-risk populations all contributed to the increased interest in genetic counseling. Development of the technique of amniocentesis, which detects both chromosomal and biochemical defects in fetal cells, led to the increased specialization of genetic counseling. By the 1970's, training of genetic counselors focused on addressing patients' psychosocial as well as medical needs. Genetic counseling thus became a voluntary social service intended exclusively for the benefit of the particular family involved.

Genetic counselors provide information and support to families who have members with genetic disorders, individuals who themselves are affected with a genetic condition, and families who may be at risk for a variety of inherited genetic conditions, including Huntington's chorea, cystic fibrosis, and Tay-Sachs disease. The counselor obtains the family medical history and medical records in order to interpret information about the inherited genetic abnormality. Genetic counselors analyze inheritance patterns, review risks of recurrence, and offer available options for the genetic condition. Other functions of genetic counselors include discussing genetic risks with blood-related couples considering marriage, contacting parents during the crisis following fetal or neonatal death, preparing a community for a genetic population screening program, and informing couples about genetically related

causes of their infertility. A patient is most commonly referred to a genetic counselor by an obstetrician because of advanced maternal age (thirty-five years or older).

In addition to obtaining accurate diagnosis of the genetic abnormality, genetic counselors strive to explain the genetic information as clearly as possible, making sure that the individual or family understands the information fully and accurately. The genetic counselor must evaluate the reliability of the diagnosis and the risk of occurrence of the genetic disease. Because the reliability of various tests will affect a patient's decision about genetic testing and abortion, the counselor must give the patient a realistic understanding of the meaning and inherent ambiguity of test results. Most genetic counselors practice the principle of nondirectiveness and value patient autonomy. They present information on the benefits, limitations, and risks of diagnostic procedures without recommending a course of action, encouraging patients to reach their own decisions based on their personal beliefs and values. This attitude reflects the historical shift of genetic counseling away from eugenics toward a focus on the individual family. The code of ethics of the National Society of Genetic Counselors states that its members strive to "respect their clients' beliefs, cultural traditions, inclinations, circumstances, and feelings as well as provide the means for their clients to make informed independent decisions, free of coercion, by providing or illuminating the necessary facts and clarifying the alternatives and anticipated consequences."

Diagnosis of Genetic Abnormalities

In the latter half of the twentieth century, discoveries in genetics and developments in reproductive technology contributed to the advancements in prenatal diagnosis and genetic counseling. Prenatal diagnostic procedures eventually became an established part of obstetrical practice with the development of amniocentesis in the 1960's, followed by ultrasound, chorionic villus sampling (CVS), and fetal blood sampling. Amniocentesis, CVS, and fetal blood sampling are ways to obtain fetal cells for analysis and detection of various types of diseases. Amniocentesis, a cytogenetic analysis of the cells within the fluid surrounding the fetus, is performed between the fifteenth and twentieth weeks of gestation and detects possible chromosomal abnormalities such as Down syndrome and trisomy 18. The information obtained from CVS is similar to that obtained from amniocentesis, except the testing can be performed earlier in the pregnancy (during the tenth to twelfth weeks of gestation). Fetal blood sampling can be performed safely only after eighteen weeks of pregnancy. An ultrasound, offered to all pregnant women, uses high-frequency sound waves to create a visual image of the fetus and detects anatomical defects such as spina bifida, cleft lip, or certain heart malformations. Pedigree analysis may also be used for diagnostic purposes and to determine the risk of passing a genetic abnormality on to future generations. A pedigree of the family history is constructed, listing the sex, age, and state of health of the patient's close relatives; from that, recurrent miscarriages, stillbirths, and infant deaths are explored.

Prenatal diagnostic techniques are used to identify many structural birth defects, chromosomal abnormalities, and over five hundred specific disorders. Genetic counselors who believe that their client is at risk for passing on a particular disease may suggest several genetic tests, depending on the risk the patient may face. Screening of populations with high frequencies of certain hereditary conditions, such as Tay-Sachs disease among Ashkenazi Jews, is encouraged so that high-risk couples can be identified and their pregnancies monitored for affected fetuses. Pregnant women may also be advised to undergo prenatal testing if an abnormality has been found by the doctor, the mother will be thirty-five years of age or older at the time of delivery, the couple has a family history of a particular genetic abnormality, the mother has a history of stillbirths or miscarriages, or the mother is a carrier of metabolic disorders that can be passed from mothers to their sons (for example, hemophilia).

The Human Genome Project is expected to have a dramatic impact on presymptomatic diagnosis of individuals carrying specific diseases, identification of multigene defects involved in common diseases such as heart disease and diabetes, and identification of individual susceptibility and environmental factors that interact with genes to produce diseases. The isolation and sequencing of genes associated with genetic abnormalities such as cystic fibrosis, kidney disease, Alzheimer's disease, and Huntington's chorea allow for individuals to be tested for those specific conditions. Between 1991 and 1995, more than fifty new genetic tests were developed so that the detection of genetic conditions could be made earlier and with more precision.

Ethical Aspects of Genetic Counseling

With advancements in human genetics and reproductive technology, fundamental moral and ethical questions may arise during difficult decision-making processes involving genetic abnormalities for which families may be unprepared. Diagnosis of a particular genetic disease may allow individuals or families to make future plans and financial arrangements. However, improvements in the capability to diagnose numerous hereditary diseases often exceed the ability to treat such diseases. The awareness that a disease with no known cure exists may lead to traumatic anxiety and depression. The psychological aspects of genetic counseling and genetic centers must therefore continue to be explored in genetic centers throughout the world.

Questions about who should have access to the data containing patients' genetic makeup must also be considered as the ability to screen for genetic diseases increases. Violating patients' privacy could potentially have devastating consequences such as genetic discrimination in job hiring and availability of health coverage. Employers and insurance companies have already denied individuals such opportunities based on information found through genetic testing. Disclosure of genetic information not only contributes to acts of discrimination but also may result in physical and psychological harm to individuals.

With data derived from the Human Genome Project increasing rapidly, problems arising from the application of new genetic knowledge in clinical practice must be addressed. The norm of nondirective counseling will be challenged, raising questions of who provides and who receives information and how it is given. Many believe that genetic counseling is beneficial to those faced with genetic abnormalities, while others fear that genetic counseling is a form of negative eugenics, an attempt to "improve" humanity as a whole. Since most genetic conditions can be neither treated nor modified in pregnancy, abortion is often the preventive measure used. Thus, ethical issues concerning the respect for autonomy of the unborn child must also be considered.

—Jamalynne Stuck
—Doug McElroy

See Also: Amniocentesis and Chorionic Villus Sampling; Genetic Screening; Genetic Testing; Heredity Diseases; Linkage Maps.

Further Reading: *Prescribing Our Future: Ethical Challenges in Genetic Counseling*, written by Bonnie Leroy and edited by Dianne M. Bartels and Arthur L. Caplan (1993), offers ethical insights into the implications of genetic counseling. An accurate and precise history on the establishment of genetic counseling is presented in Sheldon C. Reed, "A Short History of Genetic Counseling," *Social Biology* 21 (1974). Barbara Katz Rothman's *The Tentative Pregnancy* (1984) provides a discussion of situations faced by patients who seek genetic counseling. In *Mapping Fate* (1995), Alice Wexler describes her personal quest to discover the genetic basis for Huntington's chorea.

Genetic Engineering

Field of study: Molecular genetics

Significance: *The development of the tools of recombinant deoxyribonucleic acid (DNA) technology used in genetic engineering has generated unprecedented inquiry into the nature of the living system and has revolutionized the study of genetics. The implications of this research are far-reaching, ranging from a vastly greater understanding of basic biological principles and molecular mechanisms to pharmacological, diagnostic, and therapeutic applications that offer promise in the prevention and treatment of a wide range of genetic diseases.*

Key terms

CLONING: the process by which large amounts of a single gene or genome (the entire genetic content of a cell) are produced

VECTOR: a segment of DNA usually derived from virus, bacteria, or yeast genes that contains regulatory sequences of DNA that permit the amplification of single genes or genetic segments to high copy number

DNA SEQUENCE ANALYSIS: chemical methods that permit the identification of each of the four nucleotide bases of DNA (adenine, guanine, cytosine, and thymine) so that the linear sequence of nucleotide bases composing individual genes can be determined

PROBE HYBRIDIZATION: a method that permits the identification of a unique sequence of DNA bases using a single-stranded DNA segment complementary to the unique sequence and carrying a molecular tag allowing identification

COMPLEMENTARY BASE PAIRING: hydrogen bond formation that only occurs between adenine and thymine or cytosine and guanine

GENOMIC LIBRARY: a segment of DNA representing the entire genome of a single species ligated to vector DNA to permit amplification of individual genes, which can be identified using specific probes

BIOTECHNOLOGY: the use of recombinant DNA technology to produce gene products in large quantities for pharmaceutical applications

TRANSGENIC ANIMALS: species in which the genome has been modified by the insertion of genes obtained from another species (often human)

Key Discoveries in Genetic Engineering

Many of the methods used in genetic engineering represent adaptations of naturally occurring genetic processes. One of the earliest and most significant discoveries in this field was the identification of a family of deoxyribonucleic acid (DNA) enzymes called restriction enzymes. Restriction enzymes are DNA-modifying enzymes produced by microorganisms; their uniqueness and utility in recombinant DNA technology resides in their ability to cleave DNA at precise recognition sites based on DNA sequence specificity. Several hundred restriction enzymes have been isolated, and many recognize unique DNA segments and initiate DNA cleavage only at these sites. The site-specific cleavages generated by restriction enzymes can be used to produce a unique set of DNA segments that can be used to "map" individual genes and distinguish them from all other genes. This type of genetic analysis, based on differences in the sizes of DNA segments from different genes or different individuals when cleaved with restriction enzymes, is referred to as restriction fragment length polymorphism (RFLP) analysis.

If genes or DNA segments from different sources or species are cleaved with the same restriction enzyme, the DNA segments produced, though genetically unrelated, can be mixed together to produce recombinant DNA. This occurs because the restriction enzymes produce complementary, linear, single-stranded DNA ends that can join together. An additional enzyme called DNA ligase is used to form linkages between the unrelated DNA segments, generating recombinant DNA. This procedure, developed in the 1970's, is at the core of recombinant DNA technology and can be used to analyze the structure and function of the genome at the molecular level.

Another key development in genetic engineering involves the use of vectors to amplify genetic segments for analytical purposes. Vectors are genetic segments derived from viruses, bacteria, or other microorganisms such as yeast that contain regulatory sequences permitting the amplification, or increase in the number of copies, of individual genes in the appropriate host system. There are many different vectors in use, and each represents a cloning system with specific applications. Plasmids are small, circular DNAs that have been isolated from many species of bacteria. These naturally occurring molecules often encode antibiotic resistance genes that can be transferred from one bacterial cell to another in a process called transformation. In the laboratory, plasmids can be used as vectors in the amplification of genes inserted by restriction enzyme treatment of both vector and insert DNA, followed by DNA

ligation to produce recombinant molecules. The recombinant DNA is then inserted into host bacterial cells by DNA transformation, a process by which treatment of bacterial cells with calcium chloride facilitates the uptake of DNA. Once inside the host cell, the recombinant plasmid will amplify to high copy number as the bacterial cells reproduce rapidly to generate colonies of cells, each containing large amounts of recombinant plasmid DNA containing the inserted gene. This process, which results in the amplification of the inserted gene to high copy number, is termed gene cloning. The cloned DNA can then be isolated from the bacterial cells and characterized with respect to structure and function. Plasmids are useful for cloning small genes or genetic segments; larger genes can be cloned using viral vectors such as the bacterial virus (bacteriophage) lambda. This virus can infect bacterial cells and reproduce to high copy number. If nonessential viral genes are removed, recombinant viruses containing genes of interest can be produced and used to clone larger genes. Synthetic recombinant vectors incorporating bacterial and viral components, called cosmids, have also been developed for cloning purposes. In addition, synthetic minichromosomes called yeast artificial chromosomes (YACs), which incorporate large segments of chromosomal DNA and which are capable of replication in bacterial or eukaryotic systems, have been developed.

A further key discovery in genetic engineering involves the development of chemical methods of DNA sequence analysis. These methods permit a determination of the linear sequence of nucleotide bases in DNA as a result of chemical modifications in the DNA bases that can be used to distinguish one base from another in sequence. The linear sequence of bases in DNA represent the genetic information, which functions as a code specifying the structural formation of proteins within the cell. Thus DNA sequence analysis permits a direct determination of gene structure with respect to regulatory and protein coding regions and facilitates an analysis of the structure of proteins encoded by specific genes.

There are many important applications of the basic principles of genetic engineering.

Notable examples include the analysis of the structure of individual genes and the organization of the genome by gene-cloning methods, the identification and characterization of human disease genes, the production of large amounts of proteins for therapeutic or industrial purposes, the creation of genetically engineered plants that are disease resistant and show higher productivity, the creation of genetically engineered microorganisms that can serve as vaccines or fight the effects of pollution, and the treatment of genetic disorders using gene therapy.

Gene Cloning

The genetic engineering methods fundamental to the development of recombinant DNA technology have facilitated the mapping of human chromosomes and the detailed analysis of gene structure and function. By means of gene cloning and DNA sequence analysis, the structure and function of many genes have been determined. Information has also been generated on the general structural organization of the human genome and the genomes of other species. The Human Genome Project, an international effort to elucidate the structure of the entire human genome, offers the promise of greatly increasing the understanding of the genes responsible for inherited single-gene disorders as well as the involvement of specific genes in multifactorial disorders such as coronary heart disease.

The methods of genetic engineering have facilitated the molecular analysis of the structural differences between normal and disease genes responsible for inherited single-gene disorders. The genetic mutations that disrupt the functional activities of normal cellular genes and convert them into disease-causing genes have been identified for diseases such as sickle-cell anemia (which results from a single nucleotide base substitution in the hemoglobin gene), Duchenne muscular dystrophy (which results from deletions within the gene encoding the muscle protein dystrophin), and cystic fibrosis (which results from multiple mutations within a gene encoding chloride channel conductance protein). The identification of these disease genes has permitted the design of diagnostic genetic tests for some genetic disorders. In addition, analyses of the functions of the genes implicated in specific genetic diseases have led to the development of therapeutic strategies, including attempts to replace defective genes by gene therapy approaches.

Gene cloning methods have also permitted the analysis of gene function by a process called site-directed mutagenesis, which involves the targeted insertion of mutations or altered DNA sequences into cloned genes. Mutant genes constructed in this way can then be inserted into in vitro cellular systems to analyze the physiological consequences of mutations in specific genes at the cellular level. Alternatively, the mutant genes can be inserted into host animals in order to explore the effects of targeted gene mutations on developmental processes.

Transgenic Organisms

The process by which foreign genes are introduced into host animals is called germ-line transformation and involves the insertion of individual genes into fertilized eggs, which, after implantation into foster mothers, may produce offspring that have incorporated the foreign gene into all the cells of the body, including the reproductive cells. This represents a heritable genetic alteration that modifies the germline or reproductive cells of the species, generating what is termed a transgenic animal.

Many methods of gene transfer have been developed to facilitate the efficient insertion of exogenous genetic material into recipient cells. All of these methods take advantage of the naturally occurring processes of DNA transformation involving the exchange of genetic information between bacterial cells or viral infection, which results in the uptake and expression of viral genes in the infected host cell. Genetically engineered viruses such as retroviruses may insert the viral gene into the recipient cell following viral infection. Exogenous genes may be incorporated into lipid membranes to form liposomes, which can then bind to the target cell and insert the gene. Chemical methods of gene transfer include the use of calcium phosphate or dextran sulfate to gener-

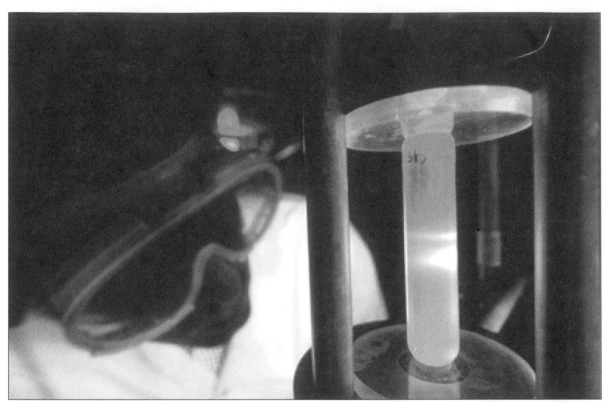

A genetic engineer examines glowing bands of DNA ready for splicing. (Dan McCoy/Rainbow)

ate pores in the recipient cell membrane through which the exogenous DNA enters the cell. Microinjection involves the use of microscopic needles to insert foreign DNA directly into the nucleus of the target cell and is often used to insert genetic material into fertilized eggs. Electroporation involves the use of an electric current to open pores in the cell membrane, permitting DNA uptake by the recipient cell. Finally, particle bombardment represents a method of gene transfer in which metal pellets coated with DNA are transferred into target cells under high pressure using "gene guns." This method is particularly useful for inserting genes into plant cells that are resistant to DNA uptake because of the presence of a thick cell wall.

Genetically engineered transgenic species have many biological uses. Transgenic animals have been used to analyze the functions of specific genes in development and to generate animal models of human diseases. For example, a transgenic mouse strain incorporating a

human breast cancer gene has been developed to explore the mechanisms by which this disease occurs. In addition, transgenic mice have been used in research to analyze the functions of specific genes in normal physiology and in disease-causing processes using methods of gene targeting to generate "knockout" mice whose genomes contain these targeted mutations. This technology, developed by Mario Capecchi, involves precise genetic replacement of specific normal cellular genes by their mutant counterparts by a process called homologous recombination, in which only complementary nucleotide base pairs carry out the genetic exchange within the host chromosome. By this method of targeted insertion, the effects of the inserted gene, or transgene, on development and physiology can be examined. Knockout mice have provided useful models for human disease processes. For example, knockout mice lacking a functional adenosine deaminase (ADA) gene show disease characteristics comparable to humans with severe combined immunodeficiency disorder (SCID). The SCID mice can be used to evaluate the efficacy of novel treatment modalities including gene therapy protocols in the restoration of immune system function.

Transgenic animals have also been developed for biotechnological purposes to produce therapeutic gene products in large quantities for the treatment of human diseases. For example, transgenic sheep have been developed that secrete the human gene product alpha-1 antitrypsin (AAT) in their milk. AAT is used to treat an inherited form of emphysema. The process involves the microinjection of fertilized sheep eggs with the human AAT gene linked to regulatory genetic sequences that allow the gene product to be produced in large amounts in the mammary tissue of the transgenic sheep. Although the process of generating transgenic animals is inefficient, individual transgenic animals can produce tremendous amounts of gene products that can be readily purified from the milk in a cost-efficient manner. Additional transgenic livestock have been genetically engineered to produce tissue plasminogen activator (used in the treatment of blood clots), hemoglobin (used as a blood substitute), erythropoietin (used to stimulate red blood cell formation in kidney dialysis patients), human growth hormone (used to treat pituitary dwarfism), and factor VIII (used to treat hemophilia).

Recombinant DNA technology has also been used to create genetically engineered plants, using the plant vector Ti (tumor inducing) plasmid. This plasmid is found naturally in the bacterium *Agrobacterium tumefaciens* as well as in plant cell viruses. The Ti plasmid has been used to transfer genes that destroy insect pests into plant cells, thereby avoiding the use of polluting pesticides.

Biotechnology

One of the most exciting applications of genetic engineering principles involves its use in biotechnology in the development of therapeutic gene products for the treatment of human diseases. Biotechnology can be defined as the production of useful gene products using the methods of recombinant DNA technology.

The most basic application of these principles involves the insertion of a gene whose product is desired in large quantities into a suitable vector using the methods of restriction enzyme digestion, DNA hybridization, and ligation to produce recombinant DNA. The recombinant vector is then inserted into a recipient host cell using one of several methods of DNA transfer. Once inside the host cell, the inserted gene replicates or reproduces every time the host cell divides, and regulatory genetic elements allow the expression of the insert gene to produce large amount of gene products that can be purified from the host system.

One of the earliest developments in this field of research involved the use of genetically engineered bacteria to produce human insulin for the treatment of diabetes. The technology involved the cloning of the human insulin gene and its insertion into bacterial expression vectors that would permit expression of the cloned gene to produce large amounts of insulin for therapeutic purposes. Subsequently, many gene products have been produced by genetically engineered microorganisms for clinical purposes, including clotting factors (used in the treatment of hemophilia), growth factors such as epidermal growth factor (used to accelerate wound healing) and colony stimulating factors (used to stimulate blood cell formation in the bone marrow), and interferons (used in the treatment of immune system disorders and certain types of cancer). The advantages of using genetically engineered products are enormous, since the therapeutic proteins or hormones can be produced in much larger amounts than could be obtained from tissue isolation and since the genetically engineered products are free of viruses or other contaminants.

Genetically Engineered Viruses

An additional medical application involves the use of genetically engineered viruses in the treatment of genetic diseases. Retroviruses are the most important group of viruses used for these purposes, since the life cycle of the virus involves the incorporation of the viral genetic material into host chromosomes, which results in the stable integration and expression of inserted genes. The genetic engineering procedure involves the removal of most of the structural genes of the virus, leaving only the regulatory viral genes that are ligated to the therapeutic gene. The recombinant retrovirus is no longer capable of causing infection since the structural genes have been removed; however, it can enter the cell and become inserted into the host cell genome from which it directs the expression of the therapeutic gene insert. Successful clinical applications include the use of genetically engineered retroviruses in the treatment of severe combined immunodeficiency disorder (SCID) by inserting a functional copy of the adenosine deaminase gene into human blood cells.

Genetic engineering approaches have also been used to produce altered viruses in the design of recombinant vaccines. In this application of recombinant DNA technology, viruses that have been used successfully as preventive vaccines, such as vaccinia virus, are genetically manipulated by the insertion of genes from other viruses to generate a recombinant viral genome. During the process of infection, the recombinant vaccinia virus produces gene products from the inserted foreign viral gene, which can serve as an antigen eliciting an immune response following vaccination. The immune system recognition of this single viral antigen blocks subsequent infection upon exposure to the infectious virus. This strategy is particularly useful in the development of vaccines against viruses that are highly pathogenic such as the human immunodeficiency virus (HIV), in which it is not possible to use a whole killed or attenuated (weakened) live viral vaccine used in traditional vaccine formulations because of the risk of infection by vaccination. Genetically engineered viruses may also be useful in the treatment of diseases such as cancer since they could be designed to target specific cells with abnormal cell surface components. Recombinant adenoviruses containing a single gene mutation have been engineered that are capable of lethal infection in cancer cells but not in normal tissues of the body.

Impact and Applications

The methods of recombinant DNA technology have revolutionized the understanding of the molecular basis of life and have led to applications affecting the quality of life and humanity's capacity to manipulate the environment. Some of the most important discoveries have involved an increased understanding of the molecular basis of disease processes, which has led to new methods of diagnosis and treatment. Genetically engineered animals can be used to produce unlimited amounts of therapeutic gene products and can also serve as genetic models to enhance understanding of the physiological basis of disease. Plants can be genetically engineered for increased productivity and disease resistance. Genetically engineered viruses have been developed as vaccines against infectious disease. The methods of recombinant DNA technology were originally developed from natural products and processes that occur within the living system. The ultimate goals of this research must involve applications that preserve the integrity and continuity of the living system. Within that context, the potential for genetic engineering approaches to increase understanding of the molecular basis of life, to prevent and treat human diseases, and to solve environmental problems is profound.

—*Sarah Crawford Martinelli*

See Also: Biotechnology; Cloning; Genetic Engineering: Agricultural Applications; Genetic Engineering: Historical Development; Genetic Engineering: Industrial Applications; Genetic Engineering: Medical Applications; Genetic Engineering: Social and Ethical Issues; Transgenic Organisms.

Further Reading: William Velander et al., "Transgenic Livestock as Drug Factories," *Scientific American* 276 (January, 1997), explains how genetic engineering methods have resulted in the production of farm animals whose milk contains large amounts of medicinal proteins. Theodore Friedmann, "Overcoming the Obstacle to Gene Therapy," *Scientific American* 276 (June, 1997), discusses the problems associated with the development of efficacious gene therapy approaches. Steve Mirsky and John Rennie, "What Cloning Means for Gene Therapy," *Sci-*

entific American 276 (June, 1997), explains how methods of gene cloning play an important role in gene therapy approaches. Mario Capecchi, "Targeted Gene Replacement," *Scientific American* 270 (March, 1994), describes a method for gene replacement involving homologous recombination. *Recombinant DNA* (1995), by James Watson, provides an in-depth explanation of the methods and applications of recombinant DNA technology. *Molecular Cloning: A Laboratory Manual* (1989), by Jay Sambrook et al., is a practical guide to gene cloning methods and analyses. *Mapping Fate* (1995), by Alice Wexler, describes the fascinating story of the discovery of the gene that causes Huntington's chorea. Tim Beardsley, "Vital Data," *Scientific American* 274 (March, 1996), discusses ethical issues related to genetic information obtained by genetic screening methods.

Genetic Engineering: Agricultural Applications

Field of study: Genetic engineering and biotechnology

Significance: *Genetic engineering is being used in many ways to improve the quality and quantity of plant and animal products used for food, oil, and fiber as an adjunct to traditional plant and animal breeding methods. The techniques of genetic engineering allow for the transfer of genes without the need for sexual reproduction. Considerable controversy over the safety and ethics of these procedures exists.*

Key terms

DEOXYRIBONUCLEIC ACID (DNA): the genetic material for most organisms

CLONE: an identical copy of a whole organism or a piece of DNA

TRANSGENIC ORGANISMS: organisms into which cloned DNA has been transferred

RECOMBINANT DNA TECHNOLOGY: techniques for joining together DNA molecules in the test tube and then inserting them into living organisms for replication

Overview

Humans have been breeding plants and animals for improved agricultural traits since agriculture began. The techniques of genetic engineering are now being used along with traditional breeding techniques in order to produce improved plants and animals for human use. Typical traits of agricultural interest such as fertility, yield, and disease resistance are being investigated, as is the production of unique products from agricultural organisms. Genetic engineering involves the use of recombinant deoxyribonucleic acid (DNA) technology to transfer cloned pieces of DNA into an organism. This transfer of genetic material can take place regardless of the source of the cloned DNA. Not all organisms can be genetically engineered, but the list of plants and animals that can be manipulated through this technology has increased steadily.

Plant Genetic Engineering

The goal of plant breeding is to increase the yield and quality of a particular crop. The overall genetic basis of yield is complex and strongly influenced by environmental conditions, and some crops may not be good candidates for genetic engineering. However, some components of yield and quality are being examined individually with some success, including disease resistance, levels of solids, and nitrogen uptake.

One of the greatest limitations to current agricultural productivity is the susceptibility of crop plants to disease. Chemical control (pesticides and fungicides, for example) is effective in controlling some diseases but not others. The use of chemical controls may also have undesirable environmental consequences. Genes exist in some plants that confer resistance to specific diseases. These disease-resistance genes have been used by plant breeders for many years to protect crops by locating them and transferring them by crossing into desirable cultivars. The usefulness of these genes was somewhat limited by the plants into which they could be transferred. For example, a resistance gene cloned from corn could not, using classical plant-breeding techniques, be transferred into wheat. The first resistance gene to be located and cloned was the *Pto* gene for resistance to bacterial speck disease in tomatoes, by Gregory Martin, Steven Tanksley, and colleagues in 1993. Now experiments can be done in which these resistance genes are inserted into plants that could not normally be crossed to the resistant plant. Additional cloned resistance genes include the *N* gene for tobacco mosaic virus resistance in tobacco, cloned by Barbara Baker and colleagues, and the *Cf* gene for resistance to a fungal disease of tomatoes, cloned by Jonathan Jones and colleagues.

Another strategy has also been used to engineer resistance to viral pathogens. In these experiments, initiated by Roger Beachy and colleagues in the 1980's, a gene from the tobacco mosaic virus was inserted into tobacco through genetic engineering techniques. Through means that are not well understood, this provided some level of resistance to the virus. Similar work has been done on different viruses, with varying results. Results are unpredictable, with increasing numbers of transferred genes not necessarily being correlated with increasing resistance to the virus.

Genes related to fruit quality (genes for high levels of solids and ripening-related genes) have been cloned in tomatoes. Tanksley started transformation of tomatoes with combinations of the high solids genes in 1997 in order to determine their individual and combined effects. Increasing the amount of solids in tomato fruit would lead directly to increased market value for the fruit. Plants transgenic for ripening-related genes have also been made, particularly for the fruit-softening enzyme polygalacturonase (PG). Control of PG may allow for fruits that can be transported unripe to avoid bruising but that will ripen at the point of delivery. Calgene received approval from the Food and Drug Administration (FDA) to release a genetically engineered tomato, "Flavr Savr," to the marketplace in 1992. This was the first example of a genetically engineered food receiving approval for release in the United States.

A bull designated as having "ideal" breeding potential. (Dan McCoy/Rainbow)

The pathways for nitrogen-fixing bacterium to obtain nitrogen directly from the air and provide it to their host plants (peas and beans) are being studied with the hope of being able to engineer the ability to fix nitrogen into additional plants. This would provide a means of obtaining nitrogen, a necessary nutrient for the plant, with a much lower application of chemical fertilizer to the field. In addition, other microorganisms are being engineered to enhance agricultural productivity. A bacterial strain was produced and named "Ice-Minus" because spraying it on plants inhibited the crystallization of water on the leaf surfaces, preventing frost damage at lower temperatures. The field trial of the use of this bacterium on strawberry plants in 1986 was one of the first allowed releases of genetically engineered organisms into the environment, and it generated intense controversy because of potential safety concerns over the uncontrolled spread of the "Ice-Minus" bacterium.

Animal Genetic Engineering

Improvement of livestock species has focused on traditional breeding practices, the cryostorage of sperm from superior males, and the use of artificial insemination techniques. In the 1980's, improved techniques in reproductive biology and cell culture allowed for the transformation of embryonic cells in some species, but not from adult cells. This changed in 1997, however, when the first mammal cloned from an adult cell, the sheep Dolly, was reported by Ian Wilmut and colleagues. Cloning has also been done in mice, cows, and monkeys. The development of cloning techniques in mammals opens the door for many advances in animal improvement and the use of animals to produce genetically engineered products. It also started a firestorm of controversy, including a renewed discussion about the ethical implications of human cloning. In 1998, Richard Seed announced that he would clone humans to provide to childless couples by the end of 1999. Many critics dismissed his announcement because of the potential technical difficulties, and many others discussed the difficult ethical questions posed by such work. The potential for the production of medically useful compounds from agricultural animals is an exciting force driving this research. Wilmut, Angelika Schnieke, and colleagues sparked this research in 1997, engineering sheep that produced human blood clotting factor IX.

Impact and Applications

The use of molecular genetic techniques to create new genotypes of plants and animals for agriculture is controversial, particularly in organisms designed to produce foodstuffs. Much controversy erupted over the approval of the "Flavr Savr" tomato for human consumption, although no human health risks were ever identified. Calgene voluntarily labelled its tomatoes as "genetically engineered," and they did not do well in the marketplace (although this may have been because of other factors such as flavor). Subsequent releases of genetically engineered foods have not generated nearly as much controversy.

Another important issue in agricultural genetic engineering is intellectual property protection. Can a gene that is present in nature be patented if it has been cloned? Can a cloned DNA sequence of unknown function be patented? Can a transgenic organism be patented? The U.S. Patent and Trademark Office (PTO) and Congress will continue to wrestle with these issues as new genes are cloned and new products are produced. The Plant Variety Protection Act is providing some direction for the protection of plant genotypes produced through traditional breeding techniques and genetic engineering. Genetic engineering and cloning of plants and microbes seems to be more acceptable to the general public than cloning of mammals. The cloning of mammals hits close to home and raises many questions: What is an individual? Should humans be cloned? If humans are eventually cloned, how will people guard against abuses? These fascinating questions will stimulate discussion and research for many years.

—*James P. Prince*

See Also: Genomics; High-Yield Crops; Sheep Cloning; Transgenic Organisms.

Further Reading: *The Frankenstein Syndrome: Issues in the Genetic Engineering of Animals* (1995), by Bernard E. Rollin, addresses moral and ethical questions surrounding the cloning of animals. Charles S. Gasser and Robert T. Fraley, "Transgenic Crops," *Scientific American* 266 (June, 1992), provides a nice review of early transgenic plant work. Jeremy Rifkin, *Algeny* (1983), discusses some of the potential dangers of genetic engineering but lacks scientific rigor.

Genetic Engineering: Historical Development

Field of study: Genetic engineering and biotechnology

Significance: *Genetic engineering, or biotechnology, is the use of biology, genetics, and biochemistry to manipulate genes and genetic materials in a highly controlled fashion. It has led to major advancements in the understanding of the molecular organization, function, and manipulation of genes. The methods have been used to identify causes and solutions to many different human genetic diseases and have led to the development of many new medicines, vaccines, plants, foods, animals, and environmental cleanup techniques.*

Key terms

CLONE: a group of genetically identical cells

PLASMIDS: small rings of deoxyribonucleic acid (DNA) found naturally in bacteria and some other organisms

RECOMBINANT DNA: a DNA molecule made up of sequences that are not normally joined together

Foundations of Genetic Engineering

Microbial genetics, which emerged in the mid-1940's, was based upon the principles of heredity that were originally discovered by Gregor Mendel in the 1880's and the resulting elucidation of the principles of inheritance and genetic mapping during the first forty years of the twentieth century. During the mid-1940's to the early 1950's, the role of deoxyribonucleic acid (DNA) as a genetic material became firmly established, and great advances occurred in understanding the mechanisms of gene transfer between bacteria. A broad base of knowledge accumulated from which later developments in genetic engineering would emerge.

The discovery of the structure of DNA by James Watson and Francis Crick in 1953 provided the stimulus for the development of genetics at the molecular level, and, for the next few years, a period of intense activity and excitement evolved as the main features of the gene and its expression were determined. This work culminated with the establishment of the complete genetic code in 1966, which set the stage for advancements in genetic engineering.

Initially, the term "genetic engineering" included any of a wide range of techniques for the manipulation or artificial modification of organisms through the processes of heredity and reproduction, including artificial selection, control of sex type through sperm selection, extrauterine development of an embryo, and development of whole organisms from cultured cells. However, during the early 1970's, the term came to be used to denote the narrower field of molecular genetics involving the manipulation, modification, synthesis, and artificial replication of DNA in order to modify the characteristics of an individual organism or a population of organisms.

The Development of Genetic Engineering

Molecular genetics originated during the late 1960's and early 1970's in experiments with bacteria, viruses, and free-floating rings of DNA found in bacteria known as plasmids. In 1967, the enzyme DNA ligase was isolated. This enzyme can join two strands of DNA together, acting like a molecular glue. It is the prerequisite for the construction of recombinant DNA molecules, which are DNA molecules that are made up of sequences that are not normally joined together in nature.

The next major step in the development of genetic engineering came in 1970 when researchers discovered that bacteria make special enzymes known as restriction enzymes. Restriction enzymes recognize particular sequences of nucleotides arranged in a specific order and cut the DNA only at those specific sites, like a pair of molecular scissors. Whenever a particular restriction enzyme or set of restriction enzymes is used, the DNA is cut into the same number of pieces of the same length and composition. With a molecular tool kit that included isolated enzymes of molecular glue (ligase) and molecular scissors (restriction enzymes), it became possible to remove a piece of DNA from one organism's chromosome and insert it into another organism's chromosome in order to produce new combinations of genes (recombinant DNA) that may not exist in nature. For example, a bacterial gene could be inserted into a plant, or a human gene could be inserted into a bacterium.

The first recombinant DNA molecules were generated by Paul Berg at Stanford University in 1971, and the methodology was extended in 1973 when DNA fragments were joined to *Escherichia coli* (*E. coli*) plasmids. These recombinant molecules could replicate when introduced into *E. coli* cells, and a colony of identical cells, or clones, could be grown on agar plates.

Paul Berg, who generated the first recombinant DNA molecules in the early 1970's, accepting the 1980 Nobel Prize in Chemistry, which he shared with Walter Gilbert and Frederick Sanger. (AP/Wide World Photos)

This development marked the beginning of the technology that has come to be known as gene cloning, and the discoveries of 1972 and 1973 triggered what became known as "the new genetics." The use of the new technology spread very quickly, and a sense of urgency and excitement prevailed. However, because of rising concerns about the morality of manipulating the genetic material of living organisms, as well as the fear that potentially harmful organisms might accidentally be produced, U.S. biologists called for a moratorium on recombinant DNA experiments in 1974, and the National Institutes of Health (NIH) issued safety guidelines in 1976 to control laboratory procedures for gene manipulation.

In 1977, the pioneer genetic engineering company Genentech produced the human brain hormone somatostatin, and, in 1978, they produced human insulin in *E. coli* by the plasmid method of recombinant DNA. Human insulin was the first genetically engineered product to be approved for human use. By 1979, small quantities of human somatostatin, insulin, and interferon were being produced from bacteria by using recombinant DNA methods. Because such research was proven to be safe, the NIH gradually relaxed their guidelines on gene splicing between 1978 to 1982. The 1978 Nobel Prize in Physiology or Medicine was shared by Hamilton O. Smith, the discoverer of restriction enzymes, and Daniel Nathans and Werner Arber, the first people to use these enzymes to analyze the genetic material of a virus.

By the early 1980's, genetic engineering techniques could be used to produce some biotechnical substances on a large scale. In December, 1980, the first genetically engineered product was used in medical practice when a diabetic patient was injected with human insulin generated in bacteria; in 1982, the Food and Drug Administration (FDA) approved the general use of insulin produced from bacteria by recombinant DNA procedures for the treatment of people with diabetes. During the same time period, genetically engineered interferon was tested against more than ten different cancers. Methods for adding genes to higher organisms were also developed

in the early 1980's, and genetic researchers succeeded in inserting a human growth hormone gene into mice, which resulted in the mice growing to twice their normal size. By 1982, geneticists had proven that genes can be transferred between plant species to improve nutritional quality, growth, and resistance to disease.

In 1985, experimental guidelines were approved by the NIH for treating hereditary defects in humans by using transplanted genes; the more efficient polymerase chain reaction cloning procedure for genes, which produces two double helixes in vitro that are identical in composition to the original DNA sample, was also developed. The following year, the first patent for a plant produced by genetic engineering, a variety of corn with increased nutritional value, was granted by the U.S. Patent and Trademark Office. In 1987, a committee of the National Academy of Sciences concluded that no serious environmental hazards were posed by transferring genes between species of organisms, and this action was followed in 1988 by the U.S. Patent and Trademark Office issuing its first patent for a genetically engineered higher animal, a mouse that was developed for use in cancer research.

Impact and Applications

The application of genetic engineering to gene therapy (the science of replacing defective genes with sound genes to prevent disease) took off in 1990. On September 14 of that year, genetically engineered cells were infused into a four-year-old girl to treat her adenosine deaminase (ADA) deficiency, an inherited, life-threatening immune deficiency. In January, 1991, gene therapy was used to treat skin cancer in two patients. In 1992, small plants were genetically engineered to produce small amounts of a biodegradable plastic, and other plants were manufactured to produce antibodies for use in medicines.

By the end of 1995, mutant genes responsible for common diseases, including forms of schizophrenia, Alzheimer's disease, breast cancer, and prostate cancer, were mapped, and experimental treatments were developed for either replacing the defective genes with working copies or adding genes that allow the cells to fight the disease. In February, 1997, a lamb named Dolly was cloned from the DNA of an adult sheep's mammary gland cell; it was the first time scientists successfully cloned a fully developed mammal. By the end of 1997, approximately fifty genetically engineered products were being sold commercially, including human insulin, human growth hormone, alpha interferon, hepatitis B vaccine, and tissue-plasminogen activators for treating heart attacks. In 1998, strong emphasis was placed on research involving gene therapy solutions for specific defects that cause cancer, as well as on a genetically engineered hormone that can help people with bad hearts grow their own bypass vessels to carry blood around blockages.

—*Alvin K. Benson*

See Also: Cloning; Gene Therapy; Genetic Engineering; Genetic Engineering: Medical Applications; Restriction Enzymes.

Further Reading: *Genetic Engineering* (1995), by Jenny Bryan, is an excellent introductory text with numerous clear illustrations about the basic concepts of genetic engineering. *Genetic Engineering* (1996), by Carol Wekesser, provides good coverage of the applications of gene technology and assesses the moral and ethical aspects associated with such applications. *An Introduction to Genetic Engineering* (1994), by D. S. Nicholl, presents a basic introduction to the world of genetic engineering and concisely describes the full range of technologies available.

Genetic Engineering: Industrial Applications

Field of study: Genetic engineering and biotechnology

Significance: *Genetic engineering involves the alteration of the genetic makeup of an organism to make it more suitable for a specific use. Such technology may offer the best hope for improving public health, from increasing longevity to enhancing the quality of life, and may ultimately become a primary source of food, fuel, fiber, chemical feedstock, and pharmaceuticals. Research in the field of genetic engineering will also provide valuable information about the function of genes and the means by which they are controlled.*

Key terms

BIOSYNTHESIS: a process by which gene coding for a particular product or function is isolated, inserted into another organism (usually bacteria), and later expressed in that organism in a final product form

DEOXYRIBONUCLEIC ACID (DNA): the genetic "blueprint" for most living organisms

GENE CLONING: the development of a line of organisms that may contain identical copies of the same gene or DNA fragment

GENE THERAPY: the insertion of a working gene or genes into a cell or an organism to correct a genetic abnormality

RESTRICTION ENZYMES: a group of enzymes that recognize and sever DNA at very specific sites

Introduction

The modern era of genetic engineering began in the early 1970's when geneticists Stanley Cohen and Herbert Boyer successfully recombined ends of bacterial deoxyribonucleic acid (DNA) after splicing a toad gene in between them. They used a group of enzymes first identified and isolated by researchers in 1971 called "restriction enzymes." Genetic engineering has revolutionized all fields of biology and has spread beyond the research laboratory to many other fields. It is now used to help produce medicine, milk, and food, and to treat various diseases that are untreatable with conventional procedures.

The first industrial application of genetic engineering was the production of insulin by the Eli Lilly company in 1982. It provided an unlimited quantity of insulin at a reasonable cost, and the engineered insulin did not cause allergic reactions in patients. Other industrial applications of genetic engineering include agriculture and environmental protection. Genetic engineering is a rapidly expanding industry in many different aspects. For example, many pharmaceutical manufacturers are expanding their development of cancer therapeutics through genetic engineering and other related biotechnology.

Applications of Biosynthesis

Biosynthesis involves the insertion of a human (or animal) gene that codes for a therapeutically important product (insulin, for example) into a suitable vector that is then cloned into another organism (mostly bacteria) and later expressed in that organism. By cultivating bacteria, large quantities of the gene product may be harvested and purified. This approach has proven to be less expensive and safer than the usual laborious alternative of extraction and purification from animal or human tissues. A number of products of therapeutic importance have already been successfully synthesized, and more are yet to be commercialized. Products released or under development include anticancer agents, antiaging substances, food ingredients, cosmetics, monoclonal antibodies for diagnostic measures or treatment of various infectious diseases, and vaccines for human immunodeficiency virus (HIV) and other contagious pathogens. Since their discovery in the mid-1970's, "pure" antibodies have been used to make diagnostic kits to screen for certain diseases and to manufacture "magic bullets" that seek out and destroy specific tumor cells. Other developments include drugs to treat AIDS-related Kaposi's sarcoma, cystic fibrosis, Gaucher's disease, multiple sclerosis, renal cancer, and reactions to transplantation and chemotherapy. Most monoclonal antibodies target cancer cells, and many have the capability to serve as both medicine and in vivo diagnostic agents. Research also continues into the production of protein-based medicines using transgenic livestock, such as cows, goats, pigs, and sheep. The products could be very helpful to hemophiliacs, people undergoing transplants, and those facing joint replacement.

Genetic engineering has also opened the gate for correction of genetic abnormalities through gene therapy. Numerous genes for diseases have already been identified, including those for Alzheimer's disease; Huntington's chorea; breast, prostate, and colon cancers; cystic fibrosis; and sickle-cell anemia. The pharmaceutical industry has used these genes in active pursuit strategies, such as modifying a patient's own tumor cells or creating synthetic mimics of cancer-cell antigens. Gene therapy is arguably the most dramatic approach for treating genetic diseases. The first attempt was approved and used in July, 1990, for the treatment of an inherited immune disorder called adenosine deaminase (ADA) deficiency. The basic strategy was to introduce copies of normal genes into the patient's blood system to achieve a temporary cure. Though somewhat successful, such treatment does not prevent affected people from passing the defective genes on to their offspring. Germ-line therapy, which corrects defective genes once and for all and prevents them from passing to offspring, is illegal in the United States.

Agriculture and Environmental Protection

Genetic engineering research and product development is revolutionizing the farm industry and leading to the production of transgenic plants (such as the tomato) with a longer shelf life, pest-resistant crops, plants that are able to fix nitrogen and require less chemical fertilizer, and herbicide-resistant crops. The first "green" revolution in crop production occurred in the 1960's and resulted in increased world grain output through the use of semidwarf varieties, irrigation, and fertilizers. Genetic engineering has driven the next wave of agricultural innovation. Genetically engineered seeds have been produced on an increasingly large scale for plants such as soybeans resistant to herbicides, canola and cotton resistant to herbicide glyphosate, corn resistant to corn borers and herbicides, and potatoes resistant to pests. The development of herbicide-resistant crops has helped achieve weed control with less cultivation, less erosion, and less compaction of soil than required with conventional means.

Even though some consumer groups are resistant to genetically engineered foods, the commercial release of the Roundup Ready soybean by the Monsanto corporation was successful. In 1996, one million acres were planted with this soybean. In 1997, seeds were sown on nine million acres. Part of the reason for success is that a complex, costly, and time-consuming testing protocol is used by federal agencies such as the United States Department of Agriculture (USDA) and the Environmental Protection Agency (EPA) for the approval of any genetically altered crops. Such crops must be repeatedly analyzed, and their safety for consumption by humans and farm animals must be established. Once federally approved, sales of seeds can expand very rapidly.

The application of genetic engineering to agriculture is expanding to other areas as well, including the improvement of the quality of fats in foods, improvement of the quality of protein by balancing amino acids, and reduction of the natural poison in food.

In the environmental arena, genetically modified bacteria may potentially be used to convert organic wastes into sugar, alcohol, and methane, which may provide alternative fuel sources and ease dependence on petroleum. Other industrial products of this kind include genetically modified bacteria that can remove heavy metals from the environment and act as biopesticides by infesting insects and other pests. Genetically engineered microbes may also help to clean oil spills.

—Ming Y. Zheng

See Also: Biopharmaceuticals; Genetic Engineering; Genetic Engineering: Agricultural Applications; Genetic Engineering: Medical Applications.

Further Reading: *An Introduction to Recombinant DNA in Medicine* (1995), by Alan Emery and Sue Malcolm, provides a straightforward outline of general principles and medical applications of biotechnology. Carol Potera, "Therapies Seek Control of Apoptosis," *Genetic Engineering News* 18 (March 15, 1998), is an easy-to-understand discussion on designing drugs for autoimmune disorders, cancer, and heart disease. *Biotechnology* (1998), by Susan Barnum, contains a basic overview of biotechnology and its applications.

Genetic Engineering: Medical Applications

Field of study: Genetic engineering and biotechnology

Significance: *The rapidly developing field of genetic engineering has produced the means of preventing or controlling many hereditary diseases and offers hope that the major hereditary diseases might eventually be eliminated.*

Key terms

DEOXYRIBONUCLEIC ACID (DNA): varieties of nucleic acids that are found in cells, each of which contains the genetic blueprint of the organism of which it is a part

DNA RESTRICTION: DNA purified from an organism after it has been purified and cut into smaller molecules

GENE: the physical unit of heredity; a segment of genetic material that determines chemical actions and reactions within cells

RIBONUCLEIC ACID (RNA): a form of genetic material, most often single-stranded, that consists of units called "ribonucleotides"

TRANSFORMATION: introduction of foreign DNA into recipient cells

The Development of Genetic Engineering

Biotechnology, of which genetic engineering is a subfield, uses living organisms to make products. Genetic engineering focuses on the characteristics of the genes found in every cell. These genes serve specific functions, determining such characteristics as gender, body build, eye color, and predisposition to diseases such as diabetes, cystic fibrosis, hemophilia, and sickle-cell anemia. For example, the human growth hormone, produced by specific genes, determines how bones grow and how weight is regulated. This hormone is a protein produced by the pituitary gland. In cases where the pituitary gland is deficient in producing this hormone, the missing protein can be introduced into the body by injections, or the genetic structure of the cells can be manipulated so that the pituitary gland produces the missing protein at a rate sufficient to assure normal growth. Genetic engineering is vitally concerned with manipulating abnormal genes to make them function as they should. It is also concerned with producing synthetic substitutes for transformation into the bodies of organisms that require such substitutes in order to function normally or, in the cases of agriculture and animal husbandry, to increase function and thereby increase production.

A watershed year in the development of genetic engineering was 1953, in which James

Watson and Francis Crick identified the molecular structure of deoxyribonucleic acid (DNA) as a double spiral helical chain. This discovery altered the course of genetic research, the history of which dates back several centuries. In 1543, Belgian anatomist Andreas Vesalius produced the first anatomical map of the human body. In the absence of powerful electronic microscopes that were to make the study of genes and chromosomes possible, anatomists had little knowledge of genes. In fact, the term "gene" was not employed until 1903, when the Dutch geneticist Wilhelm Ludwig Johannsen used it in relation to the field currently designated "genetics."

The study of cells was initiated in 1665 by British physicist Robert Hooke, who studied and described cork cells. It took another 150 years before a Scottish botanist, Robert Brown, identified the nuclei of certain cells. In 1838, German botanist Matthias Jakob Schleiden suggested that all living organisms are composed of cells whose nuclei are essential to the formation of new cells. The following year, German physiologist Theodor Schwann presented a paper in which he contended that all living things are composed of cells or substances derived from cells. He thought that cells live a life quite independent of the organism as a whole.

Connections between cells and heredity were first recognized in the 1860's by the Austrian naturalist and botanist Gregor Mendel, who detected dominant traits in plants from generation to generation. A decade later, a Swiss ophthalmologist, Johann Friedrich Horner, discovered connections between genetic abnormalities and familial aspects of color blindness, strongly suggesting a hereditary factor in such cases. Further studies by Archibald Garrod in 1909 concluded that the chemical reactions that account for cellular development are triggered by enzymes, each of which is produced by a single gene capable of producing just one enzyme. In 1910, Thomas Hunt Morgan pioneered the study of heredity in fruit flies (*Drosophila melanogaster*), the abbreviated life cycle of which makes it possible to study many generations in a relatively short time span. As early as 1944, Oswald Avery, Colin MacLeod, and Maclyn McCarty replaced a defective gene in a bacterium with DNA from another bacterium. At this point, they did not understand the structure of DNA; therefore, they could not understand its potential. Further research by Linus Pauling during the 1950's resulted in an understanding of how the amino acids necessary to cellular development function.

Pauling's studies of proteins, enzymes, and nucleic acids provided Watson and Crick with some of the fundamentals they needed to carry out the studies that led to the discovery of how DNA is structured. Watson and Crick, collaborating in their experimental work at the University of Cambridge from 1951 to 1953, concluded that DNA is a three-dimensional structure composed of two chains that form a double helix. They used metal plates from a machine shop to create a model of the DNA molecule. Two days after they completed their model, Maurice Wilkins, later to share a Nobel Prize in Physiology or Medicine with them, announced independently that he had also unraveled the mystery of DNA's structure. Without an understanding of this structure, genetic engineering would have been impossible.

Heredity and Genetic Disorders

In 1976, researchers published a catalogue that identified 1,487 genetic disorders that frequently recur in generation after generation within some families. In 1990, the list was revised to include almost five thousand such hereditary disorders. This increased understanding of how genetic characteristics are passed through families generationally, combined with an understanding of how genes determine the physical characteristics of organisms, has led to extensive experimentation in genetic manipulation. It must be remembered that the average adult human has sixty trillion cells and that each of these cells contains between 30,000 and 100,000 genes, each possessing characteristics that determine some aspect of that person's genetic composition. Except for a small number of reproductive cells in humans, every cell in the human body contains the full blueprint of the total organism of which it is a part.

The ability to alter the genetic structure of cells increased substantially during the last half of the twentieth century as electron microscopes and sophisticated laboratory instruments gave scientists access to microscopic entities that earlier scientists were unable to see. Early geneticists questioned how information found within the DNA of genes is copied for transmission to future generations of cells. With Watson and Crick's discovery of the structure of DNA, that fundamental question was answered. More significantly, implicit within the answer is the suggestion that genetic alteration might reasonably be expected to permit the treatment of diseases that were hitherto treated ineffectively through medication or left untreated because they were considered incurable.

Numerous human diseases result from the abnormal functioning of a single gene within the cells. By introducing a normal copy of such a gene into the appropriate cells of someone suffering from a disorder caused by this abnormality, the normal gene, by replicating itself, can often eliminate or alleviate the problems the defective gene has caused. Among the diseases currently being treated in this way are diabetes, some forms of cancer, cystic fibrosis, human growth hormone deficiency, severe combined immune deficiency diseases, muscular dystrophy, Huntington's chorea, hemophilia, thalassemia, and inherited predispositions to high cholesterol. The results have varied considerably because of the unique types of problems each of these diseases presents. For example, enormous progress has been made in prolonging the life spans of patients suffering from cystic fibrosis, which is characterized by accumulations of mucus in the lungs and which usually ends in death through lung infection. A study in Great Britain showed that the average life expectancy of

Maurice Wilkins accepts his Nobel Prize in 1962. (AP/Wide World Photos)

such patients before 1960 was one year; however, life expectancy had increased to five years by 1970, to ten years by 1980, to nearly twenty years by 1990, and to about twenty-five years by 1995. Nevertheless, because the cells that line the lungs are sloughed off frequently and quickly, the gene therapy that succeeds in increasing life span has to be repeated continually and has not yet resulted in an advance that promises anything like a normal life or life span to those suffering from the affliction.

Treatment of Diabetes and Cancer

Progress in the treatment of diabetes has been much more encouraging than the treatment of cystic fibrosis. Diabetes occurs when a major endocrine gland, the pancreas, fails to supply insulin in sufficient quantities to regulate the body's blood sugar. In cases in which diabetes cannot be successfully treated though diet and exercise, injections of insulin are used

to stabilize the patient's blood sugar. Initially, most of the insulin used by diabetics came from the pancreases of cows or pigs, whose insulin is similar to that produced by human pancreases but is just different enough to cause allergic reactions in some patients. Through genetic engineering, human insulin has been synthesized and is used without difficulty by those allergic to other forms of the hormone.

Genetic engineering is also creating smart genes that release insulin into the bloodstream precisely when it is needed, exactly as the pancreas releases the insulin it produces into the bloodstream when a person's blood sugar begins to rise. The use of such insulin prevents the peaks and troughs that insulin-dependent diabetics formerly suffered. Genetically engineered insulin results when DNA codes from one strand of the double helix are introduced into *Escherichia coli* (*E. coli*) bacterium while DNA from the second strand of the double

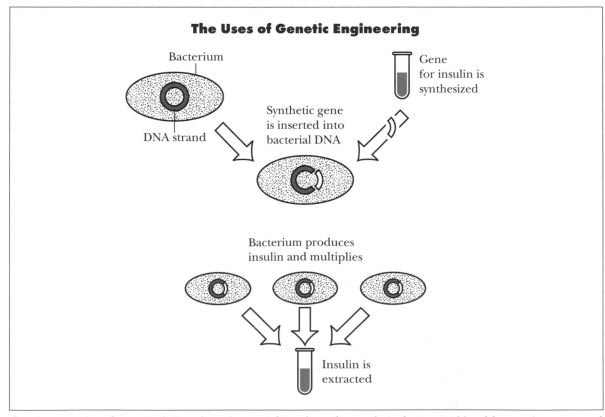

Genetic engineering, the manipulation of genetic material, can be used to synthesize large quantities of drugs or hormones, such as insulin.

helix is introduced into a different bacterium. The bacterial cells produced in this way are then gathered, combined, and treated chemically to make them join together. The product is identical to the insulin molecules the human pancreas produces.

Considerable genetic research has also been directed toward the effort to prevent or control cancer. Scientists are working on vaccines that would insulate people from developing various forms of this dreaded disease that, until the 1960's, was generally regarded as incurable unless caught very early and treated surgically. Through genetic engineering, a variety of drugs called "interferons" have been developed and offer promise in the treatment of cancer and various diseases caused by viruses, including acquired immunodeficiency syndrome (AIDS). Interferons are proteins known medically as immune regulators or lymphokines. These proteins—some twenty different varieties of them—in their natural state react early to the presence of disease cells within the body. Most humans are bombarded continually by disease cells that have the potential to cause illness, but the immune regulators are triggered automatically to attack the invading cells and kill them. Interest was aroused in the 1950's when medical researchers observed that people who contracted viral infections did not contract a second viral infection at the same time. This observation suggested that the body creates some defense against reinfection and that this defense prevents further contamination.

Before the structure and function of DNA were understood, interferons could be manufactured through blood and tissue cultures. Interferons could be obtained from blood whose white cells had been infected by a virus. Each cell, however, produced such a minute quantity of interferon that it took almost 100,000 pints of blood to produce a single gram of interferon at the prohibitive cost of about fifty million dollars. Finally, however, medical scientists found that they could use *E. coli* bacteria as a medium for cloning human interferon genes. One liter of this bacterial culture yields as much interferon as the white blood cells from one hundred pints of human blood.

Retroviruses, Prenatal Testing, and Genetic Manipulation

Retroviruses play a significant role in genetic engineering because they synthesize complementary DNA from their ribonucleic acid (RNA), then they insert this DNA into the host's DNA. Because retroviruses infect only dividing cells, they cannot be used successfully in situations in which an illness is not caused by mutations of dividing cells. The use of retroviruses, however, can result in dangerous outcomes and can even spawn cancer in otherwise healthy patients. This is because retrovirus genes insert their DNA randomly within cells. Genes placed elsewhere than in their normal places may not work effectively because, in nature, they work best when they are placed close to other genes that trigger reactions in them. The greatest danger arises from the insertion of retrovirus genes into tumor-suppressor genes, which may prevent these genes from functioning properly and which may, in some cases, cause cancer.

Several kinds of tests exist to determine the health of a fetus. Popular among these are ultrasound (high frequency sound waves) and amniocentesis. Ultrasound produces a picture of the fetus on a screen, allowing physicians watching the picture to insert a long, hollow needle into the mother's abdomen and withdraw amniotic fluid that can then be tested for defective genes. In some cases, gene repair can be performed to prevent a birth defect that amniocentesis has uncovered. Prenatal surgery has also been performed on fetuses in their mothers' wombs to repair heart defects and other life-threatening conditions revealed by amniocentesis. Many people fear, however, that despite the promise genetic manipulation offers, parents might be tempted to abort a fetus that has genetic problems or even one that is not of the sex that the parents want their child to be. As gene manipulation becomes increasingly sophisticated, parents might conceivably create made-to-order children who would possess precise characteristics that the parents deem desirable.

The Future of Genetic Engineering

Despite fears about many of the implications, gene splicing and genetic engineering technology is readily available, and progress in the field will continue. Among the reservations that have been expressed is the fear that governments might create DNA libraries in which the genetic profiles of every citizen would be stored. Such libraries would be helpful in criminal investigations, but they pose a substantial threat to the rights of privacy that most individuals wish to have preserved. A genetically defective person might be denied the same employment, insurance, and benefits that most people take for granted. A subculture of genetically impaired people might indeed spring up. Therefore, the dangers of genetic engineering must be carefully weighed against its benefits.

—*R. Baird Shuman*

See Also: Biotechnology; Genetic Engineering: Historical Development; Genetic Engineering: Social and Ethical Issues; Human Genetics; Human Genome Project.

Further Reading: For those new to the field of genetic engineering, Linda Tagliaferro's *Genetic Engineering: Progress or Peril?* (1997) offers a rudimentary overview. More advanced but still accessible is *Improving Nature? The Science and Ethics of Genetic Engineering* (1997), by Michael J. Reiss and Roger Straughan. *Biotechnology from A to Z* (1993), by William Bains, is lucid and comprehensive. Somewhat more demanding is *Genetic Engineering* (1993), by A. Ceccarelli and N. Spurr.

Genetic Engineering: Social and Ethical Issues

Field of study: Genetic engineering and biotechnology

Significance: *New technologies for manipulating the genetic makeup of living organisms raise serious questions about the social desirability of controlling genes and the moral right of humans to redesign living beings.*

Key terms

BIODIVERSITY: the presence of a wide variety of forms of life in an environment

BIOTECHNOLOGY: the technological manipulation of living organisms; genetic engineer-

ing is the most common form of biotechnology

DEOXYRIBONUCLEIC ACID (DNA): a double-helix-shaped structure made up of genes, the basic units of heredity that pass traits from one generation of a living organism to the next

ENZYMES: proteins that promote or speed up chemical reactions in a cell

RECOMBINANT DNA: a new combination of genes spliced together on a single piece of DNA; recombinant DNA is the basis of genetic engineering technology

TRANSGENIC: a plant or animal into which the DNA of another species has been inserted

Genetic Engineering as a Social and Ethical Problem

English author Mary Shelley's 1818 horror novel *Frankenstein*, about a scientist who succeeds in bringing a creature to life, expressed anxiety about the possibility of human control over the basic mysteries of existence. The novel's continuing popularity and the many films and other works based on it attest deep-seated feelings that unrestrained science may violate essential principles of nature and religion and that human powers may grow to exceed human wisdom. With the rise of genetic engineering in the 1970's, many serious philosophers and social critics began to believe that the Frankenstein story was moving from the realm of science fiction into reality.

The basic blueprint of all living beings was found in 1953, when Francis Crick and James Watson discovered the structure of DNA. A little less than two decades later, in 1970, it became possible to conceive of redesigning this blueprint when Hamilton Smith and Daniel Nathans of The Johns Hopkins University discovered a class of "restriction" enzymes that could be used as scissors to cut DNA strands at specific locations. In 1973, two researchers in California, Stanley Cohen and Herbert Boyer, spliced recombinant DNA strands into bacteria that reproduced copies of the foreign DNA. This meant that it would be possible to combine genetic characteristics of different organisms. In 1976, Genentech in San Francisco, California, became the first corporation formed to develop genetic engineering techniques for commercial purposes.

By the 1990's, genetic engineering was being used on plants, animals, and humans. The "Flavr Savr" tomato, the first genetically engineered food to be approved by the U.S. government for sale, was developed when biotechnologists inserted a gene that delayed rotting into tomatoes. Transgenic animals (containing genes from humans and other animals) became commonplace in laboratories by the middle of the 1990's. The year 1990 saw the first successful use of genetic engineering on humans, when doctors used gene therapy to treat two girls suffering from an immunodeficiency disease. The long-felt discomfort over scientific manipulation of life, the suddenness of the development of the new technology, and the application of the technology to humans all combined to make many people worry about the social and ethical implications of genetic engineering. The most serious concerns were over genetic manipulation of humans, but some also pointed out possible problems with the genetic engineering of plants and animals.

Engineering of Plants and Animals

According to a Harris Poll survey conducted for the U.S. Office of Technology in the fall of 1968, a majority of Americans were not opposed to using recombinant DNA techniques to produce hybrid agricultural plants. Some social critics, such as Jeremy Rifkin, have argued that this acceptance of the genetic engineering of plants is shortsighted. These social critics question the wisdom of intervening in the ecological balance of nature. More specifically, they maintain that manipulating the genetic structure of plants tends to lead to a reduction in the diversity of plant life, making plants less resistant to disease. It could also lead to the spread of diseases from one plant species to another, as genes of one species are implanted in another. Furthermore, new and unnatural varieties of food plants could have unforeseen health risks for human beings.

Since genetic engineering is a highly technical procedure, this gives more power over the food supply to those who control technology. Thus, both corporate power over consumers

and the power of more technologically advanced nations over less technologically advanced nations could be increased. In addition, plants that are genetically engineered to produce more often require more fertilizer and greater amounts of irrigated water than ordinary plants. The technology would therefore serve the interests of corporate agribusiness at the expense of small-scale, low-income farmers.

Many of the concerns about the genetic engineering of animals are similar to those about the engineering of plants. Loss of biodiversity, vulnerability to disease, and business control over livestock are all frequently mentioned objections to the genetic manipulation of animals. Moral issues tend to become more important, though, when opponents of genetic engineering discuss its use with animals. Many religious beliefs hold that the order of the world, including its division into different types of creatures, is divinely ordained. From the perspective of such beliefs, the relatively common experimental practice of injecting human growth genes into mice could be seen as the sacrilegious creation of monsters. Opponents of the genetic alteration of animals argue, further, that animals will suffer. They point out that selective breeding, a slow process, has led to about two hundred diseases of genetic origin in purebred dogs. Genetic engineering brings about change much faster than breeding, increasing the probability of genetic diseases.

Engineering of Humans

Some of the greatest ethical and social problems with genetic engineering involve the use of this type of technology on humans. The most common projects for the manipulation of human genetic material are medical. Gene therapy seeks to cure inherited disease by altering the defective genes that cause illnesses. Those who favor gene therapy maintain that it can be a powerful tool to overcome human misery. Those who oppose this type of medical procedure usually point out one or more of three major ethical issues. First, critics of human genetic engineering maintain that this technology raises the problem of ownership of human life. In the early 1990's, the National Institutes of Health (NIH) began filing for patents on

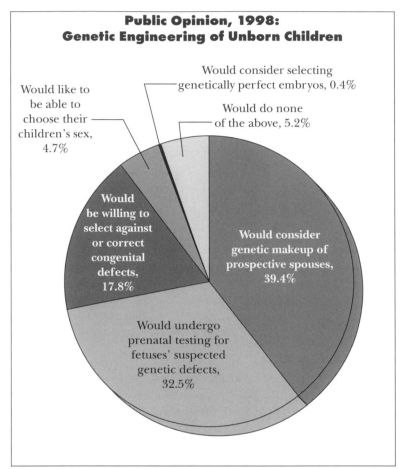

**Public Opinion, 1998:
Genetic Engineering of Unborn Children**

Would consider selecting genetically perfect embryos, 0.4%

Would like to be able to choose their children's sex, 4.7%

Would do none of the above, 5.2%

Would be willing to select against or correct congenital defects, 17.8%

Would consider genetic makeup of prospective spouses, 39.4%

Would undergo prenatal testing for fetuses' suspected genetic defects, 32.5%

An informal 1998 survey conducted on the Internet elicited the above responses to the question, "How far would you go in genetically manipulating your unborn child?" (Source: Moms Online)

human genes, meaning that the blueprints for human life could actually be owned. Because all human DNA comes from human tissue, this raises the question of whether participants in genetic experiments own their own DNA or whether it belongs to the researchers who have extracted it.

The second problem of human genetic engineering identified by critics is its eugenic implications. Eugenics is the practice of breeding human beings to produce "better" humans. If scientists can alter genes to produce humans with more desirable health characteristics, then scientists can also alter genes to produce humans with characteristics of personality or physical appearance. In this way, genetic engineering poses the risk of becoming an extreme

and highly technological form of discrimination. The third problem is related to both of the first two: the reduction of humans to objects. When human life becomes something that can be owned and redesigned at will, some ethicists claim, it will cease to be seen as a sacred mystery and will become simply another piece of biological machinery. As objects, people will gradually lose the philosophical justification for their political and moral rights.

Impact and Applications

Concerns about the social and ethical implications of genetic engineering have led to a number of attempts to limit or control the technology. The environmentalist group Greenpeace has campaigned against genetically engineered agricultural products and called for the clear labeling of all foods produced by genetic manipulation. In September, 1997, Greenpeace filed a legal petition against the U.S. Environmental Protection Agency (EPA), objecting to the EPA's approval of genetically engineered plants.

Activist Jeremy Rifkin of the Foundation on Economic Trends became one of the most outspoken opponents of all forms of genetic engineering. Rifkin and his associates called on the U.S. NIH to stop government-funded transgenic animal research. A number of organizations, such as the Boston-based Council for Responsible Genetics (CRG), lobbied to increase the legal regulation of genetic engineering. In 1990, in response to pressure from critics of genetic engineering, the Federal Republic of Germany enacted a gene law to govern the use of biotechnology. In the United States, the federal government and many state governments considered laws regarding genetic manipulation. A 1995 Oregon law, for example, granted ownership of human tissue and genetic information taken from human tissue to the person from whom the tissue was taken.

—*Carl L. Bankston III*

See Also: Biotechnology, Risks of; Cloning: Ethical Issues; Eugenics; Genetic Engineering; Genetic Engineering: Historical Development.

Further Reading: *Genetic Engineering: Opposing Viewpoints* (1996), edited by Carol Wekesser, gives both sides of major debates regarding genetic engineering. In *The Biotech Century: Harnessing the Gene and Remaking the World* (1998), Jeremy Rifkin, one of the best-known critics of biotechnology, warns that procedures such as cloning and genetic engineering could be disastrous for the gene pool and for the natural environment. Eric S. Grace's *Biotechnology Unzipped: Promises and Reality* (1997) provides a nontechnical history and explanation of biotechnology for general readers.

Genetic Medicine

Field of study: Genetic engineering and biotechnology

Significance: *The development of medicines and novel therapeutic approaches based upon an understanding of gene function holds great promise for the treatment of inherited genetic disorders, infectious diseases such as AIDS, and diseases resulting from abnormal physiology such as heart disease and cancer. The new genetic medicines can be thought of as "magic bullets" since they are designed to target abnormal gene products directly associated with specific disease processes.*

Key terms

NUCLEIC ACIDS: biomolecules in the form of deoxyribonucleic acid (DNA), which is the genetic material specifying protein structure, or ribonucleic acid (RNA), which transmits the genetic information from DNA during protein synthesis

ANTISENSE THERAPY: genetic medicine involving the production of nucleic acids, which block the formation of abnormal proteins in diseased tissues

RECOMBINANT DNA: segments of DNA from different sources that are joined together to produce hybrid molecules

RIBOZYMES: small RNAs that can recognize and destroy abnormal RNAs by enzyme action

TRANSGENIC ORGANISM: an organism whose genome has been altered by the insertion of one or more genes from another species

Tools of Genetic Medicine

Ever since Paul Ehrlich's discovery of salvarsan as a "magic bullet" for the treatment of syphilis at the beginning of the twentieth century, researchers and clinicians have dreamed of treating medical disorders by designing drugs that would target and destroy abnormal proteins responsible for specific disease processes and thereby effect a cure. With the exception of antibiotics, which exert selective toxicity against bacteria, much of this promise remained unfulfilled until the development of the tools of recombinant deoxyribonucleic acid (DNA) technology opened the door to understanding the role of specific abnormal gene products in the genesis of disease processes. This understanding has facilitated an unprecedented inquiry into methods that will allow the delivery of genetic medicines that will either destroy defective, disease-causing gene products or repair abnormal physiological pathways at the root of many complex diseases. This research also shows promise in the treatment of infectious diseases, as genetic medicines have been developed to target and block the spread of specific infections such as acquired immunodeficiency syndrome (AIDS) within the body.

The tools of recombinant DNA technology critical to the development of genetic medicines originated in methods permitting the identification of specific genes whose dysfunction contributed to the pathogenesis of disease processes. The Human Genome Project represents a worldwide effort to identify all the genes present in humans. This information has been used to identify genes responsible for inherited genetic disorders as well as genes implicated in cancer and other complex diseases. Once a gene is identified, it becomes a potential target for therapeutic strategies designed to block the function of the gene if it is behaving abnormally, as in various types of cancer, or to replace the gene product if its absence is responsible for disease, as in various types of hemophilia, a blood disorder resulting from a lack of clotting factor. The therapeutic design strategy for a genetic medicine therefore depends on the type of genetic dysfunction that results in specific disease processes.

Design Strategies for Genetic Medicines

An early approach to the design of genetic medicines involved the cloning of specific genes to obtain many copies of a gene that could then be used to produce large amounts of a gene product for therapeutic purposes. For example, human insulin prepared in this way is used to treat insulin-dependent diabetics, and various clotting-factor genes have been cloned to obtain specific clotting factors used to treat patients with hemophilia. This type of genetic medicine represents a protein-replacement therapy that can be used to treat many disorders that result from a deficiency of a single gene product. Once the gene is identified and cloned, the gene can be used to make large amounts of the therapeutic protein using cell culture systems. This method is very expensive; therefore, research efforts have been directed toward the development of transgenic animals, such as pigs and cows, that can produce large amounts of therapeutic protein in their milk. This technology involves the microinjection of therapeutic human genes into animal embryos, which are then transplanted into foster mothers. As the embryo develops, the injected human gene becomes incorporated into all the cells of the animal's body in a stable form. Specific regulatory elements called tissue-specific promoters can be attached to the gene so that it will be expressed to make human protein exclusively in the mammary tissue of the transgenic animal; secretion in milk permits the isolation of the human protein in purified form for therapeutic purposes. The use of genetically cloned animals (such as the sheep Dolly, cloned in 1997) may further enhance efforts to obtain therapeutic human products from animal sources.

Other approaches have been initiated for the treatment of diseases resulting from meta-bolic dysfunction caused by the production of abnormal gene products within the body. Treatment requires the inactivation of the abnormal gene or its product in order to alleviate the symptoms of disease. An important approach in this area is termed "antisense therapy," which involves the synthetic construction of novel nucleic acids that target and inactivate abnormally functioning genes contributing to disease processes. These therapeutic nucleic acids are called "antisense" because they are synthesized in the opposite orientation as the normal molecule and can specifically bind to it and inactivate its function. An abnormal gene can be targeted in several ways. The most direct approach involves the synthesis of specific nucleic acid sequences that will recognize and bind to a specific gene to form a three-strand structure consisting of double-strand DNA (which is the form of chromosomal DNA) bound to the synthetic DNA. In this form of "triplex DNA," as it is called, transcription, which involves the synthesis of messenger ribonucleic acid (mRNA) as a first step in protein formation, cannot occur, and expression of the abnormal gene is blocked. Additional antisense approaches involve the formation of synthetic RNAs or short DNAs called oligonucleotides, which bind to specific mRNAs to block their translation. Also in development are therapeutic RNAs called ribozymes (RNA enzymes with nuclease activity that can target and destroy RNAs that specify dysfunctional proteins).

Impact and Applications

While the full potential of genetic medicines to act as "magic bullets" in the treatment of human diseases has not yet been realized, many significant steps in this direction have been taken. The first important step has involved the identification of specific disease genes and abnormal genes responsible for many human diseases. This step was made possible by the development of recombinant DNA technology and has led to the development of therapeutic strategies based on an understanding of the molecular basis of disease. Many human gene products are available for replacement therapy, including insulin, clotting factors, and special-

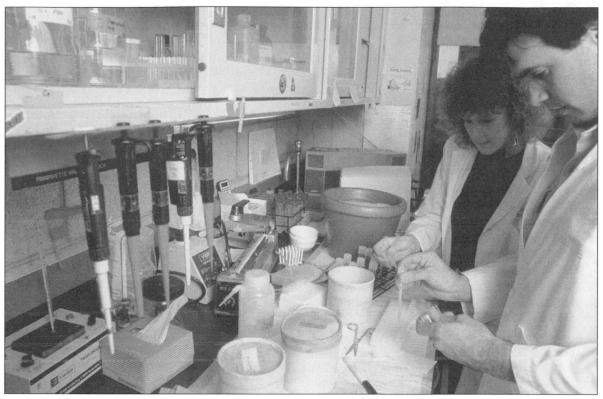

The discoverers of a gene related to Alzheimer's disease and Down syndrome at work. (Dan McCoy/Rainbow)

ized growth factors such as erythropoietin; many others are in clinical trials or in development. The use of transgenic animals to produce these therapeutic proteins represents an important development in this field. Limitations have involved the stability of the therapeutic protein and its ability to function appropriately in the body following administration.

Antisense therapy has enormous potential to control pathogenic processes responsible for disease. Experimental studies involving the insertion of antisense genes into human tumor cells have shown that these genes can block the abnormally functioning genes within the cell and have shown therapeutic impact in clinical trials. Limitations have involved the stability of the antisense molecules, their specificity with respect to gene targets, and the degree to which gene expression is blocked by this therapy.

The Human Genome Project has provided a great deal of information about the structure of specific genes and gene products implicated in human diseases as diverse as cancer and heart disease. This information has been organized in the form of computer databases used in "rational drug design," which involves the development of drugs that are designed to target specific abnormal gene products on the basis of structure and alleviate disease processes at the level of molecular dysfunction. This approach has also been applied to the treatment of infectious disease, most notably in the form of "protease inhibitors" that target the protease gene of the human immunodeficiency virus (HIV).

—*Sarah Crawford Martinelli*

See Also: Biopharmaceuticals; Biotechnology; Genetic Engineering: Medical Applications; Gene Therapy.

Further Reading: William Haseltine, "Discovering Genes for New Medicines," *Scientific American* 276 (March, 1997), explains how the discovery of disease-causing genes represents the starting point for developing genetics-based therapeutic approaches. William Velander et al., "Transgenic Livestock as Drug Factories," *Scientific American* 276 (January, 1997), explains how transgenic animals can be generated to produce large amounts of therapeutic gene products. Tapasy Mukhopadhyay and Jack Roth, "Antisense Therapy for Cancer," *The Cancer Journal* 1 (1995), explains the principles of antisense therapy and its applications in the treatment of cancer.

Genetic Screening

Field of study: Human genetics
Significance: *Genetic screening is a preventive health measure that involves the mandatory or voluntary testing of certain individuals for the purpose of detecting genetic disorders or identifying defective genes that can be transmitted to offspring. The primary goals of genetic screening include the prevention of genetic disorders, treatment, and the option to make informed and rational decisions about conception and birth. Genetic screening also has applications in gene therapy for the treatment of genetic disorders. It has raised concerns about confidentiality, discrimination, and the right to privacy.*

Key terms

CARRIER: a healthy individual who possesses a normal gene paired with a defective form of the same gene
GENETIC DISORDER: a disorder caused by a change in a gene or chromosome

Neonatal Screening

The most widespread use of genetic screening is the testing of newborn babies. This is called neonatal screening. Every year, millions of newborn babies are tested for inborn errors of metabolism (inherited diseases that are caused by mutations in the genes that code for the synthesis of enzymes). The purpose of this kind of screening is to provide immediate treatment after birth if a defect is detected so that the newborn has a chance of having a normal life. A classic example of neonatal screening in the United States and many countries is the mandatory mass screening of newborn babies for phenylketonuria (PKU), a disorder that causes irreversible brain damage when not

treated. Individuals with PKU lack the enzyme phenylalanine hydroxylase, which converts the essential amino acid phenylalanine into another amino acid, tyrosine. When the enzyme is absent, a high level of phenylalanine accumulates in the body and is broken into a toxic substance that causes irreversible brain damage.

Newborn babies are screened for PKU using the Guthrie test, named after its inventor, Robert Guthrie. The Guthrie test detects high levels of phenylalanine in the blood of newborns. Blood samples are taken from the heels of newborn babies in the hospital nursery, placed on filter papers as dried spots, and sent off to appropriate laboratories for analysis. Newborns with positive results can be effectively treated with a diet low in phenylalanine (low-protein foods). The level of phenylalanine in the blood is regularly monitored. If treatment is not initiated within the first two months of life, irreversible brain damage will occur. Newborns must also be screened for galactosemia, an inherited disorder characterized by seizures, mental retardation, vomiting, and liver disease caused by the accumulation of galactose in the blood. Some states also screen for sickle-cell anemia, an inherited blood disorder characterized by anemia, pain in the abdomen and joints, and damage to organs.

Carrier Screening

Carrier screening is the voluntary testing of healthy individuals of reproductive age who may be carriers of defective genes that could be passed on to their children. The purpose of carrier screening is to inform couples of their risk of having a child with a genetic disorder. In the United States, screening has been limited to some ethnic groups known to have a high incidence of a specific genetic disorder. In the 1970's, for example, Tay-Sachs screening of Ashkenazi Jews of reproductive age was successfully implemented. Tay-Sachs disease is an inherited, progressive disease in infants characterized by a startle response to noise, blindness, seizures, paralysis, and death in infancy caused by the absence of an enzyme called hexosaminidase A. The Tay-Sachs screening programs involved tests that measured the level of hexosaminidase A in the blood of individuals. People with Tay-Sachs disease have no detectable level of the enzyme, while carriers have half the level of the enzyme found in the blood of normal individuals.

In the early 1970's, mandatory, large-scale screening of African American couples and some schoolchildren was implemented in an effort to identify carriers of the gene for sickle-cell anemia. Blood samples taken from individuals were tested for the presence of distorted or sicked-shaped red blood cells caused by the production of abnormal hemoglobin, the molecule that transports oxygen in the body. The laws mandating screening were later repealed amid charges of racial discrimination.

After successful identification of the cystic fibrosis gene in 1989, the scientific and medical community began debating the costs and benefits of screening millions of carriers of the gene in the United States. Cystic fibrosis is a common inherited disorder characterized by accumulation of mucus in the lungs and pancreas; it

affects Caucasian children and young adults. Some companies have begun voluntary screening for the cystic fibrosis gene in couples with a family history of the disorder.

Impact and Applications

The benefits of genetic screening include early intervention and treatment, detection of new mutations by researchers, and the education of people about genetic disorders so that they are able to make informed and responsible decisions about reproductive issues. However, screening for genetic defects has raised ethical and social issues such as confidentiality, discrimination, and the right to privacy. One example is the sickle-cell screening program of the early 1970's. Screening results were not kept in strictest confidence; consequently, many healthy African Americans who were carriers of the sickle-cell gene were stigmatized and discriminated against in terms of employment and insurance coverage. There were also charges of racial discrimination because carri-

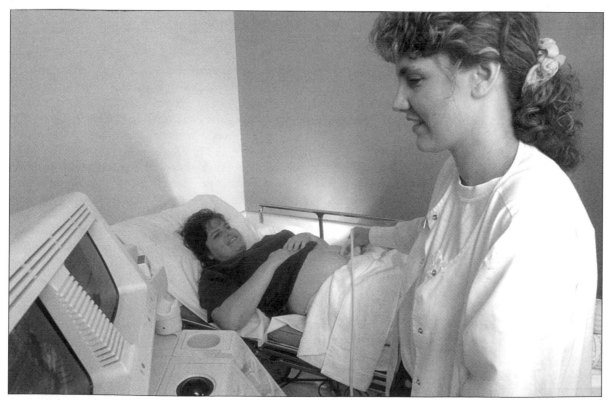

Ultrasound machines are widely used in prenatal screening for genetic disorders. (Larry Mulvehill/Rainbow)

ers were advised against bearing children. Although the sickle-cell anemia screening programs were unsuccessful, there were some successes in other screening programs. For example, the number of newborns with Tay-Sachs disease dropped dramatically as a result of carrier screening programs. The screening programs succeeded because of the tremendous effort put forth to educate Ashkenazi Jewish communities about the disorder and the consequent acceptance of the programs by these communities.

Genetic screening also has applications in the treatment of some genetic disorders. Individuals afflicted with serious genetic disorders may often be treated by gene therapy (the replacement of a defective or missing gene with a normal, functional copy of the gene). The first clinical trials of gene therapy began in 1990 for severe combined immunodeficiency, a lethal genetic disorder in which a person has no functional immunity. Much attention has been focused on gene therapy as a significant treatment option for patients with disorders such as cystic fibrosis.

—*Oluwatoyin O. Akinwunmi*

See Also: Gene Therapy; Genetic Counseling; Genetic Testing; Genetic Testing: Ethical and Economic Issues; Sickle-Cell Anemia.

Further Reading: Fritz Fuchs, "Genetic Amniocentesis," *Scientific American* 242 (June, 1980), provides an overview for the general reader. Gina Kolata, "Genetic Screening Raises Questions for Employers and Insurers," *Science* 232 (April, 1986), discusses some of the ethical dilemmas facing employers and insurers. W. French Anderson, "Gene Therapy," *Scientific American* 273 (September, 1995), provides an overview for the general reader. Leslie Roberts, "To Test or Not to Test?" *Science* 247 (January, 1990), discusses the issues surrounding cystic fibrosis screening.

Genetic Testing

Field of study: Human genetics
Significance: *Genetic testing is the use of tests and techniques to detect the presence of a genetic disor-*

der or a defective gene in a fetus, newborn, or adult. Applications of genetic testing include DNA fingerprinting for the identification of individuals and gene therapy for the treatment of genetic disorders. Genetic testing has significant implications with respect to reproductive choices, privacy, insurance coverage, and employment.

Key terms
PROTEIN: a molecule made up of building blocks called amino acids
DEOXYRIBONUCLEIC ACID (DNA): the genetic material in most organisms; a double-stranded helical molecule that contains coded instructions for the synthesis of the proteins needed in a cell
ENZYME: a protein that speeds up the rate of chemical reactions without itself being used up in the process
GENE: a unit of inheritance; a segment of DNA that codes for the synthesis of a specific protein
GENETIC DISORDER: a disorder caused by a change in a gene or chromosome
GENETIC MARKER: a distinctive DNA sequence inherited by members of a family with a certain genetic disorder but not inherited by other family members without the disorder

Prenatal Diagnosis

Prenatal diagnosis is the testing of a developing fetus in the womb, or uterus, for the presence of a genetic disorder. The purpose of this type of genetic testing is to inform a pregnant woman of the chances of having a baby with a genetic disorder. Prenatal diagnosis is limited to high-risk individuals and is usually recommended if a woman is thirty-five years of age or older, if she has had two or more spontaneous abortions, or if there is a family history of a genetic disorder. More than two hundred genetic disorders can be tested in a fetus. One of the most common genetic disorders screened for is Down syndrome, or trisomy 21, a form of mental retardation caused by three copies of chromosome 21. The incidence of Down syndrome is known to increase with maternal age.

The technique most commonly used for prenatal diagnosis is amniocentesis. It is performed between the sixteenth and eighteenth week of pregnancy. Amniocentesis involves the insertion of a hypodermic needle through the abdomen into the uterus of a pregnant woman. The insertion of the needle is guided by ultrasound, a technique that uses high-frequency sound waves to locate a developing fetus or internal organs and present a visual image on a computerized television screen. A small amount of amniotic fluid, which surrounds and protects the fetus, is withdrawn. The amniotic fluid contains fetal secretions and cells sloughed off the fetus that are analyzed for genetic abnormalities. Fetal blood can also be collected from the umbilical cord and analyzed for disorders such as sickle-cell anemia. Sickle-cell anemia is a genetic disorder caused by abnormal hemoglobin, the protein molecule that transports oxygen in the body. It is characterized by anemia, acute pain in the abdomen and joints, and blockage of blood vessels. Chromosomal disorders such as Down syndrome, Edwards' syndrome (trisomy 18), and Patau syndrome (trisomy 13) can be detected by examining the chromosome number of the fetal cells. Certain biochemical disorders such as Tay-Sachs disease, a progressive disorder characterized by a startle response to sound, blindness, paralysis, and death in infancy, can be determined by testing for the presence or absence of a specific enzyme activity in the amniotic fluid. Amniocentesis can also determine the sex of a fetus and detect common birth defects such as spina bifida (an open or exposed spinal cord) and anencephaly (partial or complete absence of the brain) by measuring levels of alpha fetoprotein in the amniotic fluid. The limitations of amniocentesis include the lack of detection of most genetic disorders, possible fetal loss, infection, and bleeding.

Chorionic villus sampling (CVS) is another technique used for prenatal diagnosis. It is performed earlier than amniocentesis (between the eighth and twelfth week of preg-

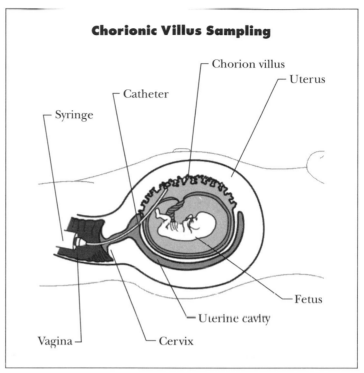

Chorionic Villus Sampling

Chorion villus

Uterus

Catheter

Syringe

Fetus

Uterine cavity

Vagina

Cervix

Chorionic villus sampling is one method of obtaining embryonic cells from a pregnant woman; examination of these cells helps physicians determine fetal irregularities or defects, which allows time to assess the problem and make recommendations for treatment.

nancy). Under the guidance of ultrasound, a catheter is inserted into the uterus via the cervix to obtain a sample of the chorionic villi. The chorionic villi are part of the fetal portion of the placenta, the organ that nourishes the fetus. The chorionic villi can be analyzed for chromosomal and biochemical disorders but not for congenital birth defects such as spina bifida and anencephaly. The limitations of this technique are inaccurate diagnosis and a slightly higher chance of fetal loss than in amniocentesis.

Neonatal Testing

The most widespread genetic testing is the mandatory testing of every newborn infant for detection of an inborn error of metabolism (a biochemical disorder caused by mutations in the genes that code for the synthesis of enzymes). The purpose of this type of testing is to initiate early treatment if an abnormality is detected. A blood sample is taken by heel prick from a newborn in the hospital nursery, placed on filter papers as dried spots, and subsequently tested in a specialized laboratory for the presence or absence of a specific protein or enzyme. For example, the Guthrie test, named after its designer Robert Guthrie, detects elevated levels of the amino acid phenylalanine in the blood of a newborn in order to determine if the infant has phenylketonuria, a disorder characterized by irreversible brain damage. Diets low in phenylalanine are promptly initiated within the first two months of life if test results are positive. Genetic disorders such as sickle-cell anemia and galactosemia (accumulation of galactose in the blood) can also be detected in newborns.

Carrier Testing

A healthy couple contemplating having children can be tested voluntarily to determine if they carry a defective gene for a disorder that runs in the family. This type of testing is known as carrier testing because it is designed for carriers (individuals who have a normal gene paired with a defective form of the same gene but have no symptoms of a genetic disorder). Carriers of the genes responsible for Tay-Sachs disease, sickle-cell anemia, cystic fibrosis (accumulation of mucus in the lungs and pancreas), Duchenne muscular dystrophy (wasting away of muscles), and hemophilia (uncontrolled bleeding caused by lack of blood clotting factor) can be detected by DNA analysis. For example, mutated genes can be detected directly by using a synthetic probe. A probe is a piece of single-stranded DNA labeled or tagged with a radioactive atom that is designed to search and bind to a normal or mutated gene.

When the gene responsible for a specific genetic disorder is unknown, the location of the gene on a chromosome can be detected indirectly by linkage analysis. Linkage analysis is a technique in which the pattern of inheritance of a mutated gene and a genetic marker are traced within large families with a history of a particular genetic disorder to determine if they are linked. If a genetic marker lies close to a gene that causes a particular genetic disorder, it is possible to trace the defective gene by looking for the genetic marker. The genetic markers used commonly for linkage analysis

are restriction fragment length polymorphisms (RFLPs). When human DNA is isolated from a blood sample and digested at specific sites with special enzymes called restriction endonucleases, RFLPs are produced. RFLPs are found scattered randomly in human DNA and are of different lengths in different people, except in identical twins. They are caused by mutations or the presence of varying numbers of repeated copies of a DNA sequence and are inherited from one parent or the other. RFLPs are separated by gel electrophoresis, a technique in which DNA fragments of varying lengths are separated in an electric field according to their sizes. The separated DNA fragments are blotted onto a nylon membrane, a process known as Southern blotting. The membrane is probed and then visualized on X-ray film. The characteristic pattern of DNA bands visible on the film is similar in appearance to the bar codes on grocery items.

Predictive Testing

High-risk individuals or families can be tested voluntarily for the presence of a mutated gene that may indicate a predisposition to a late-onset genetic disorder such as Huntington's chorea (characterized by uncontrolled movements, dementia, and ultimately death) or to other conditions such as hereditary breast, ovarian, and colon cancers. This type of testing is called predictive testing. Linkage analysis and RFLPs have been used to signal or predict the future onset of genetic disorders. For example, in 1983, James Gusella, Nancy Wexler, and Michael Conneally reported a correlation between one specific RFLP they named *G8* and Huntington's chorea. After studying numerous RFLPs of generations of an extended Venezuelan family with a history of Huntington's chorea, they discovered that *G8* was present in members afflicted with the genetic disorder and was absent in unaffected members.

Impact and Applications

Genetic testing has had a significant impact on families and society at large. It provides objective information to families about genetic disorders or birth defects and provides an analysis of the risks for genetic disorders through genetic counseling. Consequently,

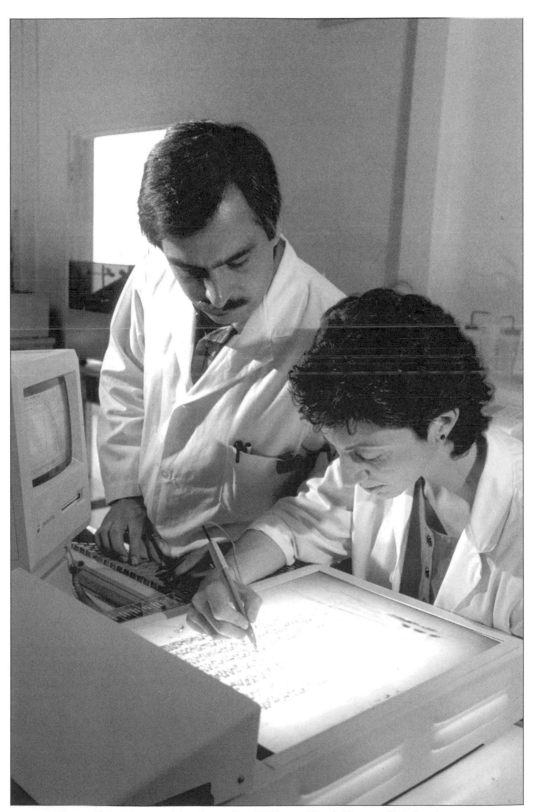

Technicians studying the results of DNA test. (Hank Morgan/Rainbow)

many prospective parents are able to make informed and responsible decisions about conception and birth. Some choose not to bear children, some terminate pregnancy after prenatal diagnosis, and some take a genetic gamble and hope for a normal child. Genetic testing can have a profound psychological impact on an individual or family. A positive genetic test could cause a person to experience depression, while a negative test result may eliminate anxiety and distress. Questions have been raised in the scientific and medical community about the reliability and astronomical costs of tests. There is concern about whether genetic tests are stringent enough to ensure that errors are not made. DNA-based diagnosis can lead to errors if DNA samples are contaminated. Such errors can be devastating to families. People at risk for late-onset disorders such as Huntington's chorea can be tested to determine if they are predisposed to developing the disease. There is, however, controversy over whether it is ethical to test for diseases for which there are no known cures or preventive therapies. The question of testing also creates a dilemma in many families. Unlike other medical tests, predictive testing involves the participation of many members of a family. Some members of a family may wish to know their genetic status, while others may not.

While there has been great enthusiasm over genetic testing, there are also social, legal, and ethical issues such as discrimination, confidentiality, reproductive choice, and abuse of genetic information. Insurance companies and employers may require prospective customers and employees to submit to genetic testing or may inquire about a person's genetic status. Individuals may be denied life and health insurance coverage because of their genetic status, or a prospective customer may be forced to pay exorbitant insurance premiums. The potential for discrimination with respect to employment and promotions also exists. For example, as a result of the sickle-cell screening programs of the early 1970's, many African Americans with sickle-cell anemia traits were denied employment and insurance coverage. Some were denied entry into the U.S. Air Force because of their carrier status. The Americans

with Disabilities Act, signed into federal law in 1990, contained provisions safeguarding employees from genetic discrimination by employers. By 1994, companies with fifteen or more employees had to comply with the law, which prohibits employment discrimination because of genetic status and also prohibits genetic testing by employers.

As genetic testing becomes standard practice, the potential for misuse of genetic tests and genetic information will become greater. Prospective parents may potentially use prenatal diagnosis as a means to ensure the birth of a "perfect" child. Restriction fragment length polymorphism analysis, used in genetic testing, has applications in DNA fingerprinting or DNA typing. DNA fingerprinting is a powerful tool for identification of individuals used to generate patterns of DNA fragments unique to each individual based on differences in the sizes of repeated DNA regions in humans. It is used to establish identity or nonidentity in immigration cases and paternity and maternity disputes; it is also used to exonerate the innocent from violent crimes and to link a suspect's DNA to body fluids or hair left at a crime scene. Several states in the United States have been collecting blood samples from a variety of sources, including newborn infants during neonatal testing and individuals convicted of violent crimes, and have been storing genetic information derived from them in DNA databases for future reference. Such information could be misused by unauthorized people.

Genetic testing also has applications in gene therapy, which involves replacing a defective gene in a cell with a normal gene with the hope of correcting a genetic disorder. There are more than four thousand documented genetic disorders, and the majority cannot be treated. Gene therapy appears to be a promising form of treatment for many genetic disorders. The first gene therapy clinical trial began in September, 1990. A four-year-old girl, Ashanti DeSilva, became the first person to be treated by gene therapy for an inherited disorder called severe combined immunodeficiency (SCID), a disorder caused by the absence of the enzyme adenosine deaminase (ADA), which is needed for the proper functioning of the immune system. Without the enzyme, a person cannot fight against infections. The implications of genetic testing are quite significant. Therefore education, knowledge, and public awareness of the ethical, legal, and social issues surrounding genetic testing are essential to prevent discrimination and abuse.

—Oluwatoyin O. Akinwunmi

See Also: Amniocentesis and Chorionic Villus Sampling; DNA Fingerprinting; Gene Therapy; Genetic Screening; Genetic Testing: Ethical and Economic Issues.

Further Reading: Fritz Fuchs, "Genetic Amniocentesis," *Scientific American* 242 (June, 1980), provides an overview for the general reader. Gina Kolata, "Genetic Screening Raises Questions for Employer and Insurers," *Science* 232 (April, 1986), discusses some of the ethical dilemmas facing employers and insurers. Rachel Nowak, "Genetic Testing Set for Takeoff," *Science* 265 (July, 1994), provides a discussion on some of the questions raised about predictive genetic testing. W. French Anderson, "Gene Therapy," *Scientific American* 273 (September, 1995), provides an overview for the general reader.

Genetic Testing: Ethical and Economic Issues

Field of study: Human genetics
Significance: *Using a suite of molecular, biochemical, and medical techniques, it is now possible to identify carriers of a number of genetic diseases and to diagnose some genetic diseases even before they display physical symptoms. In addition, numerous genes that predispose people to particular diseases such as cancer, alcoholism, and heart disease have been identified. These technologies raise important ethical questions about who should be tested, how the results of tests should be used, who should have access to the test results, and what constitutes normality.*

Key terms

RECESSIVE TRAIT: a genetically determined trait that is only expressed if a person gets the gene for the trait from both parents

DOMINANT TRAIT: a genetically determined trait that is expressed when a person gets the gene for the trait from either or both parents

The Dilemmas of Genetic Testing

Historically, it was impossible to determine whether a person was a carrier of a genetic disease or whether a fetus was affected by a genetic disease. Now both of these things and much more can be determined through genetic testing; however, although there are obvious advantages to acquiring this kind of information, there are also potential ethical problems involved. For example, if two married people are both found to be carriers of cystic fibrosis, each child born to them will have a 25 percent chance of having cystic fibrosis. Using this information, they could choose not to have any children, or, under an oppressive government desiring to improve the genetics of the population, they could be forcibly sterilized. Alternatively, they could choose to have each child tested prenatally and abort any child that tests positive for cystic fibrosis. Ethical dilemmas similar to these are destined to become increasingly common as scientists develop tests for more genetic diseases.

Another dilemma arises in the case of diseases such as Huntington's chorea, which is caused by a single dominant gene and is always lethal but which does not generally cause physical symptoms until middle age or later. A parent with such a disease has a 50 percent chance of passing it on to each child. Now that people can be tested, it is possible for a child of a parent with the disease to know whether he or she has inherited the deadly gene. If a person tests positive for the disease, he or she can then choose to remain childless or opt for prenatal testing to guard against the possibility of bringing a child into the world under a death sentence.

Tests for deadly, untreatable genetic diseases in offspring have an even darker side. If the test is negative, the person may be greatly relieved; if it is positive, however, doctors can offer no hope. Is it right to let someone know that they will die sometime around middle age or shortly thereafter if there is nothing the medical community can do to help them? The psychological trauma associated with such disclosures can sometimes be severe enough to result in suicide. Additionally, who should receive information about the test, especially if it shows positive for the disease? If the information is kept confidential, a person with a disease could buy large amounts of life insurance, to the financial advantage of beneficiaries, at the same price as an unaffected person. On the other hand, if health and life insurance companies were allowed to know the results of such tests, they might use the information to refuse insurance coverage of any kind. Lastly, none of the genetic tests are 100 percent accurate. There will be occasional false positives and false negatives. With so much at stake, how can doctors and genetic counselors help patients understand the uncertainties?

How Should Genetic Testing Information Be Used?

Scientists are now able to test for more than just specific, prominent genetic defects. Genetic tests are now available for determining potential risks for such things as cancer, alcoholism, Alzheimer's disease, and obesity. A positive result for the alcoholism gene does not mean that a person is doomed to be an alcoholic but rather that they have a genetic tendency toward behavior patterns that lead to alcoholism or other addictions. Knowing this, a person can then get counseling, as needed, to prevent alcoholism and make lifestyle decisions to help prevent alcohol abuse.

Unfortunately, a positive test for genes that predispose people to diseases such as cancer may be more ominous. It is believed that people showing a predisposition can largely prevent the eventual development of cancer with aggressive early screening (for example, breast exams and colonoscopies) and lifestyle changes. Some pre-emptive strategies, however, have come under fire. For example, some women at risk for breast cancer have chosen prophylactic mastectomies. In some cases, however, cancer still develops after a mastectomy, and some studies have shown lumpectomy and other less radical treatments to be as effective as mastectomy.

Another concern centers on who should have access to the test results. Should employers be allowed to require genetic testing as a screening tool for hiring decisions? Should insurance companies have access to the records when making policy decisions? These are especially disturbing questions considering the fact that a test for one of the breast cancer genes, for example, only predicts a significantly higher probability of developing breast cancer than is typical for the general population. Making such testing information available to employers and insurance companies would open the door to discrimination based on the probability that a prospective employee or client will become a future financial burden. A number of states have banned insurance companies from using genetic testing data for this very reason.

Impact and Applications

The long track record and accuracy of some tests, such as the tests for cystic fibrosis and Tay-Sachs disease, has led to the suggestion that they could be used to screen the general population. Although this would seem to provide positive benefits to the population at large, there is a concern about the cost of testing on such a broad scale. Would the costs of testing outweigh the benefits? What other medical needs might not get funded if such a program were started? The medical community will have to carefully consider the options before more widespread testing takes place.

As more genetic tests become available, it will eventually be possible to develop a fairly comprehensive genetic profile for each person. Such profiles could be stored on CD-ROMs or other storage devices and be used by individuals, in consultation with their personal physicians, to make lifestyle decisions that would counteract the effects of some of the defects in their genetic profiles. The information could also be used to determine a couple's genetic compatibility before they get married. When a woman becomes pregnant, a prenatal genetic profile of the fetus could be produced; if it does not match certain minimum standards, it could be aborted. The same genetic profile could be used to shape the child's life and help determine the child's profession. Although such comprehensive testing is now prohibitively expensive, the costs should drop as the tests are perfected and made more widely available.

Access to genetic profiles by employers, insurance companies, advertisers, and law enforcement agencies could result in considerable economic savings to society, allowing many decisions to be made with greater accuracy, but at what other costs? How should the information be used? How should access be limited? How much privacy should individuals have with regard to their own genetic profiles? As genetic testing becomes more widespread, these questions will need to be answered. Ultimately, the relationship between the good of society and the rights of the individual will need to be redefined.

—*Bryan Ness*

See Also: Cancer; Genetic Testing; Huntington's Chorea; Prenatal Diagnosis.

Further Reading: *Double-Edged Sword: The Promises and Risks of the Genetic Revolution* (1996), by Karl A. Drlica, provides a broad introduction to the many concerns about the genetic revolution, including the ethics and economics of genetic testing. John Rennie, "Grading the Gene Tests," *Scientific American* 273 (June, 1994), not only focuses on the accuracy and implementation of genetic tests but also considers the problems of privacy, discrimination, and eugenics inherent in genetic testing. Anne L. Finger, "How Would You Handle These Ethical Dilemmas?" *Medical Economics* (October 27, 1997), presents results of a survey in which readers were asked to settle two ethical dilemmas involving genetic testing. Karen Rothenberg et al., "Genetic Information and the Workplace: Legislative Approaches and Policy Challenges," *Science* 275 (March 21, 1997), summarizes government action designed to protect the privacy of genetic test results and outlines suggested guidelines for future legislation.

Genetically Engineered Foods

Field of study: Genetic engineering and biotechnology
Significance: *Genetic engineering (sometimes referred to as biotechnology) is the use of biology, genetics, and biochemistry to manipulate genes and genetic materials in a highly controlled fashion. The production of genetically engineered foods has helped increase plant productivity in an effort to meet the needs of a rapidly increasing world population. By transferring selected individual genes into desired plants, researchers have developed plants that provide higher yields, show greater tolerance to stress and better resistance to pests, diseases, and herbicides, and require less fertilizer.*

Key terms
GENETIC TRANSFORMATION: the transfer of extracellular deoxyribonucleic acid (DNA) among and between species
RECOMBINANT DNA: a DNA molecule made up of sequences that are not normally joined together

The Basis of Food Biotechnology
To increase the quality and amount of food production, recombinant deoxyribonucleic acid (DNA) technology is being used to introduce genes into plants or animals from strains or species with which they do not ordinarily interbreed. While conventional breeding techniques work with the whole organism and rely upon sexual means to transfer genetic material, genetic engineering operates at the cellular and molecular levels and makes it possible to bypass sexual reproduction and move desirable genes between completely unrelated organisms. Consequently, genetic engineering of food permits modification of crops and some animals with an unprecedented specificity and allows profound genetic transformations to take place in a matter of days rather than over hundreds to thousands of years.

Once a gene is isolated, it can be transferred to many different crops or animals without a lengthy breeding program. Gene transfer has produced many organisms with novel traits and improved agricultural qualities. The general steps taken in the production of genetically engineered foods are the selection of a gene of positive agricultural value, the isolation of the gene, the transfer of the isolated gene to a selected plant or animal cell, and the regeneration of a complete plant from the transformed cell or the modification of a selected animal characteristic such as fat content. The goals of genetic engineering are to produce altered animals with improved protein and amino acid composition and to produce plants that have improved photosynthetic efficiency, nitrogen fixation ability, protein quality, amino acid composition, postharvest handling, and resistance to herbicides, insects, and fungal and viral infections.

Given the complexity of the genetic code, no one can predict the effects of adding new genes to any animal or plant. Thus, genetic engineering of foods has raised special concerns that the process might result in the introduction of unfavorable and possibly dangerous traits into bacteria and viruses that will make them resistant to antibiotics, produce harmful toxins, and increase their tendency to cause and spread disease. Another concern is that since plants have been genetically engineered to be resistant to herbicides, farmers can spray higher levels of these chemicals without damaging crops, and this can result in increased contamination of food, soil, and water.

Impact and Applications

Because the somatic cells of plants can divide and reproduce the entire parent organism and because many plants have the ability to reproduce asexually in nature, giving rise to new plants from runners or even from fragments of the original plant, plants are the best candidates for genetic engineering among the higher organisms. Hundreds of genetically engineered plants have been safely tested under conditions approved by the U.S. Food and Drug Administration and the U.S. Department of Agriculture. In 1987, scientists introduced a gene from a bacterial cell into tomato plants that made the plants resistant to caterpillars. Solutions were also sought to solve the problem that tomatoes are a perishable food that requires an expensive refrigeration facility for shipping. If allowed to ripen on the vine for increased flavor, tomatoes are too soft for shipping. Consequently, most store-bought tomatoes are picked green and refrigerated while being shipped. However, special genes have

Several varieties of commercial tomatoes have been engineered for resistance to insects and prolonged shelf life. (Ben Klaffke)

been transferred into tomato plants to enable them to resist viral diseases and to produce tomatoes with increased flavor and shelf life. This gene transfer therapy has resulted in a 90 percent decrease in the softening process, so that tomatoes can now develop their natural flavor by ripening on the vine and be shipped without refrigeration, allowing distribution to poorer populations.

In the early 1990's, soybean plants resistant to herbicides and various insect pests were genetically engineered. In the mid-1990's, a number of plants with genetically engineered traits were patented, including maize (corn) plants rich in the amino acid tryptophan, potato plants resistant to various viruses, and frost-resistant potato and strawberry plants. In addition, genetically engineered food additives and enzymes have been approved and used in bread, baby foods, sugar, fruit juices, baking powder, soft drinks, corn syrup, and other processed foods. One limitation to genetic engineering of plants is the lack of efficient transformation and regeneration systems for legumes and monocots, which include the world's major cereal crops, but research is making progress on these problems.

Genetic engineering of higher animals is much less promising because the process of development is comparatively rigid and because asexual reproduction does not normally occur; however, some useful genetic modifications have been engineered by treating various animals with genetically engineered enzymes and hormones. A bovine growth hormone genetically engineered from bacteria has been injected into dairy cows to increase their yield of milk and into beef cattle to produce leaner meat. Similarly, a genetically engineered pig hormone causes hogs to grow faster and decreases the amount of fat in pork. Genetic research is underway to optimize lactose levels in milk to reduce dietary malabsorption and improve the lipid composition in cow milk to make it similar to human milk for infants.

—*Alvin K. Benson*

See Also: Biopesticides; Biotechnology, Risks of; Genetic Engineering: Agricultural Applications; High-Yield Crops; Transgenic Organisms.

Further Reading: *Plant Biotechnology* (1997), by S. Ignacimuthu, covers all the necessary and basic information regarding genetic engineering of plants. *An Introduction to Genetic Engineering* (1994), by D. S. Nicholl, contains a basic introduction to the world of genetic engineering, including gene modifications in animals and plants. *Genetic Engineering of Plants for Crop Improvement* (1993), by Rup Lal and Sukanya Lal, provides a scholarly treatment of the technologies available in the genetic engineering of plants.

Genetics, Historical Development of

Field of study: Genetic engineering and biotechnology

Significance: *Genetics is a relatively new branch of biology that explores the mechanisms of heredity. It impacts all branches of biology as well as agriculture, pharmaceuticals, and medicine. Advances in genetics may one day eliminate a wide variety of diseases and disorders and change the way that life is defined.*

Key terms

BIOTECHNOLOGY: the use of biological systems or organisms in industrial processes such as the manufacture of drugs and treatments

HEREDITY: the passing of traits from one generation to the next

GAMETE: a reproductive cell; a single sperm or egg

GENOME: all the genes carried by a single gamete

Charles Darwin

The prevailing attitude of the mid-nineteenth century was that all species were the result of a special creation and were immutable; that is, they remained unchanged over time. The work of Charles Darwin challenged that attitude. As a young man, Darwin served as a naturalist on the HMS *Beagle*, a British ship that mapped the coastline of South America from 1831 to 1836. Darwin's observations of life forms and their adaptations, especially those he encountered on the Galápagos Islands, led him to postulate that living species shared common ancestors

Many of the questions posed by Charles Darwin's theories were answered by the findings of geneticists. (Library of Congress)

but had no intention of publishing his notebooks, since he knew that his ideas would bring him into direct conflict with the society in which he lived. However, in 1858, he received a letter from a young naturalist named Alfred Russel Wallace. Wallace had done the same type of collecting in Malaysia that Darwin had done in South America, had observed the same phenomena, and had drawn the same conclusions. Wallace's letter forced Darwin to publish his findings, and in 1858, a joint paper by both men on the topic of evolution was presented at the London meeting of the Linnean Society. In 1859, Darwin reluctantly published *On the Origin of Species.* The response was immediate and largely negative. While the book became a best-seller, Darwin found himself under attack from religious leaders and other prominent scientists. In his subsequent works, he further delineated his proposals on the emergence of species, including man, but was never able to answer the pivotal question that dogged him until his death in 1882: If species are in fact mutable (capable of change over long periods of time), by what mechanism is this change possible?

with extinct species and that the pressures of nature—the availability of food and water, the ratio of predators to prey, and competition— exerted a strong influence over which species were best able to exploit a given habitat. Those best able to take advantage of an environment would survive, reproduce, and, by reproducing, pass their traits on to the next generation. He called this response to the pressures of nature "natural selection": Nature selected which species would be capable of surviving in any given environment and, by so doing, directed the development of species over time.

When Darwin returned to England, he shared his ideas with other eminent scientists

Gregor Mendel

Ironically, it was only six years later that this question was answered, and nobody noticed. Today, Gregor Mendel is considered the "father" of genetics, but, in 1865, he was an Augustinian monk in a monastery in Brunn, Austria (now Brno, Czech Republic). From 1856 to 1863, he conducted a series of experiments using the sweet pea (*Pisum sativum*), in which he cultivated over twenty-eight thousand plants and analyzed seven different physical traits. These traits included the height of the plant, the color of the seed pods and flowers, and the physical appearance of the seeds. He cross-pollinated tall plants with short plants, expecting the next generation of plants to be of medium height. Instead, all the plants produced from this cross, which he called the F_1 (first filial) generation, were tall. When he crossed plants of the F_1 generation, the next generation of plants (F_2) were both tall and short at a 3:1 ratio; that is, 75 percent of the F_2 generation of plants were tall, while 25 percent were short. This ratio held true whether he looked at one trait or multiple traits at the same time. He coined two phrases still used in genetics to describe this phenomenon: He called the trait that appeared in the F_1 generation "dominant" and the trait that vanished in the F_1 generation "recessive." While he knew absolutely nothing about chromosomes or genes, he postulated that each visible physical trait, or phenotype, was the result of two "factors" and that each parent contributed one factor for a given trait to its offspring. His research led him to formulate several statements that are now called the Mendelian principles of genetics.

Mendel's first principle is called the principle of segregation. While all body cells contain two copies of a factor (what are now called genes), gametes contain only one copy. The factors are segregated into gametes by meiosis, a specialized type of cell division that produces gametes. The principle of independent assortment states that this segregation is a random event. One factor will segregate into a gamete independently of other factors contained within the dividing cell. (It is now known that there are exceptions to this rule: Two genes carried on the same chromosome will not assort independently.)

To make sense of the data he collected from twenty-eight thousand plants, Mendel kept detailed numerical records and subjected his numbers to statistical analysis. In 1865, he presented his work. He received polite but indifferent applause. Until Mendel, scientists rarely quantified their findings; as a result, the scientists either did not understand Mendel's math or were bored by it. In either case, the scientists completely overlooked the significance of his findings. Mendel published his work in 1866. Unlike Darwin's work, it was not a best-seller. Darwin himself died unaware of Mendel's

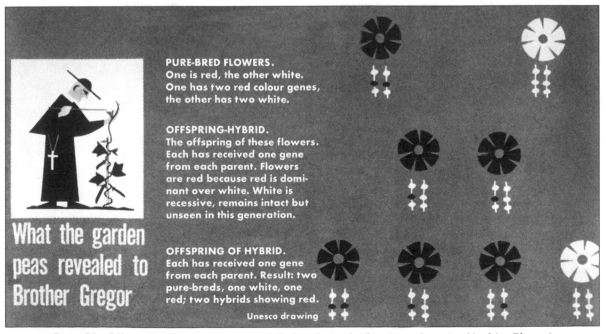

PURE-BRED FLOWERS.
One is red, the other white.
One has two red colour genes,
the other has two white.

OFFSPRING-HYBRID.
The offspring of these flowers.
Each has received one gene
from each parent. Flowers
are red because red is dominant over white. White is
recessive, remains intact but
unseen in this generation.

OFFSPRING OF HYBRID.
Each has received one gene
from each parent. Result: two
pure-breds, one white, one
red; two hybrids showing red.
Unesco drawing

What the garden peas revealed to Brother Gregor

Gregor Mendel's work with pea plants provided the foundation for the science of genetics. (Archive Photos)

work. Mendel died in 1884, two years after Darwin, with no way of knowing the eventual impact his work was to have on the scientific community. That impact began in 1900, when three botanists, working in different countries with different plants, discovered the same principles as had Mendel. Hugo De Vries, Carl Correns, and Erich Tschermak von Seysenegg rediscovered Mendel's paper, and all three cited it in their work. Sixteen years after his death, Mendel's research was given the respect it deserved, and the science of genetics was born.

Pivotal Research in Genetics

In 1877, Walter Fleming identified structures in the nuclei of cells that he called "chromosomes"; he later described the material of which chromosomes are composed as "chromatin." In 1900, William Bateson introduced the term "genetics" to the scientific vocabulary. Wilhelm Johannsen expanded the terminology the following year with the introduction of the terms "gene," "genotype," and "phenotype." In fact, 1901 was an exciting year in the history of genetics: The ABO blood group was discovered by Karl Landsteiner; the role of the X chromosome in determining gender was described by Clarence McClung; Reginald Punnett and William Bateson discovered genetic linkage; and De Vries introduced the term "mutation" to describe spontaneous changes in the genetic material. Walter Sutton suggested a relationship between genes and chromosomes in 1903. Five years later, Archibald Garrod, studying a strange clinical condition in some of his patients, determined that their disorder, called alkaptonuria, was caused by an enzyme deficiency. He introduced the concept of "inborn errors of metabolism" as a cause of certain diseases. That same year, two researchers named Godfrey Hardy and Wilhelm Weinberg published their extrapolations on the principles of population genetics.

From 1910 to 1920, Thomas Hunt Morgan, with his graduate students Alfred Sturtevant, Calvin Bridges, and Hermann Müller, conducted a series of experiments with the fruit fly *Drosophila melanogaster* that confirmed Mendel's principles of heredity and also confirmed the link between genes and chromosomes. The mapping of genes to the fruit fly chromosomes was complete by 1920. The use of research organisms such as the fruit fly became standard practice. For an organism to be suitable for this type of research, it must be small and easy to keep alive in a laboratory and must produce a great number of offspring. For this reason, bacteria (such as *Escherichia coli*), viruses (particularly those that infect bacteria, called bacteriophages), certain fungi (such as *Neurospora*), and the fruit fly have been used extensively in genetic research.

During the 1920's, Müller found that the rate at which mutations occur is increased by exposure to X-ray radiation. Frederick Griffith described "transformation," a process by which genetic alterations occur in pneumonococci bacteria. In the 1940's, Oswald Avery, Maclyn McCarty, and Colin MacLeod conducted a series of experiments that showed that the transforming agent Griffith had not been able to identify was, in fact, deoxyribonucleic acid (DNA). George Beadle and Edward Tatum proposed the concept of "one gene, one enzyme"; that is, a gene or a region of DNA that carries the information for a gene product codes for a particular enzyme. This concept was further refined to the "one gene, one protein" hypothesis and then to "one gene, one polypeptide." (A polypeptide is a string of amino acids, which is the primary structure of all proteins.)

During the 1940's, it was thought that proteins were the genetic material. Chromosomes are made of chromatin; chromatin is 65 percent protein, 30 percent DNA, and 5 percent ribonucleic acid (RNA). It was a logical conclusion that if the chromosomes were the carriers of genetic material, that material would make up the bulk of the chromosome structure. By the 1950's, however, it was fairly clear that DNA was the genetic material. Alfred Hershey and Martha Chase were able to prove in 1952 that DNA is the hereditary material in bacteriophages. From that point, the race was on to discover the structure of DNA.

For DNA or any other substance to be able to carry genetic information, it must be a stable molecule capable of self-replication. It was known that along with a five-carbon sugar and

Rosalind Franklin, whose research contributed to the determination of the structure of DNA. (Science Photo Library)

a phosphate group, DNA contains four different nitrogenous bases (adenine, thymine, cytosine, and guanine). Erwin Chargaff described the ratios of the four nitrogenous bases in what is now called Chargaff's rule: adenine in equal concentrations to thymine, and cytosine in equal concentrations to guanine. What was not known was the manner in which these constituents bonded to each other and the three-dimensional shape of the molecule. Groups of scientists all over the world were working on the DNA puzzle. A group in Cambridge, England, was the first to solve it. James Watson and Francis Crick, supported by the work of Maurice Wilkins and Rosalind Franklin, described the structure of DNA in a landmark paper in *Nature* in 1953. They described the molecule as a double helix, a kind of spiral ladder in which alter-

nating sugars and phosphate groups make up the backbone and paired nitrogenous bases make up the rungs. Arthur Kornberg created the first synthetic DNA in 1956. The structure of the molecule suggested ways in which it could self-replicate. In 1958, Matthew Meselson and Franklin Stahl proved that DNA replication is semiconservative; that is, each new DNA molecule consists of one template strand and one newly synthesized strand.

The Information Explosion

Throughout the 1950's and 1960's, genetic information grew exponentially. This period saw the description of the role of the Y chromosome in sex determination; the description of birth defects caused by chromosomal aberrations such as trisomy 21 (Down syndrome), trisomy 18 (Edwards' syndrome), and trisomy 13 (Patau syndrome); the description of operon and gene regulation by François Jacob and Jacques Monod in 1961; and the deciphering of the genetic code by Gobind Khorana, Marshall Nirenberg, and Severo Ochoa in 1966.

The discovery of restriction endonucleases (enzymes capable of splicing DNA at certain sites) led to an entirely new field within genetics called biotechnology. Mutations, such as the sickle-cell mutation, could be identified using restriction endonucleases. Use of these enzymes and DNA banding techniques led to the development of DNA fingerprinting. In 1979, human insulin and human growth hormone were synthesized in *Escherichia coli*. In 1981, the first cloning experiments were successful when the nucleus from one mouse cell was transplanted into an enucleated mouse cell. By 1990, cancer-causing genes called oncogenes had been identified, and the first attempts at human gene therapy had taken place. In 1997, researchers in England successfully cloned a living sheep. As the result of a series of conferences between 1985 and 1987, an international collaboration to map the entire human genome began in 1990. A comprehensive, high-density genetic map was published in 1994. The project plans to identify every human gene, determine the genetic information it carries, and map each gene to its chromosome.

Impact and Applications

The impact of genetics is immeasurable. In less than one hundred years, humans went from complete ignorance about the existence of genes to the development of gene therapies for certain diseases. Genes have been manipulated in certain organisms for the production of drugs, pesticides, and fungicides. Genetic analysis has identified the causes of many hereditary disorders, and genetic counseling has aided innumerable couples in making difficult decisions about their reproductive lives. DNA analysis has led to clearer understanding of the manner in which all species are linked. Techniques such as DNA fingerprinting have had a tremendous impact on law enforcement.

Advances in genetics have also given rise to a wide range of ethical questions with which humans will be struggling for some time to come. Termination of pregnancies, in vitro fertilization, and cloning are just some of the technologies that carry with them serious philosophical and ethical problems. There are fears that biotechnology will make it possible for humans to "play God" and that the use of biotechnology to manipulate human genes may have unforeseen consequences for humankind. For all the hope that biotechnology offers, it carries with it possible societal changes that are unpredictable and potentially limitless. Humans may be able to direct their own evolution; no other species has ever had that capability. How genetic technology is used and the motives behind its use will be some of the critical issues of the future.

—Kate Lapczynski

See Also: Biotechnology; Genetic Code, Cracking of; Human Genetics; Mendel, Gregor, and Mendelism; Natural Selection.

Further Reading: *The Double Helix* (1968), by James D. Watson, describes the discovery of the structure of DNA. A study of Mendel's paper, along with an interesting biography, is presented in *Mendel's Experiments on Plant Hybrids: A Guided Study* (1993), by A. Corcos and F. Monaghan. *A Dictionary of Genetics* (1997), by Robert C. King and William D. Stansfield, is a helpful resource for understanding the terminology of genetics. *Fundamentals of Genetics* (1994), by Peter J. Russell, is a textbook that thoroughly describes all the breakthrough research in genetics in understandable language.

Genomic Libraries

Field of Study: Molecular genetics

Significance: *A genomic library is a collection of relatively short pieces of the entire genome of an organism; between them, these pieces contain all of the genetic information about the organism. When scientists want to find a specific gene, they can search through such a collection to find the necessary information.*

Key terms

GENOME: all the genetic material carried by a cell

LAMBDA PHAGE: a virus that infects bacteria and then makes multiple copies of itself by taking over the infected bacteria's cellular machinery

LIGATION: the joining together of two pieces of deoxyribonucleic acid (DNA) using the enzyme ligase

What Is a Genomic Library?

Scientists often need to search through all the genetic information present in an organism to find a specific gene. It is thus convenient to have collections of genetic sequences stored so that such information is readily available. These collections are known as genomic libraries.

The library metaphor is useful in explaining both the structure and function of these information-storage centers. If one were interested in finding a specific literary phrase, one could go to a conventional library and search through the collected works. In such a library, the information is made up of letters organized in a linear fashion to form words, sentences, and chapters. It would not be useful to store this information as individual words or letters or as words collected in a random, jumbled fashion, as the information's meaning could not then be determined. The more books a library has, the closer it can come to having the complete literary collection, although no collection can guarantee that it has every piece of written word. The same is true of a genomic library. The stored pieces of genetic information cannot be individual bits but must be ordered sequences that are long enough to define a gene. The longer the string of information, the easier it is to make sense of the gene they make up, or "encode." The more pieces of genetic information a library has, the more likely it is to contain all the information present in a cell. Even a large collection of sequences, however, cannot guarantee that it contains every piece of genetic information.

How Is a Genomic Library Created?

In order for a genomic library to be practical, some method must be developed to put an entire genome into discrete units, each of which contains sufficiently large amounts of information to be useful but which are also easily replicated and studied. The method must also generate fragments that overlap one another for short stretches. The information exists in the form of chromosomes composed of millions of units known as base pairs. If the information were fragmented in a regular fashion—for example, if it were cut every ten thousand base pairs—there would be no way to identify each fragment's immediate neighbors. It would be like owning a huge multivolume novel without any numbering system: It would be almost impossible to determine with which book to start and which to proceed to next. Similarly, without some way of tracking the order of the genetic information, it would be impossible to assemble the sequence of each subfragment into the big continuum of the entire chromosome. The fragments are thus cut so that their ends overlap. With even a few hundred base pairs of overlap, the shared sequences at the end of the fragments can be used to determine the relative position of the different fragments. The different pieces can then be connected into one long unit, or sequence.

There are two common ways to fragment deoxyribonucleic acid

(DNA), the basic unit of genetic information, to generate a library. The first is to disrupt the long strands of DNA by forcing them rapidly through a narrow hypodermic needle, creating forces that tear the strands into short fragments. The advantage of this method is that the fragment ends are completely random. The disadvantage is that the sheared ends must be modified for easy joining, or ligation. The other method is to use restriction endonucleases, enzymes that recognize specific short stretches of DNA and cleave the DNA at specific positions. To create a library, scientists employ restriction enzymes that recognize four-base-pair sequences for cutting. Normally, the result of cleavage with such an enzyme would be fragments with an average size of 256 base pairs. If the amount of enzyme in the reaction is limited, however, only a limited number of sites will be cut, and much longer fragments can be generated. The ends created by this cleavage are usable for direct ligation into vectors, but the distribution of cleavage sites is not as random as that produced by shearing.

In a conventional library, information is imprinted on paper pages that can be easily replicated by a printing press and easily bound into

---

The Action of Restriction Enzymes

Part of Duplex with Bonds Broken by Chosen RE

...xxCTATAGxxxxxCTATAGxxxCTATAG...

...xxGATATCxxxxxGATATCxxxGATATC...

$\downarrow$ RE

...xxCTATA GxxxxxCTATA GxxxCTATA G...

...xxG ATATCxxxxx GATATCxxx GATATC...

DNA Fragments with Sticky Ends

A restriction enzyme (RE) breaks part of a duplex into fragments with "sticky ends." Each x denotes an unspecified base in a nucleotide unit.

a complete unit such as a book. Genetic information is stored in the form of DNA. How can the pieces of a genome be stored in such a way that they can be easily replicated and maintained in identical units? The answer is to take the DNA fragments and attach, or ligate, them into lambda phage DNA. When the phage infects a bacteria, it makes copies of itself. If the genomic fragment is inserted into the phage DNA, then it will be replicated also, making multiple exact copies (or clones) of itself.

To make an actual library, DNA is isolated from an organism and fragmented as described. Each fragment is then randomly ligated into a lambda phage. The pool of lambda phage containing the inserts is then spread onto an agar plate coated with a "lawn" or confluent layer of bacteria. Wherever a phage lands, it begins to infect and kill bacteria, leaving a clear spot, or "plaque," in the lawn. Each plaque contains millions of phages with millions of identical copies of one fragment from the original genome. If enough plaques are generated on the plate, each one containing some random piece of the genome, then the entire genome may be represented in the summation of the DNA present in all the plaques. Since the fragment generation is random, however, the completeness of the genomic library can only be estimated. It takes 800,000 plaques containing an average genomic fragment of 17,000 base pairs to give a 99 percent probability that the total will contain a specific human gene. While this may sound like a large number, it takes only fifteen teacup-sized agar plates to produce this many plaques. A genetic library pool of phage can be stored in a refrigerator and plated out onto agar petri dishes whenever needed.

How Can a Specific Gene Be Pulled out of a Library?

Once the entire genome is spread out as a collection of plaques, it is necessary to isolate the one plaque containing the specific sequences desired from the large collection. To accomplish this, a dry filter paper is laid onto the agar dish covered with plaques. As the moisture from the plate wicks into the paper, it carries with it some of the phage. An ink-dipped needle is pushed through the filter at several spots on the edge, marking the same spot on the filter and the agar. These will serve as common reference points. The filter is treated with a strong base that releases the DNA from the phage and denatures it into single-stranded form. The base is neutralized, and the filter is incubated in a salt buffer containing radioactive single-stranded DNA. The radioactive DNA, or "probe," is a short stretch of sequence from the gene to be isolated. If the full gene is present on the filter, the probe will hybridize with it and become attached to the filter. The filter is washed, removing all the radioactivity except where the probe has hybridized. The filters are exposed to film, and a dark spot develops over the location of the positive plaque. The ink spots on the filter can then be used to align the spot on the filter with the positive plaque on the plate. The plaque can be purified, and the genomic DNA can then be isolated for further study.

It may turn out that the entire gene is not contained in the fragment isolated from one phage. Since the library was designed so that the ends of one fragment overlap with the adjacent fragment, the ends can be used as a probe to isolate neighboring fragments that contain the rest of the gene. This process of increasing the amount of the genome isolated is called "genomic walking."

—*J. Aaron Cassill*

See Also: Cloning; Cloning Vectors; Restriction Enzymes.

Further Reading: *Molecular Cloning: A Laboratory Manual* (1989), by Joseph Sambrook et al., is considered by many scientists to be the "bible" of library construction and screening. *Recombinant DNA* (1992), by James D. Watson et al., covers the use of many molecular genetics techniques, including a full chapter on isolation of cloned genes, in elucidating the function of genes. *Introduction to Genetic Engineering* (1991), by William H. Sofer, and *An Introduction to Genetic Engineering* (1994), by Desmond S. T. Nicholl, cover the cloning of genes and how cloned genes can be used in science and industry.

Genomics

Field of study: Molecular genetics

Significance: *Genomics is the study of long DNA base sequences and complete genomes of biological organisms, in which the collection of genes and other genetic information is stored in cellular DNA. Genomics emphasizes computer-intensive methods of analyzing long DNA base sequences stored in large databases. Practical applications of genomics include the discovery of disease-causing bacterial and viral targets for drugs, vaccines, and other biotechnological products.*

Key terms

DEOXYRIBONUCLEIC ACID (DNA): the cellular molecule that carries genetic information

DNA BASE SEQUENCE: the linear order of bases (adenine, thymine, guanine, and cytosine) in each strand of DNA

GENOME: the total amount of DNA in an organism, emphasizing the collection of genes and other genetic information arranged linearly in chromosomes

GENE: a linear stretch of DNA bases of specific sequence, specific length, and specific location on a chromosome that usually contains information for determining a particular protein's structure and function

History of DNA Analysis

Since the discovery of the molecular structure of deoxyribonucleic acid (DNA) by James Watson and Francis Crick in 1953, scientists have understood that DNA contains genetic information. When arranged in a long, linear pattern, the chemical subunits of DNA, namely the bases adenine (A), thymine (T), guanine (G), and cytosine (C), contain information in much the same way that the twenty-six letters of the English alphabet can be arranged to provide information. The whole genome of an organism ranges from about 1 Mb (one million bases) for a simple bacterium to 3 Gb (3 billion bases) for humans. Scientists wish to know how to read the genetic language of DNA in the same way one reads a book, and they also wish to understand how the book is constructed (that is, how evolutionary forces have shaped the genome).

Although DNA can easily be extracted and purified from organisms, no analytical techniques were available to determine any DNA sequence prior to 1977. After 1977, however, very fast and accurate laboratory methods, now mostly automated, began to yield base sequences from a variety of viruses, bacteria, and other biological organisms, including humans. In 1995, the first complete DNA base sequence of an entire organism was produced by Craig Venter and scientists at the Institute for Genomic Research (TIGR). The 1,830,137 bases of the bacterium *Haemophilus influenzae* contain all the genetic information, in genes and other DNA arrangements on the bacterial chromosome, that determines everything about this organism: appearance, metabolism, behavior, inheritance, interaction with other organisms (including its potential for causing disease), and response to its environment. Scientists try to understand exactly how the cell interprets its DNA base sequence to produce these and other features of the organism.

Since 1995, many other organisms have had their DNA base sequences analyzed and deposited in DNA databases. Complete genomes of many microorganisms are known. The Human Genome Project, begun in 1990, has collected much information about the human genome and other model organisms, though the human genomic DNA sequence is incomplete. The total amount of DNA sequence data stored in computer-accessible databases exceeds 500 million bases. The science of genomics has grown rapidly with the enormous resource of raw data that these DNA sequences represent. The challenge of genomics is to interpret the biological information represented in the DNA sequences.

Genomic Analysis

Broadly speaking, genomics includes knowledge about genomes of organisms obtained from a variety of methods, such as observing the structural features of chromosomes visually through microscopes and mapping gene locations by mating organisms. "Structural genomics" emphasizes the relationship of the DNA base sequences of chromosomes to the location of genes and other regions of chromo-

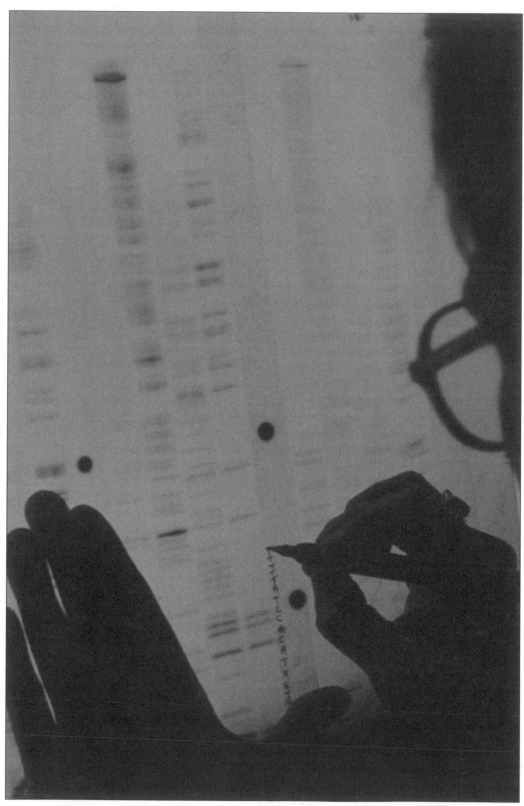

Genomics researchers seek patterns in DNA base sequences. (Dan McCoy/Rainbow)

somes measurable by other means. "Functional genomics" focuses on the expression and regulation of genes under a variety of cellular conditions and deduces properties of the cell's genome from a whole-cell analysis of gene products. "Comparative genomics" illuminates the variety of different routes and strategies evolution has taken to construct genomes. Comparing genomes between organisms leads to an understanding of evolutionary forces and strategies by which organisms respond to encode and regulate genetic information in their DNA sequences. "DNA informatics" refers to computer-intensive analysis of DNA sequences through algorithmic approaches and database searching and matching. A typical kind of DNA informatics analysis tries to find genes in very long DNA sequences. Scientists understand the general features of genes from many studies on a variety of organisms. In bacteria, a gene may be one thousand bases long. Exactly what stretch of approximately one thousand bases of a long bacterial DNA sequence represents the gene may be a difficult question to answer. Computer programs incorporating gene search algorithms are an important tool for a DNA informatics analysis.

One kind of algorithm looks for short-range patterns such as consecutive triplets of bases that code for the amino acid subunits of proteins. If a long stretch of several hundred triplets is found, this sequence is deemed an "open reading frame" (ORF), which is usually a good indication that it is contained in a gene. If the ORF is preceded by a short sequence of bases known to lie near the start positions of genes, the identification of the sequence as a gene is further strengthened. Genes can be recognized by these and other features. Note that the search for these features is greatly facilitated by computer algorithms, which can analyze DNA sequences over millions of bases quickly, accurately, efficiently, and simultaneously (for multiple sequence features) in order to locate the linear arrangements of genes.

Long-range patterns are also significant. The genome of a biological organism must contain all the genetic information needed for the full expression of all properties of the organism: appearance, life cycle, metabolism,

physiology, behavior, reproduction, and development from a single cell into a complete organism. Genetic information is contained in the genes and nearby regulatory regions of the organism's genome. However, the genome may also hold information not immediately recognizable in its genes but stored in long-range patterns in the genome. Such patterns of arrangements on the DNA base sequence may extend from a minimum size of a few thousand bases to a maximum size of hundreds of thousands or millions of bases on the chromosome. Most long-range patterns, if they truly exist, are difficult to find and, in contrast to ORFs, are poorly understood. The search for long-range patterns represents another form of DNA informatics and is especially computer intensive. The large-scale genome sequences and especially the complete genomic DNA base sequences of organisms represent a gold mine of raw data from which researchers are hoping to gain an understanding of the long-range patterns that evolution has created.

Impact and Applications

Genomics seeks to explain the structural, functional, and evolutionary features of genomes of biological organisms. Scientists know that genomes contain all the genetic information needed to specify all properties of organisms, but they do not yet know exactly how organisms interpret this information. Much, if not all, of the genetic information is stored in the linear DNA base sequence, and genomics attempts to decode the language of DNA bases over both short and long range in the genome. The large size of genomes and the enormous amount of raw DNA sequence data have led directly to computer-intensive methods of analyses.

Besides the goal of a better understanding of biological organisms, the practical applications of genomics are manifold. For instance, some organisms such as bacteria, fungi, and yeast make enzymes and other products that are useful to humans. A complete genome analysis often leads to the discovery of genes that determine further useful products. Another example pertains to disease-causing bacteria and viruses. Complete genomic analysis of these agents pinpoints gene targets for the development of vaccines, antibiotics, and other drugs to combat infections and treat humans and animals susceptible to infection. Genomics thus enhances understanding of the basic properties of biological organisms and stimulates new approaches to biotechnology.

—*R. L. Bernstein*

See Also: DNA Structure and Function; Genomic Libraries; Human Genome Project; Molecular Genetics.

Further Reading: "The Complete Genome Sequence of *Escherichia coli* K-12," by Frederick R. Blattner et al., *Science* 277 (1997), presents the summary report, with maps and tables, of the sequencing project to find the complete DNA sequence of 4,639,221 bases of *E. coli*. Anthony J. F. Griffiths et al., *An Introduction to Genetic Analysis* (1996), provides many details on the central role of DNA in molecular genetics and also contains a chapter on genomics.

Hardy-Weinberg Law

Field of study: Population genetics

Significance: *The Hardy-Weinberg law is the foundation for theories about evolution in local populations, often called "microevolution." First formulated in 1908, it continues to be the basis of practical methods for investigations in fields from plant breeding and anthropology to law and public health.*

Key terms

GENE POOL: the total set of all the genes in all individuals in an interbreeding population

ALLELE FREQUENCY: the proportion of all the genes at one chromosome location within a breeding population that are of one specific form

NATURAL SELECTION: the process by which allele and genotype frequencies are changed because of the ability of their phenotypes to survive and reproduce

GENETIC DRIFT: random changes in allele frequencies caused by chance events

GENE FLOW: movement of alleles from one population to another with different allele frequencies

Introduction

The Hardy-Weinberg law can be phrased in many ways, but its essence is that the genetic makeup of a population should not change. More important, it allows quantitative predictions about the distribution of genes and genotypes within and between generations. It may seem strange that theories about fundamental mechanisms of evolution are based on a statement that evolution should not occur. It is the nature of science that scientists must make predictions about the phenomena being studied. Without something with which to compare the results of experiments or observations, science is impossible. Newton's law of inertia plays a similar role in physics, stating that an object's motion will not change unless it is affected by an outside force.

After the rediscovery of Mendelian genetics in 1900, some scientists initially thought dominant alleles would become more common than recessive alleles, an error repeated in each generation of students. In 1908, Godfrey Hardy published his paper "Mendelian Proportions in a Mixed Population" in the journal *Science* to counteract that belief, pointing out that by themselves, sexual reproduction and Mendelian inheritance have no effect on an allele's commonness. Implicit in Hardy's paper was the idea that populations could be viewed as conglomerations of independent alleles, what has come to be called a "gene pool." Alleles randomly combine in pairs to make up the next generation. This simplification is similar to Sir Isaac Newton's view of objects as simple points with mass.

Hardy, an English mathematician, wrote only one paper in biology. Several months earlier, Wilhelm Weinberg, a German physician, independently and in more detail proposed the law that now bears both their names. In a series of papers, he made other contributions, including demonstrating Mendelian heredity in human families and developing methods for distinguishing environmental from genetic variation. Weinberg can justifiably be regarded as the father of human genetics, but his work, like Mendel's, was neglected for many years. The fact that his law was known as "Hardy's law" until the 1940's is an indictment of scientific parochialism.

The Hardy-Weinberg Paradigm

The Hardy-Weinberg "law" is actually a paradigm, a theoretical framework for studying nature. Hardy and Weinberg reenvisioned populations as collections of gametes (eggs and sperm) that each contain one copy of each gene. Most populations consist of diploid organisms that have two copies of each gene. Each generation of individuals can be regarded as a random sample of pairs of gametes from the previous generation's gamete pool. The proportion of gametes that contain a particular allele is the "frequency" of that allele.

Imagine a population of one hundred individuals having a gene with two alleles, *A* and *B*. There are three genotypes (combinations of alleles) in the population: *AA* and *BB* (homozygotes), and *AB* (heterozygotes). If the population has the following numbers of each genotype, the genotype frequencies can be computed as shown:

Genotype	Number	Genotype frequency
AA	36	36/100 = **0.36**
AB	48	48/100 = **0.48**
BB	16	16/100 = **0.16**
Total	100	1.00

The individuals of each genotype can be viewed as contributing one of each of their alleles to the gene pool, which has the following composition:

Genotype	*A* gametes	*B* gametes	Genotype contributions
AA	36 + 36 = 72		72
AB	48	48	96
BB		16 + 16 = 32	32
Total	120	80	200
Allele Frequency	120/200 = 0.6	80/200 = 0.4	200/200 = 1.0

This population can be described by the genotype ratio *AA*:*AB*:*BB* = 0.36:0.38:0.16 and the allele frequencies *A*:*B* = 0.6:0.4. Note that allele frequencies must total 1.0, as must genotype frequencies.

The Hardy-Weinberg Law and Evolution

Allele and genotype frequencies would be of little use if they only described populations. By making a Punnet square of the gametes in the population and using allele frequencies, one will get the following table of predicted genotype frequencies in the next generation:

The predicted frequencies of homozygotes are 0.36 and 0.16; the frequency of *AB* is 0.48 (adding the frequencies of *AB* and *BA*). These are the same as the previous generation.

Hardy pointed out that if the frequency of $A = p$ and the frequency of $B = q$, then $p + q = 1$. Random mating can be modeled by the equation $(p + q) \times (p + q) = 1$, or more compactly, $(p + q)^2 = 1$. This can be expanded to provide the genotype frequencies: $p^2 + 2pq + q^2 = 1$. In other words, the ratio of *AA*:*AB*:*BB* = p^2:$2pq$:q^2. Substituting 0.6 for p and 0.4 for q produces the figures shown in the preceding table, but more compactly and easily. The Hardy-Weinberg concept may also be extended to genes with more than two alleles. Therefore, three predictions may be made for a Hardy-Weinberg population: Frequencies of alleles p and q sum to 1.0 and will not change; the frequencies of genotypes *AA*, *AB*, and *BB* will be p^2:$2pq$:q^2 respectively, will sum to 1.0, and will not change (that is, they are in equilibrium); and if the genotype frequencies are not in the equilibrium ratios in one generation, they will reach equilibrium in the next.

There are within-generation and between-generation predictions. Within any one generation, the ratios of the genotypes are predictable if allele frequencies are known; if the frequency of a genotype is known, allele frequencies can be estimated. Between generations, allele and genotype frequencies should not change. However, evolution does occur, so real populations must often be out of Hardy-Weinberg equilibrium. Hardy and Weinberg made assumptions that mean it strictly applies only to diploid sexual populations when genes do not change from one allele to another (there is no muta-

Sperm	Eggs	
	A (frequency = 0.6)	*B* (frequency = 0.4)
A (frequency = 0.6)	*AA* (frequency = 0.6 x 0.6 = **0.36**)	*BA* (frequency = 0.6 x 0.4 = **0.24**)
B (frequency = 0.4)	*AB* (frequency = 0.6 x 0.4 = **0.24**)	*BB* (frequency = 0.4 x 0.4 = **0.16**)

tion); when alleles do not enter or leave a population, or adjacent populations have identical allele and genotype frequencies (there is no gene flow); when gametes are produced and combine randomly (mating is random); when the population size is infinite or at least very large (there is no genetic drift); and when all offspring survive and reproduce equally (there is no natural selection). Violations of these assumptions define the five major evolutionary forces: mutation, gene flow, nonrandom mating, genetic drift, and natural selection.

Despite its seeming limitations, the Hardy-Weinberg concept has been crucially useful in three major ways. First, its predictions of allele and genotype frequencies in the absence of evolution provide what statisticians call the "null hypothesis," which is essential for statistically rigorous hypothesis tests. If measured frequencies do not match predictions, then evolution is occurring. This redefines evolution from a vague "change in species over time" to a more useful, quantitative "change in allele or genotype frequencies." However, it is a definition that cannot be used in the domain of "macroevolution" and paleontology above the level of biological species. Similarly, Newton's definition of a moving object does not apply in quantum physics. Second, Hardy-Weinberg provides a conceptual framework for investigation. If evolution is happening, a checklist of potential causes of evolution can be examined in turn. Finally, the Hardy-Weinberg paradigm provides the foundation for mathematical models of each evolutionary force. These models help biologists determine whether a specific evolutionary force could produce observed changes.

Using the Hardy-Weinberg Law

Sickle-cell anemia is a severe disease of children characterized by reduced red blood cell number, bouts of pain, fever, gradual failure of major organs, and early death. In 1910, physicians noticed the disease and associated it with distortion ("sickling") of red blood cells. They realized that victims of the disease were almost entirely of African descent. Studies showed that the blood of about 8 percent of adult American blacks exhibited sickling, although few actually had the disease. By the 1940's, they knew sick-

ling was even more common in some populations in Africa, India, Greece, and Italy.

In 1949, James Neel proved the disease was caused by a recessive gene: Children homozygous for the sickle allele developed the disease and died. Heterozygotes showed the sickle trait but did not develop the disease. Using the Hardy-Weinberg law, Neel computed the allele frequency among American blacks as follows: Letting p = the frequency of the sickle allele, $2pq$ is the frequency of heterozygotes (8 percent of adult African Americans). Since $p + q = 1$, $q = 1 - p$ and $2p(1 - p) = 0.08$. From this he computed $p = 0.042$ (about 4 percent). From the medical literature, Neel knew the frequency of the sickle trait in several African populations and computed the sickle allele frequency to be as high as 0.10 (since then the frequency has been found to be as high as 0.20). These are extraordinarily high frequencies for a lethal recessive allele and begged the question: Why was it so common?

The Hardy-Weinberg assumptions provided a list of possibilities, including nonrandom mating (mathematical models based on Hardy-Weinberg showed nonrandom mating distorts genotype frequencies but cannot change allele frequencies), mutation (for the loss of sickle alleles via death of homozygotes to be balanced by new mutations, scientists estimated the mutation rate from normal to sickle allele would have to be about three thousand times higher than any known human mutation rate, which seemed unlikely) and gene flow (models showed gene flow reduces differences between local populations caused by other evolutionary forces; gene flow from African populations caused by slavery explained the appearance of the sickle allele in North America but not high frequencies in Africa).

Another possibility was genetic drift. Models had shown deleterious alleles could rise to high frequencies in very small populations (smaller than one thousand). It was possible the sickle allele "drifted" to a high frequency in a human population reduced to small numbers by some catastrophe (population "bottleneck") or started by a small number of founders (the "founder effect"). If so, the population had since grown far above the size at which drift is significant. Moreover, drift was random; if there had been several small populations, some would have drifted high and some low. It was unlikely that drift would maintain high frequencies of a deleterious allele in so many large populations in different locations. Therefore, the remaining possibility, natural selection, was the only reasonable possibility: The heterozygotes must have some selective advantage over the normal homozygotes.

A few years later, A. C. Allison was doing field work in Africa and noted that the incidence of the sickle-cell trait was high in areas where malaria was prevalent. A search of the literature showed this was also true in Italy and Greece. In 1954, Allison published his hypothesis: In heterozygotes, sickle-cell alleles significantly improved resistance to malaria. It has been repeatedly confirmed. Scientists have found alleles for several other blood disorders that also provide resistance to malaria in heterozygotes.

Impact and Applications

The Hardy-Weinberg law has provided scientists with a more precise definition of evolution: change in allele or genotype frequencies. It allows them to measure evolution, provides a conceptual framework for investigation, and continues to serve as the foundation for the theory of microevolution. Beyond population genetics and evolution, the Hardy-Weinberg paradigm is used in such fields as law (analysis of DNA "fingerprints"), anthropology (human migration), plant and animal breeding (maintaining endangered species), medicine (genetic counseling), and public health (implementing screening programs). In these and other disciplines, the Hardy-Weinberg law and its derivatives continue to be useful.

The Hardy-Weinberg law also has implications for social issues. In the early twentieth century, growing knowledge of genetics fueled a eugenics movement that sought to improve society genetically. Eugenicists in the 1910's and 1920's promoted laws to restrict immigration and promote sterilization of "mental defectives," criminals, and other "bad stock." The Hardy-Weinberg law is often credited with the decline of eugenics. The ratio $2pq/q^2 = 1$ makes it clear that if a recessive trait is rare (as most deleterious alleles are), most copies of a recessive allele are hidden in apparently normal heterozygotes. Selecting against affected individuals will be inefficient at best. However, a host of respected scientists championed eugenics into the 1920's and 1930's, long after the implications of Hardy-Weinberg were understood. It was really the reaction to the horrors of Nazi leader Adolf Hitler's eugenics program that made eugenics socially unacceptable. Moreover, it is premature to celebrate the end of the disturbing questions raised by eugenics. Progress in molecular biology makes it possible to detect deleterious alleles in heterozygotes, making eugenics more practical. Questions of whether genes play a major role in criminality and mental illness are still undecided. Debate about such medical and social issues may be informed by knowledge of the Hardy-Weinberg law, but decisions about what to do lie outside the domain of science.

—*Frank E. Price*

See Also: Eugenics; Inbreeding and Assortative Mating; Natural Selection; Population Genetics; Sickle-Cell Anemia.

Further Reading: William Provine provides the best overview of the history of population genetics, including the Hardy-Weinberg law, in *The Origins of Theoretical Population Genetics* (1971). Jared Diamond discusses the importance of genetic drift in human evolution in "Founding Fathers and Mothers," *Natural History* (June, 1988). Diane Paul and Hamish Spencer examine the role of Hardy-Weinberg and geneticists in the rise and fall of the eugenics movement as well as current issues in "The Hidden Science of Eugenics," *Nature* 23 (March, 1995). The November, 1994, issue of *Discover* magazine is a special edition that contains a number of articles about the genetics of race. Among other articles, Jared Diamond's "Race Without Color" discusses race from a Hardy-Weinberg perspective.

Heart Disease

Field of study: Human genetics
Significance: *The leading cause of death in the industrialized world is heart disease. While many*

lifestyle risk factors for heart disease have been identified, a relationship between family heredity and the risk of heart disease has been established for many years.

Key terms

ATHEROSCLEROSIS: the deposit of fatty materials on the inner walls of medium and large arteries

CHOLESTEROL: a steroid compound naturally found in the human body that is a risk factor for heart disease when blood levels are elevated

COAGULATION: the process of blood-clot formation

FIBRINOGEN: a protein found in blood plasma that is converted to fibrin during the blood-clotting process

HYPERTENSION: high blood pressure

Historical Research

The relationship between heart disease and genetics was first determined in the 1930's. Carl Miller of Oslo, Norway, found that family history was related to high blood cholesterol, coronary heart disease, and xanthomas (small pockets of fat deposited beneath the skin). In the 1950's, further research was conducted to determine how much lifestyle was related to heart disease in an attempt to better understand the relationships between genetics and lifestyle. The Framingham study was conducted from 1950 to 1970 and identified the following risk factors for heart disease: age, sex, blood cholesterol levels, blood pressure, diabetes, electrocardiogram abnormalities, physical activity, obesity, and cigarette smoking. Many of these risk factors are related to lifestyle, and some appear to be related to genetics. Subsequent research has attempted to identify specific relationships between risk factors and genetic determinants.

One of the strongest relationships has been found between heart disease and diabetes. Atherosclerosis is a major cause of death in individuals with diabetes. Both diseases have been found to have several risk factors in common, including high blood fat levels, high blood pressure, and obesity. However, diabetes is considered an independent risk factor for atherosclerosis. It is believed that different

genes are involved with the two diseases. Therefore, individuals who have a genetic predisposition to both diseases are more likely to have cardiovascular disease.

High blood cholesterol levels have been found to be a major risk factor for heart disease. Diet is a lifestyle variable that influences cholesterol levels in the blood. However, since a similar diet may result in a wide variation in the cholesterol levels of different individuals, a strong genetic influence is suspected. Cells have receptors for cholesterol that are needed to remove cholesterol from the blood. In some individuals, there is a mutation of the receptor gene, which results in the receptor being formed differently. This makes it difficult for the cholesterol to bind the cells and be removed from the blood. This problem results in higher blood cholesterol levels and increased risk of cholesterol deposits on the arterial walls.

The incidence of high blood pressure, another major risk factor for heart disease, has been found to be strongly connected to genetic inheritance. However, specific genes that may contribute to the disease have not been well documented. The genetic mechanism that has been found to have the strongest connection to the development of high blood pressure is sodium transport in cells. Several mechanisms of transport have been identified, and each may play a role in hypertension. Other systems in blood-pressure control have also been investigated. Proteins in these systems are good candidates for genetic influence because specific genes are responsible for manufacturing specific proteins. The proteins being considered in this research are renin, angiotensin, atrial naturetic factor, and an enzyme that converts angiotensinogen I to angiotensinogen II.

An acute heart attack is most frequently caused by the formation of a blood clot at the part of an artery where atherosclerotic narrowing has occurred. A major factor in blood clotting is a protein called fibrinogen, which forms fibrin during the coagulation process. Several studies have identified elevated blood fibrinogen as a risk factor for heart disease. Increased fibrinogen levels may be in part genetically determined. Studies have revealed that relatives of individuals who had heart attacks at an early age had higher levels of fibrinogen in their blood. Mechanisms of how fibrinogen levels are genetically determined are not understood. Clearly, other factors such as blood pressure, obesity, smoking, diabetes, and cholesterol are also important. The relationships among genetics, lifestyle, and heart disease are not well understood. Certainly both genetics and lifestyle play a major role in the development of heart disease, and some interaction between the two is significant.

Impact and Applications

The role of genetics in the development of heart disease is of great interest to researchers. Prevalent diseases such as high blood cholesterol and some types of blockages in coronary arteries are currently being studied as candidates for gene therapy, which has already been used to treat some rare, congenital heart diseases. Since heart disease tends to run in families, which indicates that heart problems are related to genes, gene therapy is a promising procedure to pursue for treatment. Furthermore, since heart disease is the leading cause of death in industrialized countries, gene therapy has the potential to improve the lives of many people.

—*Bradley R. A. Wilson*

See Also: Aging; Congenital Defects; Diabetes; Gene Therapy; Hereditary Diseases.

Further Reading: A good discussion of how gene therapy will help the fight against heart disease is presented in A. Rosenfeld, "The Medical Story of the Century," *Longevity* 4 (May, 1992). In "Gene Therapy for the Heart: When?" *Harvard Heart Letter* 6 (February, 1996), by an anonymous author, the future of gene therapy in heart disease is discussed. For general information about heart disease, including gene therapy, see American Heart Association, *Heart and Stroke A-Z Guide* (1997).

Hemophilia

Field of study: Human genetics
Significance: *Hemophilia is an inherited genetic disorder in which the blood does not clot adequately. Although incidents of hemophilia are rela-tively rare, the study of this disease has yielded important information about genetic transmission and the factors involved in blood clotting.*

Key terms

GENES: inherited cells that carry information determining physical characteristics
HEMOPHILIA A: a blood disease with a deficiency of clotting factor VIII
HEMOPHILIA B: a blood disease with a deficiency of clotting factor IX
HEMOSTASIS: the process by which blood flow is stopped at an injury site

Causes and Symptoms

When an injury occurs that involves blood loss, the body responds by a process known as hemostasis. Hemostasis involves several steps that result in the blood clotting and stopping the bleeding. With hemophilia, an essential substance is absent. For blood to clot, a series of chemical reactions must occur in a "domino effect." The reaction starts with a factor called the Hageman factor or factor XII, which cues factor XI, which in turn cues factor X and so on until factor I is activated. A particular gene on a certain chromosome determines each factor. If a gene is missing, the blood will not clot properly. Hemophilia A is the most common type, affecting over 80 percent of all hemophiliacs and resulting when clotting factor VIII is deficient. Hemophilia B (also known as Christmas disease) affects about 15 percent of hemophiliacs and results when clotting factor IX is deficient.

Hemophilia affects males almost exclusively and is transmitted as an abnormal gene on an X chromosome from the mother. Although it is possible for women to develop hemophilia, it is an extremely rare occurrence. A female has two X chromosomes, and a male has an X and Y chromosome. In order for a baby to be male, it must inherit the Y chromosome from the father. The X chromosome is inherited from the mother. The daughter of a hemophiliac father will carry the disease because she inherits one X chromosome (with the abnormal gene) from the father and one from the mother. A carrier has a 50 percent chance of having a hemophiliac son, since she will either pass on a healthy gene on the X chromosome or the abnormal gene that will result in the disease. In order for a female to have hemophilia, she would have to inherit the abnormal gene on the X chromosomes from both her mother and her father.

Hemophilia can be mild, moderate, or severe, depending on the extent of the clotting factor deficiency. Mild hemophilia may not be evident until adulthood when prolonged bleeding is observed after surgery or a major injury. The symptoms of moderate or severe hemophilia often appear early in life. These symptoms may include easy bruising, difficulty in stopping minor bleeding, bleeding into the joints, and internal bleeding without any obvious cause (spontaneous bleeding). When bleeding occurs in the joints, the person experiences severe pain, swelling, and possible deformity in the affected joint. The weight-bearing joints such as ankles and knees are usually affected. Internal bleeding requires immediate hospitalization and could result in death if the condition is severe. People who experience prolonged or abnormal bleeding are often tested for hemophilia. Testing the specific blood-clotting factors can determine the type and severity of hemophilia. Although a family history of hemophilia may help in the diagnosis, approximately 20 percent of hemophiliacs have no such history of the disease.

Impact and Applications

Hemophilia is not curable, although advances in the treatment of the disease are prolonging life and preventing crippling deformities. Symptoms of hemophilia can be reduced by replacing the deficient clotting factor. People with hemophilia A may receive antihemophilic factors to raise their blood-clotting factor above normal levels so that the blood clots appropriately. People with hemophilia B may receive clotting factor IX during bleeding episodes in order to increase the clotting factor levels. The clotting factors may be taken from plasma (the fluid part of blood), although it takes a great deal of plasma to produce a small amount of the clotting factors. Risks include infection by the hepatitis virus or human immunodeficiency virus (HIV), although advanced screening procedures have greatly re-

Alleles and Hemophilia

Father's Sperm Cells

		X	Y
Mother's Egg Cells	X	XX Normal Girl	XY Normal Boy
	X$_h$	XX$_h$ Normal Girl (carrier)	X$_h$Y Hemophiliac Boy

The daughters produced by the above union will be physically normal, but half will be carriers of hemophilia. Half the sons produced will be hemophiliacs.

duced such risks. Patients with mild hemophilia may be treated with a synthetic hormone known as desmopressin acetate (DDAVP).

Treatment with the plasma clotting factors has increased longevity and quality of life. In addition, many patients are able to treat bleeding episodes as outpatients with home infusions or self-infusions of the clotting factors. However, problems do exist with the treatment of hemophilia. Various illnesses, such as HIV, liver disease, or cardiovascular disease, have resulted from contamination of the clotting factors. Several techniques are used to reduce the risk of contamination, and most difficulties were largely eliminated by the mid-1990's. Bleeding into the joints is often controlled by the use of elastic bandages and ice. Exercise is recommended to help strengthen and protect the joint. Painkillers are used to reduce the chronic pain associated with joint swelling and inflammation, although hemophiliacs cannot use products containing aspirin or antihistamines because they prolong bleeding. Patients and their families have also benefitted from genetic education, counseling, and testing. Hemophilia centers can provide information on how the disease is transmitted, potential genetic risks, and whether a person is a carrier. This knowledge provides options for family planning as well as support in coping with the disease.

—Virginia L. Salmon

See Also: Genetic Counseling; Genetic Screening; Genetic Testing; Hereditary Diseases; Human Genetics.

Further Reading: *Everything You Need to Know About Diseases* (1996), edited by Michael Shaw, provides a general overview of hemophilia and how it is inherited. Peter Jones, *Living with Haemophilia* (1984), provides an understandable discussion of hemophilia and its transmission, symptoms, and management. "Outlook Brighter for Youngsters with Hemophilia," *FDA Consumer* 31 (July-August, 1993), includes information on treatment advances.

Hereditary Diseases

Field of study: Human genetics
Significance: *Hereditary diseases are assuming an ever-increasing proportion of the list of diseases that affect children and adults. The Human Genome Project was begun in 1990 with the goal of determining and mapping all human genes by the year 2005. As knowledge about the ways that genetics are involved in different diseases is gained, opportunities will increase for the diagnosis, prevention, and treatment of these diseases.*

Key terms

HEMIZYGOUS: characterized by being present only in a single copy, as in the case of genes on the single X chromosome in males

HUMAN GENOME PROJECT: a research project whose goal is to identify all the genes of humans

MENDELIAN TRAIT: a trait controlled by a single gene pair

MODE OF INHERITANCE: the pattern by which a trait is passed from one generation to the next

MULTIFACTORIAL TRAIT: a trait determined by one or more genes and environmental factors

The Causes and Impact of Hereditary Diseases

The twentieth century has seen an unprecedented success by medicine in conquering the infectious diseases that have plagued humankind. Elimination, control, and treatment of diseases such as smallpox, measles, diphtheria, and plague have lessened their impact as causes of infant and adult mortality. Improved prenatal and postnatal care have also contributed to a decrease in childhood mortality. Shortly after the rediscovery of Mendelism in the early 1900's, reports of genetic determination of human traits began to appear in medical and biological literature. For the first half of the twentieth century, most of these reports were regarded as interesting scientific reports of isolated clinical diseases that had no real relevance to the practice of medicine. The field of medical genetics is considered to have begun in 1956 with the first description of the correct number of chromosomes in humans (forty-six). Between 1900 and 1956, findings were accumulating in cytogenetics, Mendelian genetics, biochemical genetics, and other fields that began to draw medicine and genetics together. Victor A. McKusick describes medical genetics as "the science of human biologic variation as it relates to health and disease."

The causes of hereditary diseases may be classified into four major categories: single-gene disorders or Mendelian disorders (cystic fibrosis, Huntington's chorea, color blindness, and phenylketonuria), chromosomal disorders caused by changes in the number or alterations in the structure of chromosomes (Down syndrome, Klinefelter's syndrome, and Turner's syndrome), multifactorial disorders caused by both genetic and environmental factors (congenital hip dislocation, cleft palate, and cardiovascular disease), and mitochondrial disorders caused by defects of the genetic information in the mitochondria (Leber hereditary optic neuropathy). These four categories are relatively clear-cut. It is likely that genetic factors also play a less well-defined role in all human diseases, including susceptibility to many common diseases and degenerative disorders. Genetic factors may affect a person's health from the time before birth to the time of death.

Congenital defects are birth defects and may be caused by genetic factors, environmental factors (such as trauma, radiation, alcohol, infection, and drugs), or the interaction of genes and environmental agents. Alan Emery and David Rimoin noted that the proportion of childhood deaths with nongenetic causes was estimated to be 83.5 percent in London in 1914 but had declined to 50 percent in Edinburgh by 1976, whereas childhood deaths with genetic causes went from 16.5 percent in 1914 to 50 percent in 1976. Concerning the lifetime frequency of genetic diseases, Rimoin, J. Michael Connor, and Reed Pyeritz estimate that single gene disorders have a lifetime frequency of 20 in 1000, chromosomal disorders have a frequency of 3.8 in 1000, and multifactorial disorders have a frequency of 646 in 1000. It is evident that hereditary diseases are and will be of major concern to the health professions for some time.

Single-Gene Disorders

Single-gene disorders result from a change or mutation in a single gene and are referred to as Mendelian disorders. In 1865, Gregor Mendel described the first examples of monohybrid or unifactorial inheritance. In a trait governed by a single pair of genes, individuals inherit a member of each pair from each parent. If the alleles (or genes for the same trait) are identical, the individual is said to be homozygous for those genes and has a homozygous genotype. If the alleles are different, the individual is said to be heterozygous and has a heterozygous genotype. Single-gene disorders will typically show a family history consistent with the pattern of dominant or recessive inheritance. A dominant gene will be expressed as the trait whether it is present in two copies, as in the homozygous condition, or in a single copy, as in the heterozygous condition. A recessive gene, on the other hand, is expressed only when it is present in the homozygous condition. When it is found together in a heterozygous condition with a dominant allele, the dominant allele "hides" or masks the expression of the recessive gene. In addition to being dominant or recessive, single genes may be characterized according to their position on chromosomes, being either autosomal or sex linked.

Sex-linked genes are located on chromosomes associated with the organism's gender. Human males have an unlike pair of sex chromosomes, one designated as the X chromosome and another, smaller one designated as the Y chromosome. Females have a pair of like sex chromosomes similar to the X chromosome of the male. Technically, sex-linked inheritance refers to traits governed by genes on the X or Y sex chromosomes. However, since Y chromosomes contain few genes, "sex linked" usually refers to genes on the X chromosome; therefore, such traits might better be referred to as "X linked." Regarding X-linked genes, males are said to be hemizygous. Females have two X chromosomes and may have homozygous or heterozygous genotypes, as in the case of all other genes located on all other chromosomes. The latter are called autosomes. Males have twenty-two pairs of autosomes and one pair of unlike sex chromosomes, whereas females have twenty-two pairs of autosomes and one pair of like sex chromosomes.

It is possible to have four basic types of single-gene inheritance: autosomal dominant, autosomal recessive, X-linked dominant, and X-linked recessive. Medically important disorders for autosomal dominant inheritance include achondroplasia (a disorder of cartilage formation that results in a form of disproportionate dwarfism) and Huntington's chorea (a progressive deterioration of motor and mental skills with first symptoms usually appearing after age thirty). Autosomal recessive disorders include cystic fibrosis (a fatal disease characterized by abnormal secretions of the pancreas and other glands) and phenylketonuria, or PKU (an inborn error of metabolism that leads to severe mental retardation if not diagnosed and treated shortly after birth). Disorders passed by X-linked inheritance include Duchenne muscular dystrophy (a severe disorder leading to progressive weakness, wasting of muscles, and death usually by age twenty) and hemophilia (a "bleeders" disease characterized by the failure of blood to clot).

Chromosomal Disorders

Chromosomal disorders are a major cause of birth defects, different types of cancer, infertility, mental retardation, and other abnormalities. They are also the leading cause of spontaneous abortions. Deviations from the normal number of forty-six chromosomes or structural changes usually result in abnormalities. Variations in the number of chromosomes may involve complete sets of chromosomes resulting in cases of polyploidy. Polyploidy among live newborns is very rare, and the few polyploid babies who are born usually die within a few days of birth as a result of severe malformations. The vast majority of embryos and fetuses with polyploidy are spontaneously aborted.

Chromosomal defects may also be caused by a change that involves only one or a few chromosomes rather than a whole set. If a cell does not contain some multiple of a set, the condition is known as aneuploidy. If there are forty-five chromosomes, it is called monosomy. This means that only one member of a normal chromosome pair is present. If there are forty-seven chromosomes, the condition is called trisomy, which is three chromosomes in place of one pair. Monosomy of the autosomes usually leads to death during development. A few cases are known of individuals surviving to birth with forty-five chromosomes, but they suffered from severe multiple malformation. Most embryos and fetuses that have autosomal trisomies abort early in pregnancy. Invariably, the trisomies that are born have severe physical and mental abnormalities. The most common trisomies occur in chromosome 21 (Down syndrome), chromosome 13 (Patau syndrome), and chromosome 18 (Edwards' syndrome). Trisomies 13 and 18, both rare, have major malformations, and all children afflicted with these disorders die early. Down syndrome is the most common (about one in seven hundred births) and is the best known of the chromosomal disorders. Individuals with the disorder are short and have slanting eyes, a nose with a low bridge, and stubby hands and feet; about one-third suffer severe mental retardation. The risk of giving birth to a child with Down syndrome increases dramatically for women over thirty-five years of age.

Variations in the number of sex chromosomes are not as lethal as those of the autosomes. Turner's syndrome is the only monosomy that survives in any number, although 98 percent of them are spontaneously aborted. Patients have forty-five chromosomes consisting of twenty-two pairs of autosomes and

only one X chromosome. They are short in stature, sterile, and have underdeveloped female characteristics but normal or near-normal intelligence. Other diseases caused by variations in the number of sex chromosomes include Klinefelter's syndrome, caused by having forty-seven chromosomes, including two X and one Y chromosome (affected individuals are male with small testes and are likely to have some female secondary sex characteristics such as enlarged breasts and sparse body hair) and multiple X syndrome (affected individuals are females whose characteristics are variable;

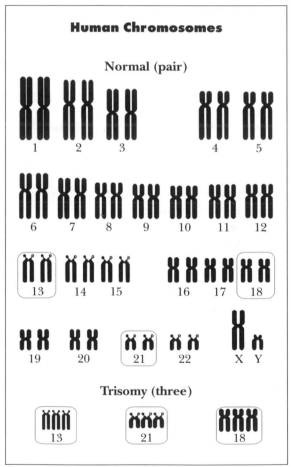

Genetic diseases are caused by defects in the number of chromosomes, in their structure, or in the genes on the chromosome (mutation). Shown here is the human complement of chromosomes (23 pairs) and three errors of chromosome number (trisomies) that lead to the genetic disorders Patau's syndrome (trisomy no. 13), Edward's syndrome (no. 18), and the more common Down syndrome (trisomy no. 21).

some are sterile or have menstrual irregularities or both).

Variations in the structure of chromosomes include added pieces (duplications), missing pieces (deletions), and transfer of a segment to a member of a different pair (translocation). Most deletions are likely to have severe effects on developing embryos, causing spontaneous abortion. Only those with small deletions are likely to survive and will have severe abnormalities. The cri du chat ("cry of the cat") syndrome produces an infant whose cry sounds like a cat's meow.

Multifactorial Traits

Multifactorial traits (sometimes referred to as complex traits) result from an interaction of one or more genes with one or more environmental factors. Sometimes the term "polygenic" is used for the case of a trait that is determined by multiple genes with small effects. Multifactorial traits do not follow any simple pattern of inheritance and do not show distinct Mendelian ratios. Specific diseases will show an increased recurrence risk within families. "Recurrence risk" refers to the likelihood of any trait showing up multiple times in a family; in general, the more closely related someone is to an affected person, the higher the risk. Recurrence risk may be complicated further by factors such as the degree and severity of expression of the trait, the sex of the affected relative, and the number of affected relatives. For example, pyloric stenosis is a disorder involving an overgrowth of muscle between the stomach and small intestine. It is the most common cause of surgery among the newborn. It has an incidence of about 0.2 percent in the general, unrelated population. Males are five times more likely to be affected than females. If a male is affected, there is a 5 percent chance the first child will be affected. If a female is affected, there is a 16 percent chance the first child will be affected.

It is necessary to develop separate risks of recurrence for each multifactorial disorder. The impact of multifactorial disorders is considerable as it is thought that they account for 50 percent of all congenital defects. In addition, they play a significant role in many adult disorders, including hypertension and other cardiovascular diseases, rheumatoid arthritis, psychosis, dyslexia, epilepsy, and mental retardation. In total, multifactorial disorders account for more genetic diseases than do single-gene and chromosome disorders combined. Because of the complexity of the interaction of genes and environmental factors, it has been difficult to deal with questions concerning prevention, diagnosis, genetic counseling, and treatment for multifactorial disorders. Application of the techniques of molecular genetics have allowed researchers to be able to locate specific genes associated with multifactorial traits.

Impact and Applications

The Human Genome Project's goal is to identify the entire human genome of some sixty thousand to seventy thousand genes and to map all of these genes to specific locations on chromosomes. The complete specifications of the genetic material on each of the twenty-two autosomes and the X and Y chromosomes will improve the understanding of the biological and molecular bases of hereditary diseases. Once the location of a gene is known, it allows for a better prediction of how that gene is transmitted within a family and of the probability that an individual will inherit a specific genetic disease.

For many hereditary diseases, the protein produced by the gene and its relation to the symptoms of the disease are not known. Locating a gene facilitates this knowledge. It becomes possible to develop new diagnostic tests and therapies. The number of hereditary disorders that can be tested prenatally and in newborns will increase dramatically. In the case of those single genes that do not produce clinical symptoms until later in life, many more of these disorders will be diagnosed before symptoms appear, opening the way for better treatments and even prevention. Possibilities will exist to develop the means of using gene therapy to repair or replace the disease-causing gene. The identification and mapping of single genes and those identified as having major effects on multifactorial disorders will greatly affect hereditary disease treatment and genetic counseling techniques. It is evident that knowledge of genes, both those that cause disease and those that govern normal functions, will begin to raise many questions about legal, ethical, and moral issues.

—*Donald J. Nash*

See Also: Congenital Defects; Down Syndrome; Human Genetics; Human Genome Project; Monohybrid Inheritance.

Further Reading: *Mendelian Inheritance in Man* (1994), compiled by Victor A. McKusick, is a comprehensive catalog of Mendelian traits in humans. Although it is filled with medical terminology and clinical descriptions, there are interesting and fascinating accounts of many traits. Benjamin A. Pierce provides an introduction to the principles of heredity and a catalog of more than one hundred human traits in the *Family Genetic Sourcebook* (1990). The Alliance of Genetic Support Group publishes the *Directory of National Genetic Voluntary Organizations and Related Resources* (1998).

Heredity and Environment

Field of study: Population genetics
Significance: *"Heredity and environment" is the modern incarnation of the age-old debate on the effects of nature versus nurture. Research in the field has implications ranging from the improvement of crop plants to the understanding of the heritability of behavioral traits and intelligence in humans.*

Key terms

PHENOTYPE: the ensemble of all measurable traits of a plant or animal

GENOTYPE: the composite of all the genes of an individual

PHENOTYPIC PLASTICITY: the ability of a genotype to produce different phenotypes when exposed to different environments

REACTION NORM: the graphic illustration of the relationship between environment and phenotype for a given genotype

HERITABILITY: a measure of the genetic variation for a quantitative trait in a population

QUANTITATIVE TRAIT LOCI (QTL) MAPPING: a molecular biology technique used to identify genes controlling quantitative traits in natural populations

Nature Versus Nurture and the Origin of Genetics

Is human behavior controlled by genes or by environmental influences? The nature-versus-nurture controversy has raged throughout human history, eventually leading to the modern antithesis between hereditarianism and environmentalism in biological research. These two schools of thought have shaped a dispute that is at once a difficult scientific problem and a thorny ethical dilemma. Many disciplines, chiefly genetics but also the cognitive sciences, have contributed to the scientific aspect of the discussion. At the same time, racist and sexist overtones have muddled the inquiry and inextricably linked it to the implementation of social policies. Nevertheless, the relative degree of influence of genes and environments in determining the characteristics of living organisms is a legitimate and important scientific question, apart from any social or ethical consideration.

At the beginning of the twentieth century, scientists rediscovered the laws of heredity first formulated by Gregor Mendel in 1865. Mendel understood a fundamental concept that underlies all genetic analyses: Each discrete trait in a living organism, such as the color of peas, is influenced by minute particles inside the body that behave according to simple and predictable patterns. Mendel did not use the term "gene" to refer to these particles, and his pioneering work remained largely unknown to the scientific community for the remainder of the nineteenth century. Immediately following the rediscovery of Mendel's laws in 1900, the Danish biologist Wilhelm Johannsen proposed the fundamental distinction between "phenotype" and "genotype." The phenotype is the ensemble of all measurable traits of a plant or animal. The composite of all the genes of an individual is its genotype. To some extent, the genotype determines the phenotype.

Reaction Norm: Environments and Genes Come Together

It was immediately clear to Johannsen that the appearance of a trait is the combined result of both the genotype and the environment, but to understand how these two factors interact took the better part of the twentieth century and is still a preeminent field of research in ecological genetics. One of the first important discoveries was that genotypes do not always produce the same phenotype but that this varies with the particular environment to which a genotype is exposed. For instance, if genetically identical fruit flies are raised at two temperatures, there will be clear distinctions in several aspects of their appearance, such as the size and shape of their wings, even though the genes present in these animals are indistinguishable.

This phenomenon can be visualized in a graph by plotting the observed phenotype on the ordinates versus the environment in which that phenotype is produced on the abscissa. A curve describing the relationship between environment and phenotype for each genotype is called a reaction norm. If the genotype is insensitive to environmental conditions, its reaction norm will be flat (parallel to the environmental axis); most genotypes, however, respond to alterations in the environment by producing distinct phenotypes. Their reaction norms are therefore characterized by a slope (or even by being nonlinear). When the latter case occurs, that genotype is said to exhibit phenotypic plasticity. One can think of plasticity as the degree of responsiveness of a given genotype to changes in its environment: The more responsive the genotype is, the more plasticity it displays.

The first biologist to fully appreciate the importance of reaction norms and phenotypic plasticity was the Russian Ivan Schmalhausen, who wrote a book on the topic in 1947. Schmalhausen understood that natural selection acts on the shape of reaction norms: By molding the genotype's response to the environment, selection can improve the ability of that genotype to survive under the range of environmental conditions it is likely to encounter in nature. For example, some butterflies are characterized by

The bright spots on the wings of many butterflies reflect seasonal genetic adaptations to their environment. (Janet Haas/Rainbow)

the existence of two seasonal forms. One form exists during the winter, when the animal's activity is low and the main objective is to avoid predators. Accordingly, the coloration of the body is dull to blend in with the surroundings. During the summer, however, the butterflies are very active, and camouflage would not be an effective strategy against predation. Therefore, the summer generation develops brightly colored "eyespots" on its wings. The function of these spots is to attract predators' attention away from vital organs, thereby affording the insect a better chance of survival. Developmental geneticist Paul Brakefield has demonstrated, in a series of works published in the 1990's, that the genotype of these butterflies codes for proteins that sense the season by using environmental cues such as photoperiod and temperature. Depending on the perceived environment, the genotype directs the butterfly developmental system to produce or not produce the eyespots.

Quantitative Genetics of Heredity and Environment

An important aspect of modern science is the description of natural phenomena in mathematical form. This allows predictions on future occurrences of such phenomena. In the 1920's, Ronald Fisher developed the field of quantitative genetics, a major component of which is a powerful statistical technique known as analysis of variance. This allows a researcher to gather data on the reaction norms of several genotypes and then mathematically partition the observed phenotypic variation (V_p) into its three fundamental constituents:

$$V_p = V_g + V_e + V_{ge}$$

where V_g is the percentage of variation caused by genes, V_e is the percentage attributable to environmental effects, and V_{ge} is a term accounting for the fact that different genotypes may respond differently to the same set of environmental circumstances. The power of this approach is in its simplicity: The relative balance among the three factors directly yields an answer to any question related to the nature-nurture conundrum. If V_g is much higher than the other two components, genes play a primary role in determining the phenotype ("nature"). If V_e prevails, the environment is the major actor ("nurture"). However, when V_{ge} is more significant, this suggests that genes and environments interact in a complex fashion so that any attempt to separate the two is meaningless. Anthony Bradshaw pointed out in 1965 that large values of V_{ge} are indeed observable in most natural populations of plants and animals.

The quantity V_g is particularly important for the debate because when it is divided by V_p, it yields the fundamental variable known as "heritability." Contrary to intuition, heritability does not measure the degree of genetic control over a given trait but only the relative amount of phenotypic "variation" in that trait that is attributable to genes. In 1974, Richard Lewontin pointed out that V_g (and therefore heritability) can change dramatically from one population to another, as well as from one environment to another. This is because V_g depends on the frequencies of the genes that are turned on (active) in the individuals of a population. Since different sets of individuals may have different sets of genes turned on, every population can have its own value of V_g for the same trait. Along similar lines, some genes are turned on or off in response to environmental changes; therefore, V_g for the same population can change depending on the environment in which that population is living. Accordingly, estimates of heritability cannot be compared between different populations or species and are only valid in one particular set of environmental conditions.

Molecular Genetics of Heredity and Environment

The modern era of the study of nature-nurture interactions relies on the developments in molecular genetics that characterized the whole of biology throughout the second half of the twentieth century. In 1993, Carl Schlichting and Massimo Pigliucci proposed that specific genetic elements known as "plasticity genes" supervise the reaction of organisms to their surroundings. A plasticity gene normally encodes a protein that functions as a receptor of environmental signals; the receptor gauges the state of a relevant environmental variable such as temperature and sends a signal that initiates a cascade of effects eventually leading to the production of the appropriate phenotype. For example, many trees shed their leaves at the onset of winter in order to save energy and water that would be wasted by maintaining structures that are not used during the winter months. The plants need a reliable cue that winter is indeed coming to best time the shedding process. Deciduous trees use photoperiod as an indicator of seasonality. A special set of receptors known as phytochromes are capable of sensing day length, and they initiate the shedding whenever day length gets short enough to signal the onset of winter. Phytochromes are, by definition, plasticity genes.

Research on plasticity genes is a very active field in both evolutionary and molecular genetics. Johanna Schmitt's group has demonstrated that the functionality of photoreceptors in plants has a direct effect on the

Deciduous trees shed their leaves at the prompting of "plasticity" genes that interpret environmental conditions. (Ben Klaffke)

fitness of the organism, thereby implying that natural selection can alter the characteristics of plasticity genes. Harry Smith and collaborators have contributed to the elucidation of the action of photoreceptors, uncovering an array of other genes that relate the receptor's signals to different tissues and cells so that the whole organism can appropriately respond to the change in environmental conditions. Similar research is ongoing on an array of other types of receptors that respond to nutrient availability, water supply, temperature, and a host of other environmental conditions.

From an evolutionary point of view, it is important not only to uncover which genes control a given type of plasticity but also to find out if and to what extent these genes are variable in natural populations. According to neo-Darwinian evolutionary theory, natural selection is effective only if populations harbor different versions of the same genes, thereby providing an ample set of possibilities from

which the most fit combinations are passed to the next generation. Thomas Mitchell-Olds has pioneered a combination of statistical and molecular techniques known as quantitative trait loci (QTL) mapping, which allows researchers to pinpoint the location in the genome of those genes that are both responsible for phenotypic plasticity and variable in natural populations. These genes are the most likely targets of natural selection for the future evolution of the species.

Heritability of Human Traits

The most important consequence of nature-nurture interactions is their application to the human condition. Humans are compelled to investigate questions related to the degree of genetic or environmental determination of complex traits such as behavior and intelligence. Unfortunately, such a quest is a potentially explosive mixture of science, philosophy, and politics, with the latter often perverting the practice of the first. For example, the original intention of intelligence quotient (IQ) testing in schools, introduced by Alfred Binet at the end of the nineteenth century, was simply to identify pupils in need of special attention in time for remedial curricula to help them. Soon, however, IQ tests became a widespread tool to support the supposed "scientific demonstration" of the innate inferiority of some races, social classes, or a particular gender (with the authors of such studies conveniently falling into the "best" race, social class, or gender). During the 1970's, ethologist Edward Wilson freely extrapolated from behavioral studies on ant colonies to reach conclusions about human nature; he proposed that genes directly control many aspects of animal and human behavior, thereby establishing the new and controversial discipline of sociobiology.

The reaction against this trend of manipulating science to advance a political agenda has, in some cases, overshot the mark. Some well-intentioned biologists have gone so far as to imply that either there are no genetic differences among human beings or that they are at least irrelevant. This goes against everything that is known about variation in natural populations of any organism. There is no reason to think that humans are exceptions: Since humans can measure genetically based differences in behavior and problem-solving ability in other species and relate these differences to fitness, the argument that such differences are somehow unimportant in humans is based on social good will rather than scientific evidence.

The problem with both positions is that they do not fully account for the fact that nature-nurture is not a dichotomy but a complex interaction. In reality, genes do not control behavior; their only function is to produce a protein, whose only function is to interact with other proteins at the cellular level. Such interactions do eventually result in what is observed as a phenotype, or a particular behavior, but this occurs only in a most indirect fashion and through plenty of environmental influences. On the other hand, plants, animals, and even humans are not infinitely pliable by environmental occurrences. Some behaviors are indeed innate, and others are the complex outcome of a genotype-environment feedback that occurs throughout the life span of an organism. In short, nature-nurture is not a matter of "either/or" but a question of how the two relate and influence each other.

As for humans, it is very likely that the precise extent of the biological basis of behavior and intelligence will actually never be determined because of insurmountable experimental difficulties. While it is technically feasible, it certainly is morally unacceptable to clone humans and study their characteristics under controlled conditions, the only route successfully pursued to experimentally disentangle nature and nurture in plants and animals. Studies of human twins help little, since even those separated at birth are usually raised in similar societal conditions with the result that the effects of heredity and environment are hopelessly confounded from a statistical standpoint. Regardless of the failure of science to answer these questions fully, the more compelling argument that has been made so far is that the actual answer should not matter to society, in that every human being is entitled to the same rights and privileges of any other one, regardless of indubitably real and sometimes profound differences in genetic makeup. Even the best science is simply the wrong tool to answer ethical questions.

—Massimo Pigliucci

See Also: Behavior; Evolutionary Biology; Mendel, Gregor, and Mendelism; Natural Selection; Quantitative Inheritance.

Further Reading: *The Mismeasure of Man* (1996), by Stephen Jay Gould, provides a fascinating account of the misuse of biology in supporting racial policies. *Sociobiology* (1975), by Edward Wilson, is the reference book for the modern reductionist approach to interpreting human behavior. A popular account of sociobiological theories is presented in *The Selfish Gene* (1989), by Richard Dawkins. A critique of sociobiology and biological determinism is found in Richard Levins and Richard C. Lewontin, *The Dialectical Biologist* (1985). A comprehensive treatment of phenotypic plasticity is found in Carl Schlichting and Massimo Pigliucci, *Phenotypic Evolution: A Reaction Norm Approach* (1998).